W0042691

THE THEORY OF THIN WINGS IN SUBSONIC FLOW

THE THEORY OF THIN WINGS IN SUBSONIC FLOW

Sergei Mikhailovich Belotserkovskii

Central Aero-Hydrodynamics Institute
Moscow, USSR

Translated from Russian

Translation Editor
Maurice Holt
Professor of Aeronautical Sciences
University of California
Berkeley, California

Springer Science+Business Media, LLC 1967

Sergei Mikhailovich Belotserkovskii, born in 1920, was graduated from the Moscow State University in 1941 and completed his studies at the N. E. Zhukov Military Aero-Engineering Academy in 1945. He is currently associated with both that institution and with the Central Aero-Hydrodynamics Institute in Moscow.

ISBN 978-1-4899-6161-7   ISBN 978-1-4899-6299-7 (eBook)
DOI 10.1007/978-1-4899-6299-7

The original Russian text, published by Nauka Press, Moscow, in 1965, has been revised by the author for the American Edition.

Сергей Михайлович Белоцерковский

Тонкая несущая поверхность в дозвуковом потоке газа

TONKAYA NESUSHCHAYA POVERKHNOST' V DOZVUKOVOM POTOKE GAZA

Library of Congress Catalog Card Number 66-17189

Plenum Press

© 1967 Springer Science+Business Media New York
Originally published by Plenum Press in 1967.
Softcover reprint of the hardcover 1st edition 1967

All rights reserved

No part of this publication may be reproduced in any form without written permission from the publisher

FOREWORD

This monograph contains a complete exposition of the theory of thin lifting wings at moderate and high subsonic speeds. The basic theory is developed for incompressible flow, and subsonic compressible flow is treated by means of the Prandtl-Glauert correction. The earlier chapters set out the formal solution to the problem of flow past a wing in terms of potential theory, which reduces the problem to an integral equation for a distribution of vortices over the wing. This equation can be solved exactly only in very special cases, and for wings of general plan form approximate methods of solution must be used. These are based on the representation of the wing either as a single horseshoe vortex (Prandtl lifting line theory) or by a series of horseshoe vortices. Alternatively the wing may be approximated by a series of local horseshoe vortices (vortex lattice theory).

These approximate methods have been developed and refined both in the Soviet Union and in Western countries during the past 30 years. The present volume gives a full description of Russian work, including the classical contribution of Zhukovskii and Chaplygin and extending to the more recent work of Kochin, Chushkin, and the author. In addition to single monoplane wings, cascades of wings, multiplanes, and annular wings are considered.

Many books are available in English dealing with Wing Theory; these include well known works by Abbott and Doenhoff, Betz (in the Durand volumes), Glauert, Milne-Thomson, Carafoli, Robinson and Laurmann, and Woods. A recent article in Handbuch der Physik by Weissinger (in German) should also be mentioned. With the exception of the works by Woods and Weissinger the main material in these books is at least 10 years old. The value of the present monograph is that it is very recently written, covers a large amount of Soviet work not readily available from Western sources, yet provides a complete treatment of the subject. It will serve as a valuable manual to the many aerodynamicists who continue to be concerned with subsonic aircraft design. It can also be used as the basis for a graduate course in Wing Theory.

Berkeley, California
March 1967

Maurice Holt

PREFACE

This book is devoted to one of the most completely developed sections of gas dynamics—wing theory. In the book we describe various types of thin lifting surfaces (monoplane wings of arbitrary shape in plan, annular wings of arbitrary aspect ratio, profile cascades), and we study most of the aerodynamic characteristics of wings on the basis of linear theory. This theory deals with small disturbances and so the problems can be greatly simplified. In the case of subsonic velocities for three-dimensional bodies, however, even with this simplification, it is difficult to obtain solutions in closed form. Therefore, numerical methods using electronic digital computers are the most effective.

The results of such calculations can be used not only in the qualitative explanation of phenomena, but also often lead to satisfactory quantitative results.

Wing theory has been studied in great detail both in the USSR and abroad. The original work was done by N. E. Zhukovskii, S. A. Chaplygin, and L. Prandtl. The ideas of N. E. Zhukovskii concerning bound vortices were especially fruitful. On the one hand these ideas supplied a basis for the explanation of the mechanism of lift in a perfect medium, and on the other hand they were used as a starting point for obtaining methods of numerical calculation.

The hypothesis of the coalescence of streamlines from the trailing edge of a lifting surface due to S. A. Chaplygin and N. E. Zhukovskii was of fundamental importance.

Even though the branch of aerodynamics described above is quite well developed, to the author's knowledge there is no sufficiently complete exposition of wing theory at subsonic speeds which is up-to-date. In practical applications it is also important to present effective methods of calculating aerodynamic characteristics which can be used with modern computers, and methods of checking the results of calculations. The author has written the present monograph in the hope that it will to some extent fill this gap in the literature.

The book can be divided into three sections. The first section (Chapters I-VI) is devoted to the general theory of lifting surfaces. The problems considered in this section were chosen mainly on the basis of the methods used in their solution. For example, we consider not only translational motion, but also steady rotation of monoplane and annular wings. This enables us to obtain rather simply a series of new results by employing the consequences of the reversibility theorem (forces and moments for reversed wings, the efficiency of attached controls, the influence of deformation of the wing surface on aerodynamic characteristics, etc.). The apparent mass of a wing is also considered, and this makes it possible to take account of the action of the medium in the case of oscillations of a wing in situ. All problems are discussed for lifting surfaces, some in the linear and some in the nonlinear formulation.

In the second part of the book (Chapters VII-IX) we describe methods for the calculation of aerodynamic characteristics of thin lifting systems of arbitrary form. Basically only one numerical method is used in the solution of such systems. It is based on the replacement of the continuously distributed vortex layer modeling the wing by a discrete-vortex system.

Thus the flow past a multiplane is constructed by using vortex filaments of infinite span; the simplest vortex system for a profile grid is an infinite chain of such filaments. The vortex surface of an annular wing is constructed from annular bound vortices with a corresponding sheet of free vortices; the surface of a rectangular wing is constructed from ordinary horseshoe vortices; the surface of a monoplane wing of very general shape in plan is constructed from a set of so-called oblique horseshoe vortices. By increasing the number of very simple systems, we can reproduce a continuous vortex layer with any desired degree of accuracy.

This type of approach leads to the replacement of one- or two-dimensional integral, differential, or integrodifferential equations by a system of algebraic equations which possesses many important useful properties. The resulting numerical method is flexible, efficient, and very convenient for use with an electronic digital computer.

In the third section (Chapters X–XIV) the use of numerical methods is explained. Numerical results for various lifting surfaces are presented and used in the analysis of the influence of geometrical parameters and the Mach number on the aerodynamic characteristics of wings of various forms. Calculated and the theoretical results are compared.

The thin lifting-surface scheme can be used to calculate the over-all characteristics of, and aerodynamic loads on, a wing for plane flow and also in three-dimensional problems (and for small angles of attack). Important differences between theoretical and experimental results are observed only for wings of very small aspect ratio; these differences are due to imperfections in the linear scheme. The effect of the thickness of a profile in a dense multiplane or a dense cascade must also not be neglected.

It should be noted that a combination of vortex singularities with distributed dipoles or sources and sinks can be used in the investigation of solid (not infinitesimally thin) lifting surfaces. This is important primarily in problems in which it is necessary to know the pressure distribution on the surface of a body. These methods are somewhat specialized, however, and they require a separate investigation.

In addition to known results, including some that have already been published by the author and his collaborators, we also present some completely new results.

This book has been written for engineers and scientific workers in the field of aerodynamics using the results obtained in this area, and also for students of universities and technological schools.

The author wishes to thank B. K. Skripan and V. G. Tabachinkov for their valuable discussions of the material in this book and for their help in the preparation of the manuscript.

S. M. Belotserkovskii

CONTENTS

BASIC NOTATION

I. Basic Notation Used for Geometrical Parameters

1. Monoplane Wings

l , wing span

b, root chord of a wing

b_k, end chord of a wing

b', chord of the wing section parallel to the root chord

b_a, mean aerodynamic chord of a wing (MAC)

S, wing area

z, coordinate of longitudinal section of span

$\bar{z} = 2z/l$, dimensionless coordinate of a wing section

$\lambda = l^2/S$, wing aspect ratio

χ_0, sweep back angle of leading edge

$\eta = b/b_k$, taper ratio of a wing

2. Annular Wings

b, wing chord

$D = 2r$, mean diameter of wing

α_0, wing aspect ratio

$\lambda = D/b$, wing aspect ratio

3. Profile Cascade and Multiplane

b, chord

t, the step equal to the distance between two congruent points of two neighboring profiles

$\bar{t} = t/b$, relative step

$\tau = b/t$, density

$\bar{f}$, relative curvature of a wing

β_Γ, geometrical stagger (the angle between a perpendicular to the front of a grid or multiplane, i.e., a straight line joining congruent points of a profile and a chord of the profile)

$\bar{x}_f = x_f/b$, relative coordinate of the maximum curvature of a profile

II. Basic Notation Used for Kinematic Characteristics

$\mathbf{U}_0$, absolute-velocity vector of the moving origin

$\boldsymbol{\Omega}$, absolute angular-velocity vector of a wing

W_x, W_y, W_z or W_x, W_r, W_φ, projections of the absolute (disturbed) velocity of the medium on the axes of a rectangular or cylindrical coordinate system

w_x, w_y, w_z or w_x, w_r, w_φ, projections of the dimensionless disturbed velocity of the medium

U_x, U_y, U_z or U_x, U_r, U_φ, projections of the velocity generated by the bound vortices

u_x, u_y, u_z or u_x, u_r, u_φ, projections of the dimensionless velocity generated by attached vortices

V_x, V_y, V_z or V_x, V_r, V_φ, projections of the velocity generated by free vortices

v_x, v_y, v_z or v_x, v_r, v_φ, projections of the dimensionless velocity generated by the free vortices

$\mathbf{W}^*, \mathbf{W}_0$, vector translational and relative velocities

$\mathbf{M}_\infty$, Mach number of the undisturbed flow

U_0, mean velocity of a profile cascade (one half the sum of the relative vector velocities of the medium at infinity in front of and behind a cascade)

$q = \rho_\infty U_0^2/2$, velocity head (dynamic pressure)

Φ, potential of a disturbed flow

α, angle of attack of a monoplane wing (the angle between the root chord and the projection of the velocity U_0 on the plane of symmetry Oxy of the wing)

α, angle of attack of an annular wing (the angle between the axis of symmetry of a wing and the velocity U_0)

α_Γ, geometrical angle of attack of a profile in a cascade (the angle between the geometrical chord of a profile and the mean velocity U_0)

ϑ, angle between the perpendicular to the front of a cascade and the mean velocity U_0, thus $\alpha_\Gamma + \beta_\Gamma = \vartheta$.

α_0, angle of zero lift of a cascade (angle between the geometrical and aerodynamic chords of a profile

α, aerodynamic angle of attack of a cascade or multiplane (angle between the aerodynamic chord of a profile and the velocity U_0)

β, angle of aerodynamic stagger of a profile cascade (angle between the aerodynamic chord of a profile and the mean velocity U_0); the following relations hold: $\alpha + \beta = \vartheta$, $\alpha_0 = \beta - \beta_\Gamma$

$\omega_x = \Omega_x b/U_0$, dimensionless angular velocity of bank of a monoplane wing with characteristic linear dimension b

$\omega_{x1} = \Omega_x l /2U_0$, dimensionless angular velocity of bank of a monoplane wing with characteristic linear dimension $l/2$

$\omega_z = \Omega_z b/U_0$, dimensionless angular velocity of pitch with characteristic linear dimension b

$\omega_{za} = \Omega_z b_a/U_0$, dimensionless angular velocity of pitch with characteristic linear dimension b_a

$\bar{x}_T = x_T/b$, center of a wing (x_T is the distance from the origin O to the leading edge of the wing)

U_1, U_2, U_3, U_4, U_5, U_6, the projections of the vectors U_0 and Ω on the Ox, Oy, and Oz axes ($U_1 = U_{0x}$, $U_2 = U_{0y}$, $U_3 = U_{0z}$, $U_4 = \Omega_x$, $U_5 = \Omega_y$, $U_6 = \Omega_z$); notation used in the investigation of apparent mass

$\gamma_{z+} = U_0(\gamma_z^\alpha \alpha + \gamma_z^{\omega z}\omega_z + \gamma_z^{\omega x}\omega_x)$, strength of the vortex layer attached to a monoplane wing

i; j; μ; ν, subscripts for the numbering of: attached vortices; points where boundary conditions are satisfied; bound vortex filaments; lines on which boundary conditions are satisfied

$\Gamma_+ = U_0 l /2N (\Gamma_\alpha \alpha + \Gamma_{\omega_z} \omega_z + \Gamma_{\omega_x} \omega_x)$, circulation of bound vortices

N, number of strips into which multipans are divided

$\gamma_+ = U_0(\gamma_0 \alpha_0 + \gamma_\alpha \alpha \cos \varphi + \gamma_{\omega_z}\omega_z \cos \varphi)$, strength of a bound vortex layer of an annular wing

i; j, subscripts for the numbering of annular bound vortices and sections $\bar{x}$ = const, at which boundary conditions are satisfied

$\Gamma_+ = U_0 r(\Gamma_0 \alpha_0 + \Gamma_\alpha \alpha \cos \varphi + \Gamma_{\omega_z}\omega_z \cos \varphi)$, circulation of bound vortices of an annular wing

$\gamma = U_0[\gamma_1 \sin (\alpha + \alpha_0) + \gamma_2 \cos (\alpha + \alpha_0)]$, strength of bound vortex layer of a profile cascade

$\Gamma_+ = 2U_0 t[\Gamma^{(1)}\sin (\alpha + \alpha_0) + \Gamma^{(2)} \cos (\alpha + \alpha_0)]$, circulation of bound vortices of a profile in a cascade

III. Basic Notation Used for Dynamic Characteristics

1. Monoplane Wings without Attached Controls

$c_y = Y/qS$, lift coefficient for characteristic dimension $c_y = c_y^\alpha \alpha + c_y^{\omega z}\omega_z$ or (for characteristic dimensions b_a) by the formula $c_y = c_y^\alpha \alpha + c_{ya}^\omega za\omega_{za}$

$m_z = M_z/qSb$, longitudinal-moment coefficient for characteristic dimension b, given by formula $m_z = m_z^\alpha \alpha + m_z^{\omega z}\omega_z$

$m_{za} = M_z/qSb_a$, longitudinal-moment coefficient relative to an axis passing through the nose of the mean aerodynamic chord for characteristic dimension b_a given by the formula $m_{za} = m_{za}^\alpha \alpha + m_{za}^\omega za\omega_{za}$

$m_x = M_x/qSb$, bank-moment coefficient for characteristic dimension b; given by the formula $m_x = m_x^{\omega x}\omega_x$

$m_{x1} = M_x/qS l$, bank-moment coefficient for characteristic dimension l (for m_{x1}) and $l/2$ (for ω_{x1}); given by the formula $m_{x1} = m_{x1}^{\omega x1}\omega_{x1}$

$\bar{x}_F = x_F/b_a$, coordinate of the focus relative to the nose of the mean aerodynamic chord in units of b_a

$c_y' = dY/qb' dz$, lift coefficient of a section; given by the formula

$$c_y' = c_y'^\alpha\, \alpha + c_y'^{\omega z}\, \omega_z + c_y'^{\omega x}\, \omega_x$$

$c_{xi} = X_i/qS$, induced drag coefficient with suction force

$\bar{c}_{xi}$, the same without suction force

$k_{22} = \lambda_{22}/\rho Sb$, $k_{44} = \lambda_{44}/\rho Sb^3$, $k_{66} = \lambda_{66}/\rho Sb^3$, $k_{26} = \lambda_{26}/\rho Sb^2$, apparent-mass coefficients

2. Monoplane-Wing Controls

$\Delta c_y = \Delta Y/qS$, increment of lift coefficient due to a deflection δ_f of a flap; given by the formula
$\Delta c_y = c_y{}^f \delta_f$

$\Delta m_z = \Delta M_z/qSb$, increment of the longitudinal-moment coefficient due to a deflection δ_f of a flap; given by the formula $\Delta m_z = m_z^{\delta f} \delta_f$

$\Delta m_x = \Delta M_x/qSb$, increment of the bank-moment coefficient due to a deflection δ_a of an aileron; given by the formula $\Delta m_x = m_x^{\delta a} \delta_a$

3. Annular Wings

$c_y = Y/qDb$, lift coefficient; given by the formula $c_y = c_y^\alpha \alpha + c_y^{\omega z} \omega_z$

$m_z = M_z/qDb^2$, longitudinal-moment coefficient; given by the formula $m_z = m_z^\alpha \alpha + m_z^{\omega z} \omega_z$

$\bar{x}_F = x_F/b$, dimensionless coordinate of the focus relative to the nose of a wing

$c_{xi} = X_i/qDb$, induced-drag coefficient with suction force

$\bar{c}_{xi}$, the same without suction force

$k_{22} = \lambda_{22}/\rho Sb$, $k_{66} = \lambda_{66}/\rho Sb^3$, $k_{26} = \lambda_{26}/\rho Sb^2$, apparent-mass coefficients

4. Profile Cascades

$c_y = Y/qb$, lift coefficient of a profile; given by the formula $c_y = c_y^\alpha \sin \alpha$

$m_z = M_z/qb^2$, longitudinal-moment coefficient relative to the nose of a profile; given by the formula $m_z = - m_{z0} - (m_z^\alpha/2) \sin 2\alpha - (m_z^{\alpha\alpha}/2) \sin^2 \alpha$

CHAPTER 1

GENERAL CONSIDERATIONS

§ 1. Coordinate Systems

In investigations of three-dimensional airfoils (monoplane and annular wings), we use a standard coordinate system fixed in the wing (Fig. 1.1). This consists of a rectangular right-handed Cartesian system with the Ox axis directed along the wing axis and the Oy axis in the plane of symmetry of the wing.

Let U_0 be the total velocity vector of the moving origin O and let Ω be the vector describing the total angular velocity of the wing. The vectors describing the kinematic characteristics of the complete motion of the wing are projected on the moving axes of the bound coordinate system:

$$U_0 = iU_{0x} + jU_{0y} + kU_{0z},$$
$$\Omega = i\Omega_x + j\Omega_y + k\Omega_z.$$

The moment of the aerodynamic forces acting on the wing, referred to the origin O, is projected on the same axes:

$$M = iM_x + jM_y + kM_z.$$

The projection M_x is usually called the transverse or rolling moment, M_y is the yawing moment, and M_z the longitudinal or pitching moment. The positive directions of the angular velocities of rotation of the wing and the aerodynamic moments are reckoned according to the right-hand screw rule and are shown in Fig. 1.1. The position of the wing in the flow field is described by the so-called angles of attack α and angle of slip β, also shown in Fig. 1.1, where

$$\left. \begin{array}{l} U_{0x} = U_0 \cos\alpha \cos\beta, \\ U_{0y} = -U_0 \sin\alpha \cos\beta, \\ U_{0z} = U_0 \sin\beta. \end{array} \right\} \qquad (1.1)$$

We project the resultant R of the aerodynamic forces on the axis of the bound coordinate system:

$$R = -iT + jN + kZ.$$

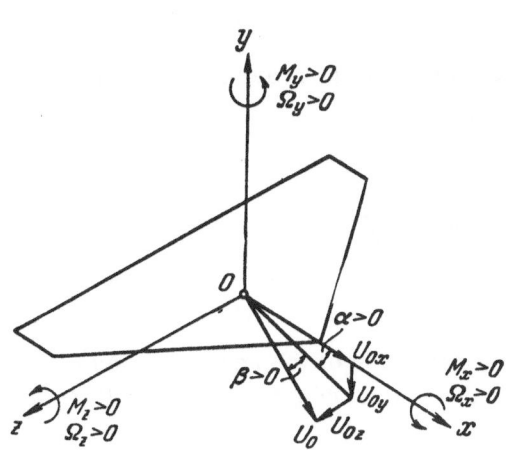

Fig. 1.1. The standard coordinate system.

1

The quantities T, N, and Z are called the tangential, normal, and transverse force components, respectively.

The force $\mathbf{R}$ is often projected on a moving coordinate system in which the x_m axis is in the direction of the velocity U_0 and the y_m axis is perpendicular to the x_m axis and in the plane of symmetry of the wing. The moving coordinate system is also a rectangular right-handed system. If we project on this set of axes we obtain

$$R = -i_m X + j_m Y + k_m Z_m;$$

the projection X is called the drag, Y is called the lift, and Z_m the lateral or side force.

The aerodynamic forces and moments depend essentially on the dimensions of the body and on experimental conditions. The dimensionless aerodynamic coefficients are much more standard; these are obtained by dividing the forces by the stagnation pressure (velocity head $q = \rho U_0^2/2$ (ρ is the mass density of the medium) and a characteristic area S, and by dividing the moments by a characteristic linear dimension as well.

We will be mainly interested in determining the drag X, the lift Y, the rolling moment M_x, the pitching moment M_z, and also the tangential and normal forces T and N. The corresponding coefficients are given by the formulas

$$c_x = \frac{X}{qS}, \qquad c_y = \frac{Y}{qS}, \qquad m_x = \frac{M_x}{qSl_1},$$

$$c_t = \frac{T}{qS}, \qquad c_n = \frac{N}{qS}, \qquad m_z = \frac{M_z}{qSl_2}.$$

We will consider below the choice of the characteristic area S and the characteristic linear dimensions l_1 and l_2.

In the most important case to be considered, when $\beta = 0$, the relations between the various aerodynamic-force coefficients in the bound and moving coordinate systems are

$$c_x = c_t \cos\alpha + c_n \sin\alpha,$$

$$c_y = c_n \cos\alpha - c_t \sin\alpha,$$

or for small α,

$$c_x = c_t + c_y\alpha, \qquad c_y = c_n. \tag{1.2}$$

Hence for small angles of attack, within an accuracy of the first order in α, there is no difference between the normal and the lift forces.

For small α and β, instead of (1.1) we have

$$U_{0x} = U_0, \qquad \alpha = -\frac{U_{0y}}{U_0}, \qquad \beta = \frac{U_{0z}}{U_0}, \tag{1.3}$$

to the first order.

In the investigation of wing oscillations and the determination of the apparent mass, it is more convenient to use a somewhat different notation. We will describe this in §8 of Chapter 1.

§2. The Monoplane Wing and its Geometrical Parameters

One of the most important forms of three-dimensional lifting surface is the monoplane wing (Fig. 1.2). The origin of the bound coordinate system Oxyz will be located in the middle of the root-chord b of the wing. The wing span will be denoted by l, the tip chord by b_k, and the general chord (in the z section) by b'. The sweepback angle of the wing is denoted by χ_0 for the leading edge, by χ_1 for the trailing edge, and by χ_Δ for the line that divides the chord in the ratio Δ (the leading edge corresponds to $\Delta = 0$, the trailing edge to $\Delta = 1$.

For the characteristic area S it is natural to use the area of the wing. For the longitudinal moment M_z we use b as the characteristic dimension, for the transverse moment M_x we use b or sometimes b and sometimes l, and so we can write

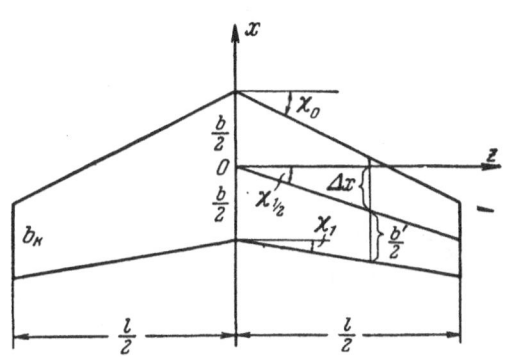

Fig. 1.2. The geometrical parameters of a monoplane wing.

$$c_x = \frac{X}{qS}, \quad c_y = \frac{Y}{qS},$$
$$m_x = \frac{M_x}{qSb}, \quad m_z = \frac{M_z}{qSb}, \quad m_{x1} = \frac{M_x}{qSl}. \qquad (1.4)$$

The shape of the wing in plan is usually characterized by dimensionless parameters. For one of these parameters we use the aspect ratio λ of the wing and for the other the taper ratio η:

$$\lambda = \frac{l^2}{S}, \qquad \eta = \frac{b}{b_k}. \qquad (1.5)$$

We mainly consider wings that are symmetrical in plan with constant leading- and trailing-edge sweepback and with parallel tip edges; such wings are called briefly, wings with straight edges. The shape of such wings in plan (with scale accuracy) is determined by λ, η, and χ_0.

Figures 1.3, 1.4, and 1.5 show how the shape of a wing varies in plan for changes in the basic geometrical parameters.

We now introduce some geometrical relations for such wings which will be useful in the sequel.

If $\Delta x(z)$ is the distance from the z axis to the mid-line of the wing in the section z = const, then

$$\Delta x = -z \, \mathrm{tg} \, \chi_{\frac{1}{2}},$$
$$b' = b - z(\mathrm{tg} \, \chi_0 - \mathrm{tg} \, \chi_1).$$

From Fig. 1.2 we obtain the relation

$$\mathrm{tg} \, \chi_\Delta = \frac{(l/2)\,\mathrm{tg}\,\chi_0 + b_k \Delta - b\Delta}{l/2}$$
$$= \mathrm{tg}\,\chi_0 - 2\frac{b}{l}\left(1 - \frac{1}{\eta}\right)\Delta,$$

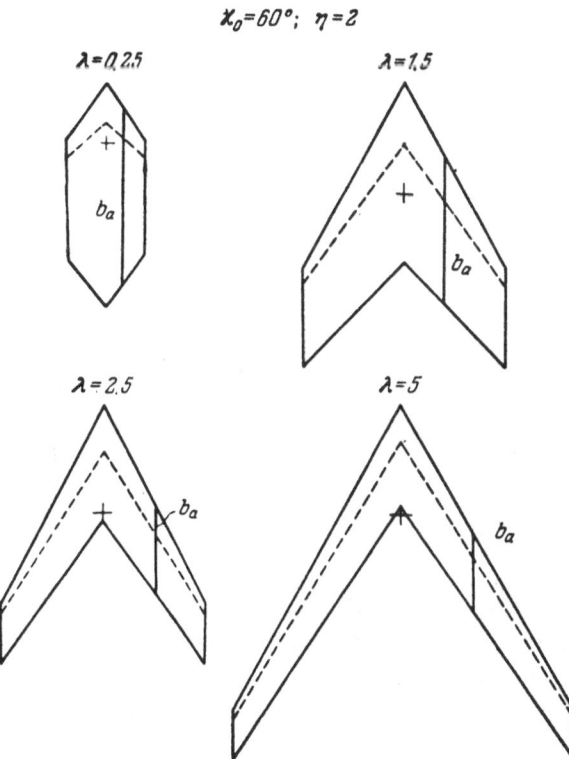

$\chi_0 = 60°; \quad \eta = 2$

Fig. 1.3. The effect of aspect ratio λ on the shape of a wing in plan. On each wing we denote the mean aerodynamic chord by b_a; the symbol "+" indicates the wing focus; --- is the focus of the sections (see Chapters IV and XII).

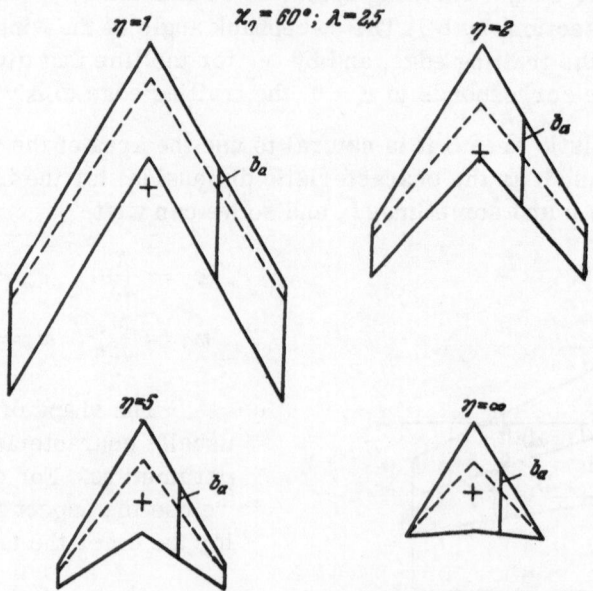

Fig. 1.4. The effect of the taper ratio η on the shape of a wing in plan. On each wing we denote the mean aerodynamic chord by b_a; the symbol "+" indicates the wing focus;--- is the focus of the sections (see Chapters IV and XII).

but since

$$S = \frac{b_\kappa + b}{2} l = \frac{1 + 1/\eta}{2} bl, \qquad \frac{l}{b} = \frac{\lambda}{2}\left(1 + \frac{1}{\eta}\right).$$

it follows that

$$\operatorname{tg} \chi_\Delta = \operatorname{tg} \chi_0 - \frac{4\Delta}{\lambda} \frac{\eta - 1}{\eta + 1}. \qquad (1.6)$$

The expressions for b' and Δx can now be written

$$\frac{b'}{b} = 1 - \bar{z}\left(1 - \frac{1}{\eta}\right), \qquad \bar{z} = \frac{2z}{l/b}, \quad \frac{l}{b} = \frac{\lambda}{2}\frac{\eta + 1}{\eta},$$

$$\frac{\Delta x}{b} = -\bar{z}\left[\frac{\lambda}{4}\left(1 + \frac{1}{\eta}\right)\operatorname{tg}\chi_0 - \frac{1}{2}\left(1 - \frac{1}{\eta}\right)\right]. \qquad (1.7)$$

§3. The Annular Wing and its Geometrical Parameters

An annular wing is an axially symmetric airfoil with a duct; it can be obtained by the rotation of a profile about the axis of symmetry of the wing (Fig. 1.6). As above, the Ox axis is directed forward along the axis of symmetry, and the origin O is located so that the yz plane divides the chord of the meridian sections into two equal parts.

The most important dimensionless geometrical parameter of an annular wing is the aspect ratio

$$\lambda = \frac{D}{b}. \qquad (1.8)$$

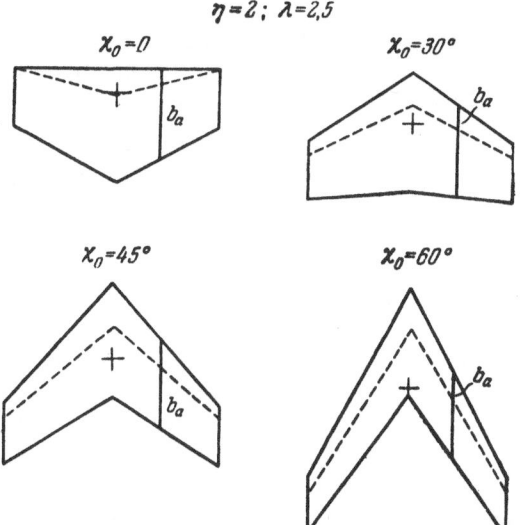

Fig. 1.5. The effect of the sweepback χ_0 on the shape of a wing in plan. On each wing we denote the mean aerodynamic chord by b_a; the symbol "+" indicates the wing focus; --- is the focus of the sections (see Chapters IV and XII).

where b is the chord and D the mean diameter of the wing. For a monoplane wing, this parameter characterizes the relation between the transverse and longitudinal dimensions of an annular wing.

The second geometrical parameter is the adjustable angle of attack α_0 of the wing sections. If the flow through the wing is axial ($\alpha = 0$), then α_0 is the angle of attack of all meridian sections.

The study of this type of wing is important both from the practical and theoretical point of view. It is a three-dimensional airfoil with a shape differing sharply from that of a monoplane wing Its characteristics can be obtained by much simpler calculations than those needed for a monoplane wing. Much useful information can be derived by comparing the aerodynamic properties of annular and rectangular wings with the same over-all dimensions (the span of the rectangular wing equal to the diameter D and the chords the same).

For the characteristic area of an annular wing, it is most convenient to use the area of the rectangular wing with the same over-all dimensions:

$$S = Db \tag{1.9}$$

and for the characteristic linear dimension, to use the chord b. Then the drag, lift, and longitudinal-moment coefficients are

$$c_x = \frac{X}{qS}, \quad c_y = \frac{Y}{qS}, \quad m_z = \frac{M_z}{qSb}. \tag{1.10}$$

If viscous forces are neglected, the transverse moment M_x of an annular wing will be zero for steady motion, i.e., $m_x = 0$.

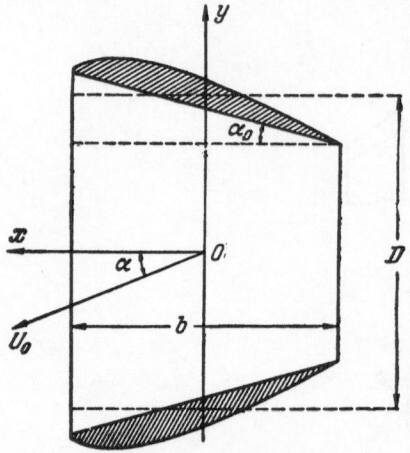

Fig. 1.6. The geometrical parameters of an annular wing.

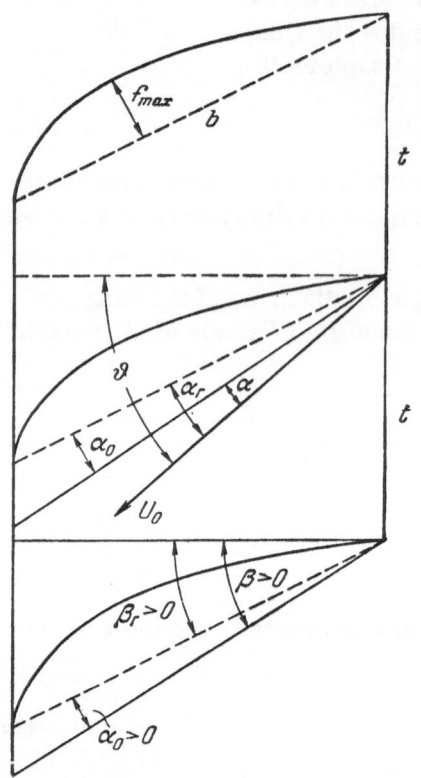

Fig. 1.7. The geometrical parameters of a profile grid.

§4. Cascade of Airfoils

We now consider cascades of airfoils, i.e., systems consisting of an infinite number of identical and similarly oriented profiles. Each of these profiles can be obtained by a displacement of the neighboring profile through a distance t along some straight line. This straight line is called the axis or front of the cascade, and the distance t is the gap (Fig. 1.7). We assume that the profile is infinitesimally thin, we denote its chord by b, and we impose no limitations on its curvature.

The angle between the geometrical chord b and the line perpendicular to the grid axis is usually denoted by β and is called the angle of geometrical stagger or the angle of the profile in the cascade. The second geometrical parameter, describing the position of a profile in the grid, is the relative gap

$$\bar{t} = \frac{t}{b} \qquad (1.11)$$

or its reciprocal

$$\tau = \frac{b}{t}, \qquad (1.12)$$

called the cascade density.

The shape of the profile itself can only be given by an equation, a graph, or by a table. Certain parameters can be given, however, which indicate the main effect of the aerodynamic characteristics. For a thin profile, there is firstly the relative concavity or profile camber

$$\bar{f} = \frac{f_{max}}{b}, \qquad (1.13)$$

where f_{max} is the maximum height of the profile relative to the chord. Secondly there is the relative position of the maximum height point of the profile

$$\bar{x}_f = \frac{x_f}{b}. \qquad (1.14)$$

The symbol x_f usually stands for the distance from the nose of the profile to the point of maximum height.

Let U_0 be the progressive velocity of the grid. The angle between this velocity and the perpendicular to the grid axis is denoted by ϑ (Fig. 1.7). In the theory of cascades the angle β of the circulation-free flow plays an important part. If $\vartheta = \beta$, then the cascade does not deflect the flow towards infinity behind it, and it has no lift. The angle of attack of a grid can be reckoned either from the geometrical chord of the profile (α_Γ), or from the direction of circulation-free flow (α). In the first case the angle of attack is usually called

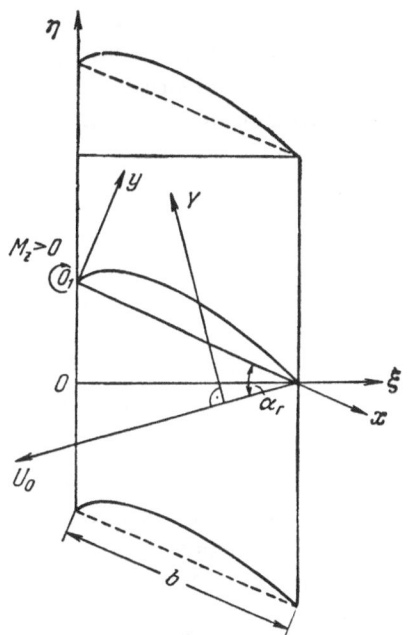

Fig. 1.8. Coordinate systems for profile cascades.

geometrical, in the second aerodynamic, and

$$\alpha = \vartheta - \beta, \qquad \alpha_\Gamma = \vartheta - \beta_\Gamma. \qquad (1.15)$$

If α_0 denotes the geometrical angle of attack corresponding to circulation-free flow past a cascade, then

$$\alpha_0 = \beta - \beta_\Gamma. \qquad (1.16)$$

Figure 1.7 shows an angle $\alpha_0 > 0$.

In the investigation of cascades, we will use the various coordinate systems shown in Fig. 1.8. The lift per unit span of the airfoil, perpendicular to U_0, will be denoted by Y. The moment on an airfoil per unit span relative to the leading edge will be denoted by M_Z. The aerodynamic lift coefficient and the longitudinal moment are introduced as follows:

$$c_y = \frac{Y}{qb}, \qquad m_z = \frac{M_z}{qb^2}. \qquad (1.17)$$

§5. Some Basic Experimental Results

In the formulation of the problem of the calculation of the flow about lifting surfaces, we must use certain information of experimental origin.

One of the basic conditions in the determination of aerodynamic characteristics is the so-called boundary condition at the wing surface. As a rule we try to design lifting surfaces so that there will be a smooth (non-reversing) flow of gas or liquid about them. This usually ensures the best possible lift properties and the least resistance. The determination of conditions for realizing smooth flow about a lift surface is not simple. Not only the theoretical but also the experimental determination of these conditions is difficult. We will, however, give some general information on this subject.

A sharp change in the direction of a subsonic flow can be accomplished by regions of separation of the boundary layer leading to the loss of smooth flow. These separated regions can be local with smooth flow again appearing behind them, but they can be more extensive with no smooth flow established behind them. The first case occurs in flow past sharp leading edges even for small angles of attack α. If flow past a leading edge were smooth, then in general, in a perfect medium, the velocity close to the leading edge would be infinite. The physical impossibility of such a flow is the cause of local separation. Extended separation of a boundary layer occurs in regions of flow with fairly high positive pressure gradients. If the wing sections have a high thickness ratio or a large camber, then for positive angles of attack the conditions for a separation will be most favorable on the upper surface behind the point of maximum thickness or maximum curvature of the section. Here there is a rapidly growing boundary layer, the separation of which can be induced by a positive pressure gradient. The most favorable conditions for smooth flow are obtained with an airfoil of intermediate thickness ratio and small curvature at low angles of attack.

For large positive angles of attack, a region appears on the upper surface of a wing where breakaway occurs. On the lower surface, smooth flow is usually maintained for large angles of attack α. A decrease in the aspect ratio of a wing strengthens the role of interaction between the effects of the lower and upper surfaces, and as is sometimes said, the flow spills over at the tips. Since in this case the pressure on the lower surface is on the whole higher and the pressure on the upper surface is lower than the pressure in the undisturbed flow, there will be a flow from the lower to the upper surface at the tips of the wing. This flow leads to an equalization of pressures on the surfaces and to an improvement of conditions for smooth flow along the upper surface. Hence a decrease in the aspect ratio of a monoplane wing leads to an increase in the range of angles of attack for which smooth flow is observed.

For a cascade of profiles, the conditions for smooth flow are better than for an isolated profile. Neighboring profiles act together to ensure smooth flow, and direct the oncoming flow correspondingly. We can say that the lower surfaces of neighboring profiles have a positive effect on the upper profile surfaces; this leads to a reinforcement of smooth flow at large angles of attack, and maintains it for high-curvature profiles. For small angles α, this effect is strengthened with decreasing gap ratio $\bar{t}$. There is a similar picture for annular wings, but here the reinforcement of smooth flow becomes stronger when the aspect ratio λ is decreased (down to a certain value λ).

In the general case, the action of flow on an airfoil can be reduced to two systems of distributed loads — the normal pressures and the tangential stresses. The second of these is due to the viscosity of the medium. In many cases we can determine the pressure and the tangential loads separately with sufficient accuracy for practical applications. In the determination of the pressure and the corresponding aerodynamic characteristics, this permits us to neglect the viscosity and to assume that we are dealing with a perfect medium.

There is, however, an important feature of the flow under consideration which influences all the aerodynamic characteristics of airfoil, and which in the final reckoning is related to the indirect introduction of viscous forces. This feature is essential in all problems concerning flow about surfaces at subsonic speeds, and is taken into account by the so-called Chaplygin-Zhukovskii hypothesis [5, 6, 3].

We consider flow about a profile or any section of a wing by the plane z = const. We can imagine three types of flow differing in the properties of the flow near the trailing edge of the body (Fig. 1.9). Experiment has shown that flows of types a and b, in which the rear stagnation point is displaced relative to the trailing edge, does not occur. When the flow is smooth there are flows close to that shown in c in which there is no flow around the trailing edge, but in which the flow coalesces behind it. In theoretical investigations we often deal with infinitely thin airfoils. Then, flow about sharp edges is accompanied by the appearance of velocities that are theoretically infinite. In this case the Chaplygin-Zhukovskii hypothesis can be considered as a condition for the finiteness of the velocity at the trailing sharp edge of a wing.

We consider further a very important experimental result which we will use later. Consider a wing of finite span (monoplane or annular), fixed at some constant angle of attack α when there is a lift on the wing. By using fine threads or special small vanes in the flow behind a wing, we can observe that the particles of the medium converging with the narrowing of the wing revolve about axes approximately parallel to the velcity of the approaching flow U_0. This is evidence of vortex motion in the trace of the wing. These vortices, converging from the wing and observed in the flow, are called free vortices.

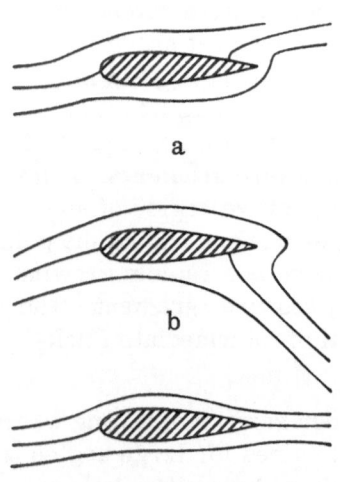

Fig. 1.9. The Chaplygin-Zhukovskii hypothesis.

It is convenient to use the idea of so-called bound vortices, which can replace a lifting surface. As we will see below these vortices, if suitably chosen, will have properties equivalent to those of an airfoil. The fact that they can be introduced is related to N. E. Zhukovskii's theorem concerning lift. Use of this approach permitted Zhukovskii to explain the Euler-d'Alembert paradox [1, 5]. He showed how to calculate the lift on a wing by using perfect-fluid theory that was of prime importance. The replacement of a lifting surface by bound vortices and corresponding conjugate free vortices is a method that is still in use. This method leads to convenient numerical methods for determining the aerodynamic characteristics of thin airfoils.

The simplest scheme is that in which a wing is replaced by a single bound vortex of a certain form and a system of free vortices [5, 10]. This method is, however, the least accurate, and in general it does not yield a sufficiently complete reproduction of the real flow pattern [26]. A much more accurate scheme is that based on the replacement of a wing by a sheet of bound vortices from which trails a sequence of free vortices (the airfoil scheme). This, however, leads to a very complicated theoretical method of investigation. In this latter method, the whole surface of a thin wing is replaced by an infinitely thin vortex layer. The inclination of each elementary vortex and the strength of the layer are not known. Behind the wing there is a sheet of free vortices with axes directed along the streamlines, in agreement with the theorems concerning the hydrodynamics of stationary motion. The vortex system must be such that the circulation is constant.

We will investigate methods based on the airfoil scheme, and in most cases related to the linearization of the problem [10, 20-22, 24-30]. The latter simplification is needed mainly because of the difficulty of taking account of the real position of the free vortices [25]. When there are no free vortices, the problem is not linearized. If a linear theory is used, then the perturbations in the velocity are assumed to be small compared to U_0, and the angle of attack α, measured in radians is assumed to be small compared to unity. In this case it can be shown that we may assume that the free vortices have the same direction as the velocity U_0, and the boundary conditions are satisfied not on the wing but on a very simple surface close to the wing (on a plane for a monoplane wing, on a circular cylinder for an annular wing). In linear theory, the vortices can be located at a position somewhat different from that indicated. For a monoplane wing in the Oxz plane (Fig. 1.1), the bound and free vortices can be located in the same plane, and it is assumed that the axes of the free vortices are parallel to the Ox axis. In the case of an annular wing, the vortices can be on a circular cylinder of diameter D with its axis parallel to Ox (Fig. 1.6), where the axes of the free vortices can be taken to be parallel to Ox. In both cases, the system of free vortices stretches behind the wing from the trailing edge to infinity. As to the vortices on the wing, it is immaterial whether we consider them all to be bound or whether we assume that some of them are considered to be free.

It should be noted that in the investigation of stationary flow about a vortex grid there will be no free vortices formed behind the grid, since the flow is plane-parallel and the bound vortices, parallel to the Ox axis (Fig. 1,8), will have a constant circulation relative to the span. This means that the problem can be solved in its nonlinear formulation with the profile-cascade curvature arbitrary.

Up to this stage, we have considered flow about a moving airfoil. Somewhat different problems are of interest when it is a question of finding the action of flow on a wing that is oscillating in a fixed position when the velocity U_0 is zero. In this case the idea of leading and trailing edges has no meaning and the conditions controlling flow around all edges with tangential faces will be identical. Such problems can usually be solved by using the hypothesis that the circulation is zero, which also ensures that the indicated conditions are satisfied for the edges. If we require that this condition be satisfied for every wing section, we find that no trail of free vortices will be formed behind the wing. In this case the lifting-surface scheme can also be used to solve the problem in its nonlinear formulation.

In the last two cases, in which the nonlinear problem can be solved, the bound vortices must be actually on the lifting surface and not on its projection; the complete boundary conditions (unlinearized) are also satisfied on this surface.

§ 6. Statement of the Problem

In the present monograph, we consider problems concerned with flow about a thin airfoil moving in an arbitrary fashion in a gas, and also the problem of arbitrary nonstationary oscillations of such an airfoil fixed in space in an incompressible medium. We assume that the shape of the airfoil and its motion are known. We must find the total and distributed aerodynamic load on the body and also, in general, the flow parameters outside the body.

If a lifting surface is replaced by a vortex system, then everywhere outside the body and the vortex trail, if one is formed, the flow will be vortex-free, which means that it will be potential flow. Hence the unknown functions determining all the characteristics of the flow will be the potential of the velocity perturbations Φ, the pressure p, and the density ρ in the gas flow.

We first consider the linear problem of the motion of an airfoil in a compressible gas. We must have three equations for the determination of the above unknown functions. One of these is the continuity equation which, for linearized stationary flow of a compressible medium, is

$$\left(1 - M_\infty^2\right) \frac{\partial^2 \Phi}{\partial x^2} + \frac{\partial^2 \Phi}{\partial y^2} + \frac{\partial^2 \Phi}{\partial z^2} = 0. \tag{1.18}$$

This equation is written in a moving coordinate system, fixed in the wing, with $M_\infty = U_0/a_\infty$, where a_∞ is the sound velocity in the undisturbed flow. In deriving this equation we assume that the dimensionless kinematic parameters characterizing the motion of the wing

$$\alpha, \ \beta, \ \omega_x = \frac{\Omega_x b}{U_0}, \quad \omega_y = \frac{\Omega_y b}{U_0}, \quad \omega_z = \frac{\Omega_z b}{U_0}, \tag{1.19}$$

which are independent of time, are small compared with unity. The squares and products of these quantities and of the disturbed velocities

$$W_x = \frac{\partial \Phi}{\partial x}, \quad W_y = \frac{\partial \Phi}{\partial y}, \quad W_z = \frac{\partial \Phi}{\partial z}, \tag{1.20}$$

are neglected. The relation between the disturbed pressure p' and the disturbed velocity W_x can be found by using the linearized Bernoulli equation

$$p' = \rho_\infty U_0 W_x, \tag{1.21}$$

where ρ_∞ is the mass density of the undisturbed medium. The disturbed density ρ' is obtained from p' by using the adiabatic equation, which in its linearized form can be written

$$p' = a_\infty^2 \rho'. \tag{1.22}$$

The gas parameters characterizing the state of the undisturbed medium p_∞, ρ_∞, a_∞ are assumed to be unknown. We also assume that the flow about the lifting surface is smooth, i.e., at each point of the flow the relative gas velocity is tangential to the surface. The boundary conditions will be considered in more detail below.

In an incompressible medium, in which the density ρ is constant, the unknowns will be the potential Φ and the pressure p. To derive the basic equations for these functions we use the continuity equations and the Cauchy-Lagrange integral,* since these relations will be needed in the investigation of both stationary and nonstationary flow. The exact (unlinearized) continuity equation in this case is the same for moving and fixed axes, since

$$\frac{\partial^2\Phi}{\partial x^2} + \frac{\partial^2\Phi}{\partial y^2} + \frac{\partial^2\Phi}{\partial z^2} = 0 \qquad (1.23)$$

for both these coordinate systems. The Cauchy-Lagrange integral in the moving system can be written

$$\frac{p}{\rho} = F(t) + \frac{W^{*2}}{2} - \frac{W_0^2}{2} - \frac{\partial\Phi}{\partial t}. \qquad (1.24)$$

Here $F(t)$ is an arbitrary function of time, W^* is the translational velocity of the frame of reference and W_0 is the relative velocity; the derivative is taken with respect to the moving axes. In linearized stationary problems, Eq. (1.24) naturally reduces to (1.21). Since (1.20) is exact, it can also be used in nonlinear problems.

One of the boundary conditions in this case is also the requirement that the flow over the surface be smooth. We assume that the parameters p_∞ and ρ_∞ of the undisturbed fluid are known.

Solutions obtained which satisfy all the above conditions for subsonic velocities are, however, not unique. In particular, we will investigate two different types of solution for an incompressible medium — flow with and flow without circulation about monoplane and annular wings. The first case corresponds to motion of a wing at very low Mach numbers M_∞, and the second to the oscillations of a fixed wing. To obtain a unique solution with subsonic velocities, we must impose some extra condition. In problems with circulation this is the Chaplygin-Zhukovskii condition, and for oscillation problems this is the requirement that the circulation be zero for every wing section.

In the following we will develop effective numerical methods for the solution of the problems described above, based on the replacement of a lifting surface by bound vortices. The strength of the bound vortices will be assumed to be unknown.

The convenience of this approach lies in the fact that the hydrodynamic singularities (the vortices) satisfy the continuity equation, and also have a series of important properties which will be described below. The strength of the free vortices is given simply in terms of the strength of the bound vortices, the relation between the strengths being based on the theorems of hydrodynamics of flows with constant circulation. Since the continuity equation will be automatically satisfied, the problem reduces to that of finding the unknown strength of the bound vortices from the boundary conditions and the Chaplygin-Zhukovskii condition, or from the condition that there be no circulation about the wing sections.

§7. Boundary Conditions

In all the problems investigated below, one boundary condition will be the requirement of smooth flow past the airfoil, which means that the normal component of the relative velocity

---

*Usually called Bernoulli's equation (M. H.).

will be zero at every point of the airfoil:

$$W_{0n} = 0. \tag{1.25}$$

Let

$$n = i \cos(n, x) + j \cos(n, y) + k \cos(n, z)$$

be the unit external normal; then the normal component of the disturbed velocity will be

$$W_n = W_x \cos(n, x) + W_y \cos(n, y) + W_z \cos(n, z). \tag{1.26}$$

The relative velocity is the vector difference between the absolute velocity and translational velocity of the frame of reference:

$$W_0 = W - W^*,$$

where for the point (x_0, y_0, z_0) we have

$$\left. \begin{array}{l} W^* = U_0 + \Omega \times r_0, \\ r_0 = ix_0 + jy_0 + kz_0. \end{array} \right\} \tag{1.27}$$

The boundary condition (1.25) can be written

$$W_n = W_n^*,$$

or, if we use (1.26) and (1.27),

$$W_x \cos(n, x) + W_y \cos(n, y) + W_z \cos(n, z) =$$
$$= U_{0x} \cos(n, x) + U_{0y} \cos(n, y) + U_{0z} \cos(n, z) + \cos(n, x)(\Omega_y z_0 - \Omega_z y_0) +$$
$$+ \cos(n, y)(\Omega_z x_0 - \Omega_x z_0) + \cos(n, z)(\Omega_x y_0 - \Omega_y x_0). \tag{1.28}$$

Using (1.1), we divide both sides of the boundary condition by U_0 and then use (1.19):

$$\frac{W_x \cos(n, x) + W_y \cos(n, y) + W_z \cos(n, z)}{U_0} = \cos \alpha \cos \beta \cos(n, x) -$$
$$- \sin \alpha \cos \beta \cos(n, y) + \sin \beta \cos(n, z) + \cos(n, x)\left(\omega_y \frac{z_0}{b} - \omega_z \frac{y_0}{b}\right) +$$
$$+ \cos(n, y)\left(\omega_z \frac{x_0}{b} - \omega_x \frac{z_0}{b}\right) + \cos(n, z)\left(\omega_x \frac{y_0}{b} - \omega_y \frac{x_0}{b}\right). \tag{1.29}$$

This is the most general form of the boundary conditions for a fixed lifting surface.

We now consider some special cases. We first consider a monoplane wing in the form of a plate of arbitrary shape in plan (Fig. 1.1); then

$$\cos(n, x) = 0, \quad \cos(n, y) = 1, \quad \cos(n, z) = 0.$$

Linearizing the boundary conditions (1.29) relative to the kinematic parameters (1.19), and in particular making the approximations $\sin \alpha \approx \alpha$, $\cos \beta \approx 1$, we obtain

$$\frac{W_y}{U_0} = -\alpha + \omega_z \frac{x_0}{b} - \omega_x \frac{z_0}{b}. \tag{1.30}$$

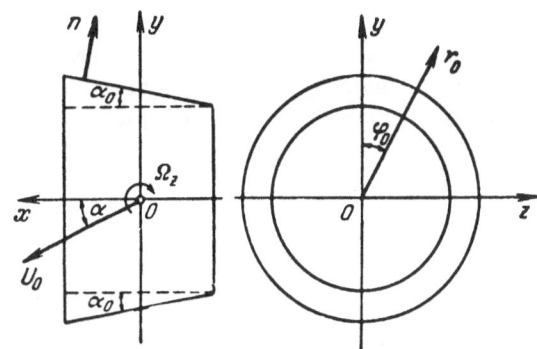

Fig. 1.10. The derivation of the boundary condition on an annular wing.

Secondly, we write the linearized boundary conditions for a conical annular wing, assuming that it is thin and the angle α_0 is small. We introduce cylindrical coordinates as illustrated in Fig. 1.10, and obtain the equation,

$$\cos(n, \; x) = -\alpha_0,$$
$$\cos(n, \; y) = \cos\varphi_0,$$
$$\cos(n, \; z) = \sin\varphi_0,$$
$$\frac{y_0}{b} = \frac{r_0}{b}\cos\varphi_0,$$
$$\frac{z_0}{b} = \frac{r_0}{b}\sin\varphi_0,$$

where $(x_0, \; r_0, \; \varphi_0)$ is the point at which the boundary conditions are to be satisfied and

$$W_r = W_y\cos\varphi_0 + W_z\sin\varphi_0.$$

Since an annular wing is axially symmetric, we can always locate the axes Oy and Oz so that $\Omega_y = 0$. In the framework of the linear theory, as will be seen below, the translational and rotational motion of the wing will be investigated separately and so we can set $\beta = 0$. Thus the boundary condition (1.29) becomes

$$\frac{W_r}{U_0} = -\alpha_0 - \alpha\cos\varphi_0 + \omega_z\frac{x_0}{b}\cos\varphi_0, \tag{1.31}$$

this condition is approximately satisfied, not on the wing, but at the corresponding points of a cylindrical surface close to the surface.

Thirdly, we consider the translational motion of a profile cascade with velocity U_0. We assume that the profile is infinitely thin but, in general, that it has a large curvature. The angle between the tangent to the surface profile at an arbitrary point and the Ox axis will be denoted by ε (see Fig. 1.11, where the angles ε and β_Γ are positive). The normal component of the disturbed velocity can be written

$$W_n = W_\eta\cos(\beta_\Gamma - \varepsilon) - W_\xi\sin(\beta_\Gamma - \varepsilon),$$

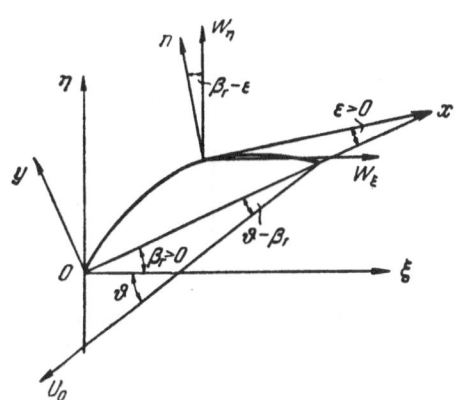

Fig. 1.11. The derivation of the boundary condition on a profile cascade.

and the normal component of the translational velocity U_0 is

$$U_n = -U_0\sin(\vartheta - \beta_\Gamma + \varepsilon),$$

or

$$U_n = -U_0[\sin(\vartheta - \beta_\Gamma)\cos\varepsilon + \cos(\vartheta - \beta_\Gamma)\sin\varepsilon].$$

The boundary condition

$$W_n = U_n$$

can now be written

$$W_n = -U_0[\sin(\vartheta - \beta_\Gamma)\cos\varepsilon + \cos(\vartheta - \beta_\Gamma)\sin\varepsilon], \tag{1.32}$$

where

$$W_n = W_\xi\sin(\varepsilon - \beta_\Gamma) + W_\eta\cos(\varepsilon - \beta_\Gamma). \tag{1.33}$$

§ 8. Apparent Mass of a Wing

Nonstationary aerodynamic forces and moments are sometimes determined approximately by a method in which circulation-free flow of a perfect, incompressible medium is used. This method can be used in studying the time history of the motion of a fixed wing, and also in several cases of nonstationary motion of an elongated body in water and low-velocity flight in air [38, 39]. In this scheme, the flow everywhere outside the body is potential and it is assumed that there are no vortices behind the body. Then the aerodynamic action of the flow on the body can be determined by using apparent masses [11].

If we write the six component equations of motion of the body resolved along the coordinate axes (three for the force and three for the moments), then these contain the mass and the corresponding moments of inertia which completely characterize the inertial properties of the body in a vacuum. In the framework of the above scheme, the equations of motion of a body in a perfect incompressible medium will be of the same form as for a vacuum, except that the mass properties will vary by amounts identified as the apparent masses.

Disturbances caused in the medium by a body of given shape will depend not only on the body shape and the medium density but also on the motion of the body. Thus, for example, the apparent masses for translational motion of the body along the different axes will be different only if the body is not symmetric.

In the case under consideration, the action of the medium on the body is reduced to a resultant force $\mathbf{R}$ applied at the origin of the coordinate system fixed to the body, and to a resultant moment $\mathbf{M}$, and these quantities are calculated from the formulas [11]

$$R = -\frac{\partial B}{\partial t} - \Omega \times B, \quad M = -\frac{\partial J}{\partial t} - \Omega \times J - U \times B, \tag{1.34}$$

where

$$B = iB_x + jB_y + kB_z, \quad J = iJ_x + jJ_y + kJ_z. \tag{1.35}$$

If, as above, the potential of the disturbed velocities is denoted by Φ, then

$$\left.\begin{array}{l} B_x = -\rho \iint\limits_{\Sigma} \Phi \cos(n, x)\, d\sigma, \quad B_y = -\rho \iint\limits_{\Sigma} \Phi \cos(n, y)\, d\sigma, \\[2ex] B_z = -\rho \iint\limits_{\Sigma} \Phi \cos(n, z)\, d\sigma, \\[2ex] J_x = -\rho \iint\limits_{\Sigma} \Phi \left[y_0 \cos(n, z) - z_0 \cos(n, y) \right] d\sigma, \\[2ex] J_y = -\iint\limits_{\Sigma} \Phi \left[z_0 \cos(n, x) - x_0 \cos(n, z) \right] d\sigma, \\[2ex] J_z = -\rho \iint\limits_{\Sigma} \Phi \left[x_0 \cos(n, y) - y_0 \cos(n, x) \right] d\sigma, \end{array}\right\} \tag{1.36}$$

where $\cos(n, x)$, $\cos(n, y)$, and $\cos(n, z)$ are the direction cosines of the external normal n to the surface of the body. The integrations in (1.36) are taken over the whole surface Σ of the body. For brevity in the investigation of apparent masses, it is convenient in addition to (1.1) to use the relations

$$U_0 = iU_1 + jU_2 + kU_3, \quad \Omega = iU_4 + jU_5 + kU_6. \tag{1.37}$$

For general circulation-free nonstationary motion, the potential can be expressed in the form (t is the time)

$$\Phi = \sum_{i=1}^{6} U_i \Phi_i, \qquad \Phi_i = \Phi_i(x, y, z), \qquad U_i = U_i(t). \tag{1.38}$$

If we know the potential Φ_i for the special type of motion of the body for $U_i = 1$, we can find the apparent masses:

$$\lambda_{ik} = - \rho \int\!\!\int_{\Sigma} \Phi_i \frac{\partial \Phi_k}{\partial n}\, d\sigma = - \rho \int\!\!\int_{\Sigma} \Phi_k \frac{\partial \Phi_i}{\partial n}\, d\sigma. \tag{1.39}$$

From the boundary condition (1.28) and the relations (1.20), (1.36), (1.38), and (1.39), we obtain

$$\left. \begin{array}{lll} B_x = \sum\limits_{i=1}^{6} U_i \lambda_{i1}, & B_y = \sum\limits_{i=1}^{6} U_i \lambda_{i2}, & B_z = \sum\limits_{i=1}^{6} U_i \lambda_{i3}, \\[2mm] J_x = \sum\limits_{i=1}^{6} U_i \lambda_{i4}, & J_y = \sum\limits_{i=1}^{6} U_i \lambda_{i5}, & J_z = \sum\limits_{i=1}^{6} U_i \lambda_{i6}. \end{array} \right\} \tag{1.40}$$

We substitute the expressions for $\mathbf{U}$, Ω, $\mathbf{B}$, and $\mathbf{J}$ in (1.34) and project the resulting vector relations on the axes of the Oxyz coordinate system. We now write

$$R = iX + jY + kZ, \qquad M = iM_x + jM_y + kM_z, \tag{1.41}$$

and the forces and moments acting on an arbitrary body in a circulation-free flow are given in terms of the apparent masses by the formulas

$$\left. \begin{array}{l} X = - \sum\limits_{i=1}^{6} \frac{dU_i}{dt} \lambda_{i1} - \sum\limits_{i=1}^{6} U_5 U_i \lambda_{i3} + \sum\limits_{i=1}^{6} U_6 U_i \lambda_{i2}, \\[3mm] Y = - \sum\limits_{i=1}^{6} \frac{dU_i}{dt} \lambda_{i2} - \sum\limits_{i=1}^{6} U_6 U_i \lambda_{i1} + \sum\limits_{i=1}^{6} U_4 U_i \lambda_{i3}, \\[3mm] Z = - \sum\limits_{i=1}^{6} \frac{dU_i}{dt} \lambda_{i3} - \sum\limits_{i=1}^{6} U_4 U_i \lambda_{i2} + \sum\limits_{i=1}^{6} U_5 U_i \lambda_{i1}, \\[3mm] M_x = - \sum\limits_{i=1}^{6} \frac{dU_i}{dt} \lambda_{i4} - \sum\limits_{i=1}^{6} U_5 U_i \lambda_{i6} + \sum\limits_{i=1}^{6} U_6 U_i \lambda_{i5} - \sum\limits_{i=1}^{6} U_2 U_i \lambda_{i3} + \sum\limits_{i=1}^{6} U_3 U_i \lambda_{i2}, \end{array} \right\} \tag{1.42'}$$

$$\left. \begin{array}{l} M_y = - \sum\limits_{i=1}^{6} \frac{dU_i}{dt} \lambda_{i5} - \sum\limits_{i=1}^{6} U_6 U_i \lambda_{i4} + \sum\limits_{i=1}^{6} U_4 U_i \lambda_{i6} - \sum\limits_{i=1}^{6} U_3 U_i \lambda_{i1} + \sum\limits_{i=1}^{6} U_1 U_i \lambda_{i3}, \\[3mm] M_z = - \sum\limits_{i=1}^{6} \frac{dU_i}{dt} \lambda_{i6} - \sum\limits_{i=1}^{6} U_4 U_i \lambda_{i5} + \sum\limits_{i=1}^{6} U_5 U_i \lambda_{i4} - \sum\limits_{i=1}^{6} U_1 U_i \lambda_{i2} + \sum\limits_{i=1}^{6} U_2 U_i \lambda_{i1}. \end{array} \right\} \tag{1.42''}$$

We now consider a monoplane wing – a flat plate of arbitrary plan form (Fig. 1.1). Since. the motion of a wing corresponding to the velocities U_1, U_3, and U_5 does not disturb the medium, we have

$$\Phi_1 = 0, \quad \Phi_3 = 0, \quad \Phi_5 = 0, \quad \Phi = U_2\Phi_2 + U_4\Phi_4 + U_6\Phi_6. \tag{1.43}$$

Equations (1.39) for the apparent masses can also be written so that the integrations are carried out, not over the upper and lower surfaces of the wing, but over the area S. Using the subscript "+" for all quantities referred to the upper surface ($y > 0$, $y \to 0$), and "−" for ($y < 0$, $y \to 0$), we can write $\Phi(x, y, z) = -\Phi(x, -y, z)$, and

$$\Phi_+ = -\Phi_-, \quad \left(\frac{\partial\Phi}{\partial n}\right)_+ = -\left(\frac{\partial\Phi}{\partial n}\right)_- = \frac{\partial\Phi_+}{\partial y}.$$

Hence for a monoplane wing

$$\lambda_{ik} = -2\rho \int\!\!\int_S \Phi_{+i}\frac{\partial\Phi_{+k}}{\partial y}\,ds = -2\rho \int\!\!\int_S \Phi_{+k}\frac{\partial\Phi_{+i}}{\partial y}\,ds.$$

$$i = 2,\ 4,\ 6; \quad k = 2,\ 4,\ 6; \quad \lambda_{ik} = \lambda_{ki}. \tag{1.44}$$

with the integration taken over the area S of the wing.

For a wing that is symmetric in plan relative to the Oxy plane, the number of nonzero apparent masses is still further decreased. In fact Φ_2 and Φ_6 will be even and Φ_4 odd functions of z, and the same is true of $\partial\Phi_2/\partial y$, $\partial\Phi_6/\partial y$ and $\partial\Phi_4/\partial y$. Hence the apparent masses λ_{24} and λ_{64} are zero and the only nonzero masses will be

$$\left.\begin{aligned}
\lambda_{22} &= -2\rho \int\!\!\int_S \Phi_{+2}\frac{\partial\Phi_{+2}}{\partial y}\,ds, \quad &\lambda_{44} &= -2\rho \int\!\!\int_S \Phi_{+4}\frac{\partial\Phi_{+4}}{\partial y}\,ds, \\
\lambda_{66} &= -2\rho \int\!\!\int_S \Phi_{+6}\frac{\partial\Phi_{+6}}{\partial y}\,ds, \quad &\lambda_{26} &= -2\rho \int\!\!\int_S \Phi_{+2}\frac{\partial\Phi_{+6}}{\partial y}\,ds.
\end{aligned}\right\} \tag{1.45}$$

The relations (1.42) become

$$\left.\begin{aligned}
X &= U_6 U_2 \lambda_{22} + U_6^2 \lambda_{62}, \\
Y &= -\frac{dU_2}{dt}\lambda_{22} - \frac{dU_6}{dt}\lambda_{62}, \\
Z &= -U_4 U_2 \lambda_{22} - U_4 U_6 \lambda_{62}, \\
M_x &= -\frac{dU_4}{dt}\lambda_{44} - U_5 U_6 \lambda_{66} - U_5 U_2 \lambda_{26} + U_3 U_2 \lambda_{22} + U_3 U_6 \lambda_{26}, \\
M_y &= -U_6 U_4 \lambda_{44} + U_4 U_6 \lambda_{66} + U_4 U_2 \lambda_{26}, \\
M_z &= -\frac{dU_6}{dt}\lambda_{66} - \frac{dU_2}{dt}\lambda_{26} + U_5 U_4 \lambda_{44} - U_1 U_2 \lambda_{22} - U_1 U_6 \lambda_{26}.
\end{aligned}\right\} \tag{1.46}$$

We now consider an annular wing (Fig. 1.6), and for simplicity we confine ourselves to the case of a cylindrical wing $\alpha_0 = 0$ with an infinitely thin profile. Motion of the wing corresponding to velocities U_1 and U_4 does not disturb the medium, and so

$$\Phi_1 = 0, \quad \Phi_4 = 0.$$

An investigation of the motion of an annular wing characterized by the kinematic parameters U_2 and U_6 leads to the determination of all the apparent masses, where here

$$\Phi = U_2\Phi_2 + U_6\Phi_6.$$
(1.47)

The only nonzero masses are

$$\lambda_{lk} = -\rho \int_\Sigma \int \Phi_l \frac{\partial \Phi_k}{\partial n}\, d\sigma, \quad \lambda_{kl} = \lambda_{lk} \qquad (l = 2,\ 6;\ k = 2,\ 6).$$
(1.48)

The aerodynamic forces and moments acting on an annular wing in the stationary motion under consideration are

$$\left.\begin{aligned}
X &= U_6 U_2 \lambda_{22} + U_6^2 \lambda_{26}, \\
Y &= -\frac{dU_2}{dt}\lambda_{22} - \frac{dU_6}{dt}\lambda_{26}, \\
M_z &= -\frac{dU_2}{dt}\lambda_{26} - \frac{dU_6}{dt}\lambda_{66} - U_1 U_2 \lambda_{22} - U_1 U_6 \lambda_{26}.
\end{aligned}\right\}$$
(1.49)

We note that for a monoplane wing the boundary condition (1.28) can be written

$$W_y = U_2 - U_4 z_0 + U_6 x_0.$$
(1.50)

and for a cylindrical annular wing, instead of (1.31), we have

$$W_r = (U_2 + U_6 x_0)\cos\varphi_0.$$
(1.51)

CHAPTER 2

SOME GENERAL PROPERTIES OF INDUCED VELOCITIES

§ 1. General Relations

In preparation for subsequent investigations, we must study the properties of the velocity field generated (induced) by vortex sheets. In this chapter we prove some auxiliary theorems that will be used later. We establish the basic properties of the velocity perturbations caused by a vortex sheet at points close to the surface, and we also find the order of the rate at which these disturbances decay at large distances from the vortex sheet. By a vortex sheet we mean some surface Σ, on which there is a continuously distributed vortex layer such that the axis of each elementary vortex of the layer is in the plane tangent to the surface. Our analysis is valid, not only for steady but also for unsteady motion. In the latter case the time t will be assumed to be fixed.

The properties of potentials of simple and double layers are investigated and described in detail in [42]. To shorten our description, we assume that all these properties are known.

We consider an element dl of vortex filament with strength Γ. Then the velocity dW induced by this elementary vortex filament at the point (x_0, y_0, z_0) will be given by the Biot-Savart formula [11]

$$dW = \frac{\Gamma}{4\pi R^2} \sin \psi \, dl. \tag{2.1}$$

Here R is the distance from the vortex to the point (x_0, y_0, z_0), and ψ is the angle between the vortex axis and the radius vector $\mathbf{R}$ (Fig. 2.1). The velocity dW is perpendicular to the plane passing through the element dl and the radius vector $\mathbf{R}$.

We now write the expressions for the velocity perturbations caused by the vortex sheet Σ. We customarily call one side of the surface Σ, which we assume is sufficiently smooth, the lower side ("−" in Fig. 2.2) and the other the upper side ("+" in Fig. 2.2). The strength of the vortex layer will be denoted by γ; the coordinates of the current point on the surface by ξ, η, and ζ; and the coordinates of the point at which the velocity is calculated by (x_0, y_0, z_0). We introduce Cartesian coordinates with origin O at an arbitrary point on the surface Σ, and with the Oy axis directed along the normal passing from the lower to the upper side of the surface. The Ox and Oy axes are in the plane tangent to Σ at O, so that the Oz axis is tangent and the Ox axis normal to the vortex line l at O (Fig. 2.2). The element of length in the direction of the vortex line will be dl, and in the perpendicular direction $d\tau$. Then from known formulas [11] we can obtain the following expressions for the velocity per-

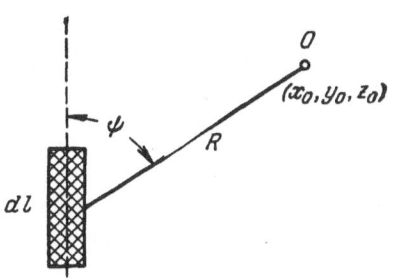

Fig. 2.1. Illustration of Biot-Savart law.

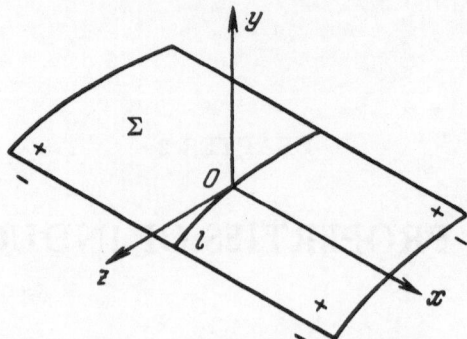

Fig. 2.2. A thin vortex sheet Σ.

turbations at (x_0, y_0, z_0):

$$W_x = \frac{1}{4\pi} \int\int_\Sigma \int \frac{\gamma}{R^2}\left[\frac{\eta - y_0}{R}\cos(l, z) - \frac{\zeta - z_0}{R}\cos(l, y)\right]d\tau\, dl,$$

$$W_y = \frac{1}{4\pi} \int\int_\Sigma \int \frac{\gamma}{R^2}\left[\frac{\zeta - z_0}{R}\cos(l, x) - \frac{\xi - x_0}{R}\cos(l, z)\right]d\tau\, dl, \qquad (2.2)$$

$$W_z = \frac{1}{4\pi} \int\int_\Sigma \int \frac{\gamma}{R^2}\left[\frac{\xi - x_0}{R}\cos(l, y) - \frac{\eta - y_0}{R}\cos(l, x)\right]d\tau\, dl,$$

where

$$R = \sqrt{(x_0 - \xi)^2 + (y_0 - \eta)^2 + (z_0 - \zeta)^2}. \qquad (2.3)$$

§2. Disturbed Velocities Close to a Vortex Sheet

We now study the properties of velocity perturbations close to any point O on the vortex sheet Σ. We assume that this surface is smooth in the Lyapunov sense [42]. We also assume that a vortex line l is sufficiently smooth. That is, if P_1 and P_2 are two points on one line l, and l_1 and l_2 are the unit tangent vectors to l at these points, then $l_1 - l_2$ satisfies the inequality

$$|l_1 - l_2| < BR^\delta,$$

where R is the distance P_1 and P_2, and B and δ are positive constants with

$$0 < \delta \leqslant 1.$$

We note that a Lyapunov surface has the following properties which will be used below.

If Q_1 and Q_2 are two points of such a surface and $\mathbf{n}_1$ and $\mathbf{n}_2$ unit vector normals at these points, then

$$|\mathbf{n}_1 - \mathbf{n}_2| \leqslant AR^\delta,$$

where A is a constant and R the distance between Q_1 and Q_2.

Moreover, close to any point, the equation of a sufficiently small part of the surface can be solved for a unique value of the coordinate η. We thus have

$$\eta = \eta(\xi, \zeta).$$

We now state some theorems concerning the velocities W_X and W_Z. The limiting values of these velocities, obtained when the point (x_0, y_0, z_0) approaches Σ, will be denoted by:

W_{X+} and W_{Z+} for the upper side of the surface;
W_{X-} and W_{Z-} for the lower side of the surface;
W_{X0} and W_{Z0} for the point O on the surface.

For the proofs of the theorems of this section, we assume that

$$|\gamma| < M = \text{const.}$$

Theorem 1. The velocities $W_{X0} = W_X(0, 0, 0)$ and $W_{Z0} = W_Z(0, 0, 0)$ as obtained from Eqs. (2.2) for a point O on Σ do exist.

We prove the convergence of the corresponding integrals in Eqs. (2.2). To do this, we consider a part Σ_1 of the surface Σ, so small that the equation of Σ_1 can be written

$$\eta = \eta(\xi, \zeta).$$

From the mean value theorem we have

$$\eta(\xi, \zeta) = \xi\left[\frac{\partial\eta}{\partial\xi}\right]_{\bar{\xi},\bar{\zeta}} + \zeta\left[\frac{\partial\eta}{\partial\zeta}\right]_{\bar{\xi},\bar{\zeta}}, \tag{2.4}$$

where

$$0 \leqslant \bar{\xi} \leqslant \xi, \quad 0 \leqslant \bar{\zeta} \leqslant \zeta.$$

Let $\mathbf{i}$, $\mathbf{j}$, and $\mathbf{k}$ be unit vectors in the direction of the axes Ox, Oy, and Oz, and let $\boldsymbol{l}$ and n be unit vectors tangent to $\boldsymbol{l}$ and normal to Σ respectively.

Since at O we have

$$\boldsymbol{l}_0 = \boldsymbol{k}.$$

it follows that

$$\cos(l, x) = \boldsymbol{l}\boldsymbol{i} = (\boldsymbol{l} - \boldsymbol{l}_0)\boldsymbol{i},$$
$$\cos(l, y) = \boldsymbol{l}\boldsymbol{j} = (\boldsymbol{l} - \boldsymbol{l}_0)\boldsymbol{j}.$$

Hence

$$\left.\begin{array}{l} |\cos(l, x)| = |\boldsymbol{l} - \boldsymbol{l}_0| < BR^{\delta}, \\ |\cos(l, y)| = |\boldsymbol{l} - \boldsymbol{l}_0| < BR^{\delta}. \end{array}\right\} \tag{2.5}$$

At O we also have $\mathbf{n}_0 = \mathbf{j}$, and so

$$\cos(n, x) = \boldsymbol{n}\boldsymbol{i} = (\boldsymbol{n} - \boldsymbol{n}_0)\boldsymbol{i},$$
$$\cos(n, y) = \boldsymbol{n}\boldsymbol{j} = (\boldsymbol{n} - \boldsymbol{n}_0)\boldsymbol{j} + 1,$$
$$\cos(n, z) = \boldsymbol{n}\boldsymbol{k} = (\boldsymbol{n} - \boldsymbol{n}_0)\boldsymbol{k}.$$

Now using the condition

$$|\boldsymbol{n} - \boldsymbol{n}_0| < AR^{\delta},$$

for the smoothness of the surface, where A is a constant, we obtain the inequalities

$$\left.\begin{array}{l} |\cos(n,\ x)| = |n - n_0| < AR^\delta, \\ |\cos(n,\ y)| = |(n - n_0)\,j + 1| \geqslant 1 - |n - n_0| > 1 - AR^\delta, \\ |\cos(n,\ z)| = |n - n_0| < AR^\delta. \end{array}\right\} \tag{2.6}$$

From the formulas

$$\frac{\partial\eta}{\partial\xi} = -\frac{\cos(n,\ x)}{\cos(n,\ y)},$$

$$\frac{\partial\eta}{\partial\zeta} = -\frac{\cos(n,\ z)}{\cos(n,\ y)}$$

and the inequalities (2.6) we obtain

$$\left|\frac{\partial\eta}{\partial\xi}\right| < \frac{AR^\delta}{1 - AR^\delta},$$

$$\left|\frac{\partial\eta}{\partial\zeta}\right| < \frac{AR^\delta}{1 - AR^\delta}.$$

The relation (2.4) now yields

$$|\eta| < \frac{2AR^\delta \rho_1}{1 - AR^\delta},$$

where

$$\rho_1 = \sqrt{\xi^2 + \zeta^2}.$$

For all points in the sphere

$$AR^\delta < \frac{1}{2},$$

we have

$$|\eta| < 4AR^\delta \rho_1. \tag{2.7}$$

The inequalities (2.5) and (2.7) yield

$$\left|\frac{\gamma}{R^2}\left[\frac{\eta}{R}\cos(l,\ z) - \frac{\zeta}{R}\cos(l,\ y)\right]\right| < \frac{C_1\rho_1}{R^{3-\delta}} \leqslant \frac{C_1}{\rho_1^{2-\delta}},$$

$$\left|\frac{\gamma}{R^2}\left[\frac{\xi}{R}\cos(l,\ y) - \frac{\eta}{R}\cos(l,\ x)\right]\right| < \frac{C_2\rho_1}{R^{3-\delta}} \leqslant \frac{C_2}{\rho_1^{2-\delta}},$$

where C_1 and C_2 are constants. Hence the integrals in (2.2) exist at O.

Similar estimates for W_y yield the conclusion that here the integral can have a meaning only in the Cauchy principal-value sense.

Theorem 2. The velocity $W_z(x_0,\ y_0,\ z_0)$ varies continuously when the point $(x_0,\ y_0,\ z_0)$ passes through Σ.

We will prove that the integral in the expression for $W_z(x_0,\ y_0,\ z_0)$ in (2.2) converges uniformly at O. We write

$$\rho_2 = \sqrt{(\xi - x_0)^2 + (\zeta - z_0)^2}.$$

Then from (2.5) and the inequalities

$$|\eta - y_0| \leqslant R, \qquad \rho_2 \leqslant R,$$
$$|\xi - x_0| \leqslant R,$$

we obtain the inequality

$$\left| \frac{\gamma}{R^2} \left[\frac{\xi - x_0}{R} \cos(l, y) - \frac{\eta - y_0}{R} \cos(l, x) \right] \right| < \frac{C_3}{\rho_2^{2-\delta}},$$

where C_3 = const. This means that the integral converges uniformly at O, and is thus a continuous function at this point [42].

Theorem 3. The velocity $W_x(x_0, y_0, z_0)$ has a discontinuity at points on Σ, and

$$\left. \begin{array}{l} W_{x_+} = -\frac{\gamma}{2} + W_{x0}, \\[2mm] W_{x_-} = \frac{\gamma}{2} + W_{x0}. \end{array} \right\} \tag{2.8}$$

We consider the potential $f(x_0, y_0, z_0)$ of a double layer with a layer density γ. This potential can be given by the formula

$$f(x_0, y_0, z_0) = \frac{1}{4\pi} \int_{\Sigma} \int \left[\frac{\xi - x_0}{R} \cos(n, x) + \frac{\eta - y_0}{R} \cos(n, y) + \frac{\zeta - z_0}{R} \cos(n, z) \right] \frac{\gamma}{R^2} \, d\tau \, dl.$$

The potential $f(x_0, y_0, z_0)$ for the case when there are distributed sources on the upper side of the surface Σ (see Fig. 2.2) has the known properties [42]

$$f_+ = -\frac{\gamma}{2} + f_0, \quad f_- = \frac{\gamma}{2} + f_0, \quad f_0 = f(0, 0, 0). \tag{2.9}$$

We form the difference

$$W_x(x_0, y_0, z_0) - f(x_0, y_0, z_0) = \frac{1}{4\pi} \int_{\Sigma} \int \left\{ \frac{\eta - y_0}{R} [\cos(l, z) - \cos(n, y)] - \right.$$
$$\left. - \frac{\zeta - z_0}{R} [\cos(l, y) + \cos(n, z)] - \frac{\xi - x_0}{R} \cos(n, x) \right\} \frac{\gamma}{R^2} \, d\tau \, dl.$$

We now find an upper bound for the absolute value of the integrand on the right-hand side of this equation. We note that

$$\cos(l, z) - \cos(n, y) = lk - nj = (l - l_0)k - (n - n_0)j,$$

so that

$$l_0 k = 1, \quad n_0 j = 1.$$

Hence

$$|\cos(l, z) - \cos(n, y)| \leqslant |l - l_0| + |n - n_0| < (A + B)R^{\delta}. \tag{2.10}$$

Recalling the inequalities

$$|\eta - y_0| \leqslant R, \qquad |\zeta - z_0| \leqslant R,$$
$$|\xi - x_0| \leqslant R, \qquad \rho_2 \leqslant R,$$

and also (2.5), (2.6), and (2.10), we easily obtain

$$\left| \frac{\gamma}{R^2} \left\{ \frac{\eta - y_0}{R} \left[\cos(l, z) - \cos(n, y) \right] - \frac{\zeta - z_0}{R} \left[\cos(l, y) + \cos(n, z) \right] - \frac{\xi - x_0}{R} \cos(n, x) \right\} \right| < \frac{C_4}{\rho_2^{2-\delta}},$$

where C_4 is a constant. Thus the integral in the expression for the difference $W_x(x_0, y_0, z_0) - f(x_0, y_0, z_0)$ converges uniformly at O, and the difference is continuous at this point. Hence

$$W_{x_+} - f_+ = W_{x_-} - f_- = W_{x0} - f_0,$$

and so by using (2.9) we obtain (2.8)

§3. Disturbed Velocities at Infinity

$\underline{\text{Theorem 4.}}$ If the vortex sheet Σ is bounded and the strength of the vortex layer is such that

$$\iint\limits_{\Sigma} |\gamma|\, d\Sigma \leqslant M_1 = \text{const},$$

then the velocities generated by the vortex sheet decrease like $1/R_0^2$ at infinity, where

$$R_0 = \sqrt{x_0^2 + y_0^2 + z_0^2}.$$

In fact for sufficiently large R_0 we have

$$R = R_0 \frac{R}{R_0} > \frac{R_0}{2},$$

and so

$$|W_x(x_0, y_0, z_0)| \leqslant \frac{1}{4\pi} \int\limits_{\Sigma} \int \left| \frac{\eta - y_0}{R} \cos(l, z) - \right.$$
$$\left. - \frac{\zeta - z_0}{R} \cos(l, y) \right| \left| \frac{\gamma}{R^2} \right| d\tau\, dl < \frac{1}{4\pi} \cdot \frac{4}{R_0^2} \cdot 2 \int\limits_{\Sigma} \int |\gamma|\, d\Sigma < \frac{M_2}{R_0^2},$$

where

$$M_2 = \text{const}.$$

Similarly we obtain

$$|W_y(x_0, y_0, z_0)| < \frac{M_2}{R_0^2}, \qquad |W_z(x_0, y_0, z_0)| < \frac{M_2}{R_0^2}.$$

$\underline{\text{Theorem 5.}}$ Let the infinite vortex sheet Σ have the following properties (Fig. 2.3):

1) Σ is cylindrical and its generators are parallel to the Ox axis $(-\infty < x < +\infty)$;
2) the cross section of Σ is finite;
3) the vortex axes are parallel to the Ox axis;
4) the strength γ of the vortices is independent of x;
5) the total strength of the vortex layer is zero:

$$\int\limits_0^{l_0} \gamma(l)\, dl = 0;$$

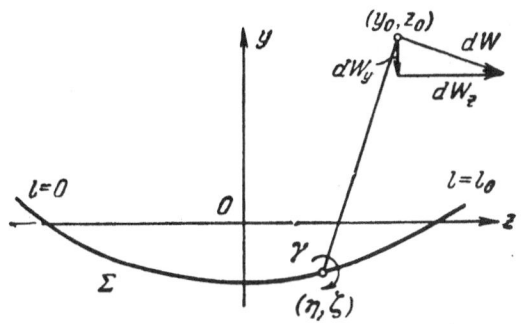

Fig. 2.3. The cylindrical vortex surface of Theorems 5 and 6.

6) the integral of $|\gamma(l)|$ is bounded:

$$\int_0^{l_0} |\gamma(l)|\,dl \leqslant M_1 = \text{const.}$$

Then the velocities $W_y(y_0, z_0)$ generated by the vortex surface Σ decrease towards infinity like $1/r_0^2$, where $r_0 = \sqrt{y_0^2 + z_0^2}$.

On the basis of Fig. 2.3 we have

$$W_y(y_0, z_0) = -\frac{1}{2\pi}\int_0^{l_0}\frac{\gamma(z_0-\zeta)}{r^2}\,dl, \quad W_z(y_0, z_0) = \frac{1}{2\pi}\int_0^{l_0}\frac{\gamma(y_0-\eta)}{r^2}\,dl, \tag{2.11}$$

where

$$r = \sqrt{(y_0-\eta)^2 + (z_0-\zeta)^2}.$$

But in the neighborhood of infinity ($y_0 \to \infty$, $z_0 \to \infty$) we have the expansions

$$\left.\begin{aligned}\frac{z_0-\zeta}{(y_0-\eta)^2+(z_0-\zeta)^2} &= \frac{z_0}{r_0^2} - \zeta\frac{y_0^2-z_0^2}{r_0^4} + \eta\frac{2y_0z_0}{r_0^4} + \cdots\\[2mm]\frac{y_0-\eta}{(y_0-\eta)^2+(z_0-\zeta)^2} &= \frac{y_0}{r_0^2} + \zeta\frac{2y_0z_0}{r_0^4} + \eta\frac{y_0^2-z_0^2}{r_0^4} + \cdots\end{aligned}\right\} \tag{2.12}$$

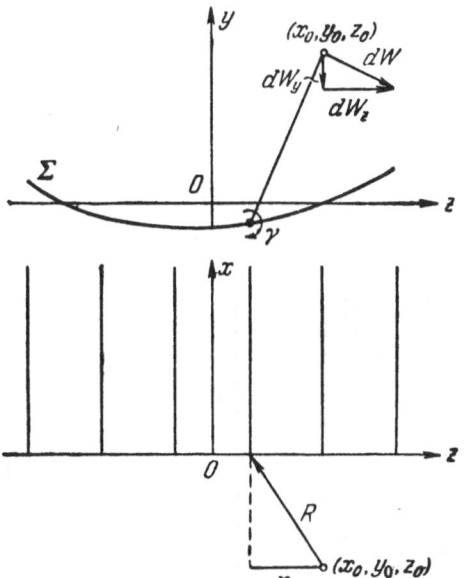

Fig. 2.4. The calculation of velocities generated by a vortex sheet Σ.

Hence

$$\left.\begin{aligned}W_y(y_0, z_0) &= \frac{1}{2\pi}\frac{y_0^2-z_0^2}{r_0^4}\int_0^{l_0}\gamma\zeta\,dl - \frac{1}{\pi}\frac{y_0z_0}{r_0^4}\int_0^{l_0}\gamma\eta\,dl + \cdots\\[2mm]W_z(y_0, z_0) &= \frac{1}{2\pi}\frac{y_0^2-z_0^2}{r_0^4}\int_0^{l_0}\gamma\eta\,dl + \frac{1}{\pi}\frac{y_0z_0}{r_0^4}\int_0^{l_0}\gamma\zeta\,dl +\end{aligned}\right\} \tag{2.13}$$

From the expansions (2.13) it follows that when the point (y_0, z_0) recedes from the vortex surface, the velocities $W_y(y_0, z_0)$ and $W_z(y_0, z_0)$ decreases like $1/r_0^2$.

Theorem 6. Let the infinite vortex sheet Σ have the following properties (Fig. 2.4):

1) Σ is cylindrical and its generators are parallel to the Ox axis ($0 \leq x < \infty$);

2) the cross section of Σ is finite;

3) the vortex axes are parallel to Ox;

4) the strength γ of the vortices is independent of x;

5) the total strength of the vortex layer is zero:

$$\int_0^{l_0} \gamma(l)\,dl = 0;$$

6) the integral of $|\gamma(l)|$ is bounded:

$$\int_0^{l_0} |\gamma(l)|\,dl \leqslant M_1 = \text{const.}$$

Then the velocities $W_y(x_0, y_0, z_0)$ and $W_z(x_0, y_0, z_0)$ generated by the vortex sheet Σ, decrease towards infinity in the half-space x < 0 like $1/R_0^2$, where

$$R_0 = \sqrt{x_0^2 + y_0^2 + z_0^2}.$$

and in the half-space x > 0 like $1/r_0^2$.

The velocity element dW at (x_0, y_0, z_0) due to the vortex filament $\gamma\,dl$ parallel to Ox will be perpendicular to the plane passing through the axis of the filament and the point (x_0, y_0, z_0), and so

$$\left.\begin{aligned}
W_y(x_0, y_0, z_0) &= \frac{1}{4\pi}\int_0^{l_0} \frac{\gamma(\zeta - z_0)}{r^2}\left[1 + \frac{x_0}{\sqrt{x_0^2 + (y_0 - \eta)^2 + (z_0 - \zeta)^2}}\right]dl, \\[2ex]
W_z(x_0, y_0, z_0) &= \frac{1}{4\pi}\int_0^{l_0} \frac{\gamma(y_0 - \eta)}{r^2}\left[1 + \frac{x_0}{\sqrt{x_0^2 + (y_0 - \eta)^2 + (z_0 - \zeta)^2}}\right]dl.
\end{aligned}\right\} \tag{2.14}$$

Using the expansion

$$\frac{x_0}{\sqrt{x_0^2 + (y_0 - \eta)^2 + (z_0 - \zeta)^2}} = \frac{x_0}{R_0} + \eta\,\frac{x_0 y_0}{R_0^3} + \zeta\,\frac{x_0 z_0}{R_0^3} + \dots.$$

and (2.12), after some transformations we obtain

$$\left.\begin{aligned}
W_y(x_0, y_0, z_0) &= \frac{1}{4\pi}\int_0^{l_0} \gamma\eta\,dl\left[\frac{2y_0 z_0}{r_0^2}\cdot\frac{1}{R_0(R_0 - x_0)} - \frac{x_0 y_0 z_0}{r_0^2 R_0^3}\right] + \\[1ex]
&\quad + \frac{1}{4\pi}\int_0^{l_0} \gamma\zeta\,dl\left[\frac{y_0^2 - z_0^2}{r_0^2}\cdot\frac{1}{R_0(R_0 - x_0)} - \frac{x_0 z_0^2}{r_0^2 R_0^3}\right] + \dots, \\[1ex]
W_z(x_0, y_0, z_0) &= \frac{1}{4\pi}\int_0^{l_0} \gamma\eta\,dl\left[\frac{y_0^2 - z_0^2}{r_0^2}\cdot\frac{1}{R_0(R_0 - x_0)} + \frac{x_0 y_0 z_0}{r_0^2 R_0^3}\right] + \\[1ex]
&\quad + \frac{1}{4\pi}\int_0^{l_0} \gamma\zeta\,dl\left[\frac{2y_0 z_0}{r_0^2}\cdot\frac{1}{R_0(R_0 - x_0)} + \frac{x_0 y_0 z_0}{r_0^2 R_0^3}\right] + \dots
\end{aligned}\right\} \tag{2.15}$$

Since

$$r_0^2 \geqslant 2y_0 z_0,$$
$$r_0^2 \geqslant y_0^2 - z_0^2,$$
$$R_0^2 \geqslant x_0^2,$$
$$R_0 - x_0 \geqslant R_0 \quad \text{for} \quad x_0 \leqslant 0,$$

we conclude from (2.15) that, for $x_0 \leq 0$, the perturbed velocities decrease like $1/R_0^2$ towards infinity.

Thus the inequalities

$$\frac{1}{R_0(R_0 - x_0)} = \frac{R_0 + x_0}{R_0(R_0^2 - x_0^2)} = \left(1 + \frac{x_0}{R_0}\right)\frac{1}{r_0^2} \leqslant \frac{2}{r_0^2}$$

and

$$\frac{1}{R_0^2} \leqslant \frac{1}{r_0^2},$$

show that, for $x_0 \geq 0$, the velocities $W_y(x_0, y_0, z_0)$ and $W_z(x_0, y_0, z_0)$ decrease towards infinity like $1/r_0^2$.

CHAPTER 3

SOME GENERALIZATIONS OF THE ZHUKOVSKII LIFT THEOREM. INDUCED DRAG

§ 1. The Theorem

The theorem of N. E. Zhukovskii concerning the lift of a wing, which he proved in 1905 and published in 1906 [1], has played an extremely important role in the development of aerodynamics. Zhukovskii was the first to give an explanation of the mechanism of the formation of lift. He showed that the lift on a cylindrical body in a plane-parallel flow is due to a circulation Γ of the velocity about a closed contour containing a wing section. He also gave an explanation of the Euler-d'Alembert paradox concerning the lack of any reaction due to the flow of perfect incompressible fluid, on a body in steady uniform motion. It was shown that this reaction would actually be zero if the circulation Γ were zero. For nonzero circulation, however, even for plane-parallel motion of a cylindrical body through a perfect incompressible fluid, there will be a lift

$$Y = \rho U_0 \Gamma l,$$

where l is the span of the body.

More than half a century has passed since Zhukovskii obtained this remarkable result. The problems aerodynamicists are called on to solve are now much more complex. In spite of this, the Zhukovskii theorem still plays an essential role in contemporary investigations in subsonic aerodynamics, since it can be used to obtain a simple method of determining aerodynamic loads in very complex problems. This theorem is at present still a powerful tool in the solution of many aerodynamic problems. There have been a great many investigations of possible generalizations of this theorem [2, 12, 14, 15, 16, 19, 43].

We obtain below formulas for the calculation of aerodynamic loads on a thin lifting surface of arbitrary shape in generally three-dimensional nonstationary motion. Two types of flow will be studied — flow with and flow without circulation, depending on the type of problem under discussion. For thin lifting surfaces, a very general form of Zhukovskii's theorem holds for flow with circulation if the instantaneous value of the lift is calculated for a surface element and if, for the velocity of the oncoming flow, we take the velocity of the fluid at a given time at the point of the lifting surface under consideration.

This result can be thought of as the Zhukovskii theorem "in the small," since it refers to an element of the lifting surface. The corresponding formula for flow without circulation has a more complex form, since it contains the derivative of the circulation with respect to the time. This formula, in analogy with the previous formula, will be called the Zhukovskii theorem "in the small" for flow without circulation.

29

§2. The Zhukovskii Theorem "in the Small" for Flow with Circulation

We will prove the Zhukovskii theorem "in the small" for arbitrary three-dimensional non-stationary motion. Let the infinitely thin lifting surface Σ be replaced by a vortex sheet with strength γ, continuously distributed over this surface. We will assume that the medium is perfect and incompressible and that the surface Σ is sufficiently smooth and not closed, so that the fluid encloses it on both sides. The flow about the lifting surface will be assumed to be smooth, and in connection with the hydrodynamics theorem concerning the constancy of the circulation, we will assume that a change with time of the strength of the bound vortices can only be caused by the formation and shedding of free vortices. The velocity of drift of free vortices at every point will be equal in magnitude and direction to the velocity of the fluid at this point at any given time.

Close to the surface Σ outside the vortex trail, the absolute flow of the fluid will be potential, with the absolute velocity

$$W = \operatorname{grad} \Phi.$$

The pressure p at any point can be obtained from Bernoulli's theorem (1.24)

When a point at which the relative velocity is calculated to be W_0 passes through the lifting surface Σ, this velocity has a discontinuity. Three values of the relative velocity should be distinguished: the limiting value from above (W_{0+}) which is obtained when the point approaches the upper side of the surface, the limiting value from below W_{0-}, and the velocity at the point s_0 on the surface (W_{00}).

The difference between the pressures on the lower and upper surfaces is, from (1.24),

$$\frac{\Delta p}{\rho} = \frac{\partial(\Phi_+ - \Phi_-)}{\partial t} + \frac{W_{0+}^2 - W_{0-}^2}{2}, \qquad (3.1)$$

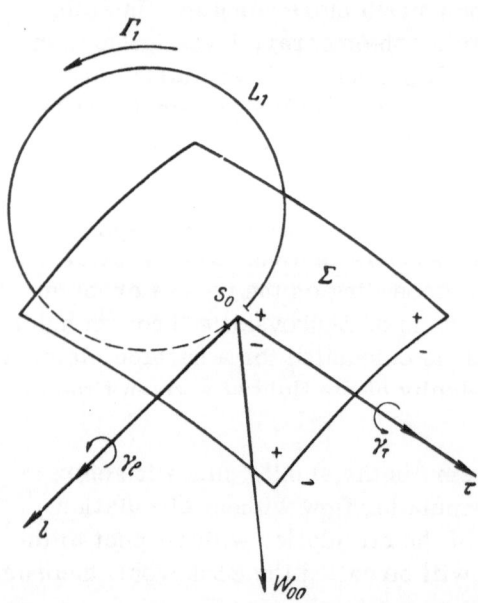

Fig. 3.1. The relation between the velocity circulation and the difference between the velocity potentials.

where Φ_+ and Φ_- are the limits of the potential of perturbed velocities obtained by approaching an arbitrary point s_0 of the lifting surface from above and from below.

We describe a closed contour L_1, intersecting the surface Σ at the single point s_0 (Fig. 3.1). Let the circulation of the disturbed velocity about the contour L_1 be Γ_1; then

$$\Gamma_1 = \int_{L_1} \overline{W}\, d\bar{s} = \int_{L_1} \frac{\partial \Phi}{\partial s}\, ds = \Phi_- - \Phi_+.$$

and so

$$\frac{\partial(\Phi_+ - \Phi_-)}{\partial t} = -\frac{\partial \Gamma_1}{\partial t}. \qquad (3.2)$$

To find the derivative $\partial\Gamma_1/\partial t$, we consider the variation during the time Δt in the circulation Γ_1 around the contour L_1 for a fixed point s_0. A change in the circulation Γ_1 by an amount ΔL_1 can

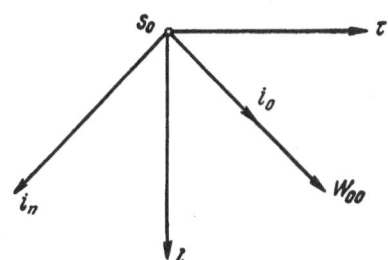

Fig. 3.2 The determination of the circulation carried away by free vortices.

be produced only by the departure of free vortices, some of which, surrounding the contour L_1, are carried away from the boundaries of this contour. These free vortices are of a completely nonstationary nature and have axes parallel to the axes of those bound vortices from which they were shed. In stationary motion there will be no free vortices of this type.

Let $\mathbf{W}_{00}$ be the relative flow velocity at the point s_0 which, in agreement with the assumption concerning the smoothness of the flow about the lifting surface, is tangential to Σ. The drift velocity of the free vortices at each point of the lifting surface Σ will be W_{00}. The total strength of the free vortices carried away during the time Δt from the limits of the contour L_1 will be $W_{00}\Delta t\,\gamma_{n-}$, where γ_{n-} is the component of the free-vortex strength in the direction of the unit vector $\mathbf{i}_n$ perpendicular to $\mathbf{W}_{00}$ (Fig. 3.2). Hence the axes of the free vortices γ_{n-} are perpendicular to the velocity $\mathbf{W}_{00}$.

At the point s_0 under consideration, we introduce the rectangular coordinate system (τ, s_0, l) in the plane tangent to the surface Σ. For l we take the direction of the bound vortex at the point s_0. The positive directions of γ_l and γ_τ are shown in Fig. 3.1. The total strength of the vortex sheet Σ will be calculated from the strengths of the bound vortices γ_+ and the free vortices γ_-, where

$$\gamma_l = \gamma_{l+} + \gamma_{l-}, \qquad \gamma_\tau = \gamma_{\tau-}. \tag{3.3}$$

If $\mathbf{i}_0$ is the unit vector in the direction of $\mathbf{W}_{00}$, then writing W_{00l} and $W_{00\tau}$ for the projections of $\mathbf{W}_{00}$ on l and τ, we obtain

$$\mathbf{i}_0 = \mathbf{i}_l \frac{W_{00l}}{W_{00}} + \mathbf{j}_\tau \frac{W_{00\tau}}{W_{00}},$$

where $\mathbf{i}_l$ and $\mathbf{j}_\tau$ are unit vectors in the directions of l and τ.

The unit vector $\mathbf{i}_n$, perpendicular to $\mathbf{i}_0$ and corresponding to the positive direction of Γ_1 (Fig. 3.1), will be given by the formula

$$\mathbf{i}_n = \mathbf{i}_l \frac{W_{00\tau}}{W_{00}} - \mathbf{j}_\tau \frac{W_{00l}}{W_{00}},$$

and so

$$\gamma_{n-} = \gamma_{l-} \frac{W_{00\tau}}{W_{00}} - \gamma_{\tau-} \frac{W_{00l}}{W_{00}}$$

and

$$W_{00}\gamma_{n-} = \gamma_{l-} W_{00\tau} - \gamma_{\tau-} W_{00l}.$$

For the change $\Delta\Gamma_1$ in the circulation during the time Δt we can write

$$\Delta\Gamma_1 = -(\gamma_{l-} W_{00\tau} - \gamma_{\tau-} W_{00l})\Delta t,$$

and so

$$\frac{\partial\Gamma_1}{\partial t} = -(\gamma_{l-} W_{00\tau} - \gamma_{\tau-} W_{00l}). \tag{3.4}$$

To find the term $(W_{0+}^2 - W_{0-}^2)/2$ in (3.1 we use the relations (2.8), which in the present case can be written (Fig. 3.1)

$$W_{0l+} - W_{0l-} = \gamma_\tau, \qquad W_{0\tau+} - W_{0\tau-} = -\gamma_l,$$

$$\frac{W_{0l+} + W_{0l-}}{2} = W_{00l}, \qquad \frac{W_{0\tau+} + W_{0\tau-}}{2} = W_{00\tau}; \tag{3.5}$$

thus

$$\frac{W_{0+}^2 - W_{0-}^2}{2} = \gamma_{\tau-} W_{00l} - \gamma_{l-} W_{00\tau} - \gamma_{l+} W_{00\tau}. \tag{3.6}$$

From (3.1), (3.2), (3.4), and (3.6), we now obtain

$$p_- - p_+ = \Delta p = -\rho \gamma_{l+} W_{00\tau}. \tag{3.7}$$

where the positive direction of γ_{l+} is shown in Fig. 3.1. It is clear from (3.7) that the pressure difference generated on an element of the upper surface is determined by the linear strength of the bound vortices only. Hence in particular it follows that on the vortex sheets consisting of free vortices which are formed behind a body of finite span both in steady and in unsteady motion there are no pressure drops.

Equation (3.7) is a generalization of Zhukovskii's theorem "in the small" to the case of arbitrary nonstationary motion. It is applicable to any thin isolated lifting surface, to systems of surfaces, to combinations of surfaces and bodies, etc.

The relation (3.7), among others, shows that any distribution of the load Δp can be reproduced by a vortex sheet.

§3. The Zhukovskii Theorem "in the Small" for Irrotational Flow

As in the previous case, we consider three-dimensional nonstationary motion of an infinitely thin lifting surface Σ. We assume that the medium is perfect and incompressible and that the surface Σ is sufficiently smooth and not closed, so that the fluid can flow smoothly over both of its sides.

In contrast to Section 2, however, we investigate here the case of irrotational flow about the lifting surface. In this case no vortex trail will be formed behind the body, and the theorem concerning the constancy of the circulation about a closed fluid contour holds because the circulation about any contour containing the lifting surface will be zero. Hence the vortex layer replacing the lifting surface can be thought of as a vortex system formed by a combination of closed vortex lines (elementary strips) of constant strength along the extent of the line. This whole vortex system is fixed in the surface Σ, and it must be considered as a system of bound vortices. In flow without circulation about a lifting surface, we may thus assume that no free vortices are formed either on the surface Σ or behind it.

We now determine the pressure difference Δp at an arbitrary point s_0 of the lifting surface Σ (Fig. 3.3). We retain the same notation for the limiting relative velocities: W_{0-} is the limit on the lower surface; W_{0+} is the limit on the upper surface; W_{00} is the relative velocity at the point s_0.

With s_0 as origin, we define arbitrary rectangular-coordinate axes $s_0\tau$ and $s_0 l$ in the plane tangent to Σ. The pressure difference Δp between the lower and upper surfaces is, in agreement with Bernoulli's theorem (1.24), given by (3.1). We express the derivative in terms

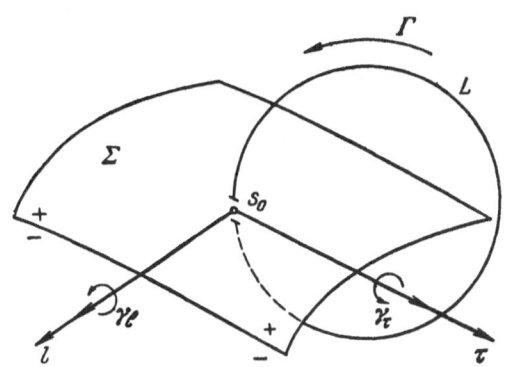

Fig. 3.3. The derivation of the Zhukovskii theorem for irrotational flow.

of the strengths of the vortices which, in the problems studied below, are determined by the solution of these problems. It is convenient to consider the contour L shown in Fig. 3.3; then

$$\Phi_- - \Phi_+ = -\Gamma,$$

where Γ is the circulation about the contour L which intersects the lifting surface at s_0. The circulation Γ can be obtained as the sum of the strengths of vortices between the trailing edge and the point s_0. We can thus write

$$\frac{\partial(\Phi_+ - \Phi_-)}{\partial t} = \frac{\partial \Gamma}{\partial t}, \tag{3.8}$$

and, from (3.6), we have

$$\frac{W_{0+}^2 - W_{0-}^2}{2} = \gamma_\tau W_{00l} - \gamma_l W_{00\tau}.$$

The use of this relation and (3.8) in (3.1) yields

$$p_- - p_+ = \Delta p = \rho \left(W_{00l} \gamma_\tau - W_{00\tau} \gamma_l + \frac{\partial \Gamma}{\partial t} \right). \tag{3.9}$$

This formula expresses Zhukovskii's theorem "in the small" for irrotational flow. A comparison of (3.7) and (3.9) easily shows that the expression for the pressure difference in the first case, for rotational flow, is simpler than that in the other case.

This same theorem could also be obtained from the relations of the preceding section by noting that

$$\gamma_{l-} = 0, \qquad \gamma_{l+} = \gamma_l,$$
$$\gamma_{\tau-} = \gamma_\tau, \qquad \Gamma = -\Gamma_1.$$

It should be stressed that (3.7) and (3.9) can only be used in the determination of the normal components of the aerodynamic loads, and that they cannot be used to obtain the so-called suction force [14, 16].

In a perfect incompressible medium, the suction forces are generated by flow around sharp edges of the lifting surface, and are in the direction of the tangent to the surface at the point of the sharp edge under consideration. Their existence is due to the infinitely high velocities and rarefactions in the flow of a perfect fluid around sharp edges.

§ 4. The Zhukovskii Theorem for a Profile in a Cascade

The lift of a cascade of airfoils without suction forces can be determined on the basis of Zhukovskii's theorem "in the small" for rotational flow. However, if we use the momentum theorem, it is not difficult in the present case to find an expression for the lift with suction force, and also to obtain certain other useful relations [2, 7, 8, 9, 12, 16]. We will investigate this problem, assuming a perfect incompressible medium and stationary flow.

We take a cascade of arbitrary profiles (not necessarily thin) with the $O\eta$ axis directed towards the front of the grid and the $O\xi$ axis perpendicular to the front (Fig. 3.4). We reverse

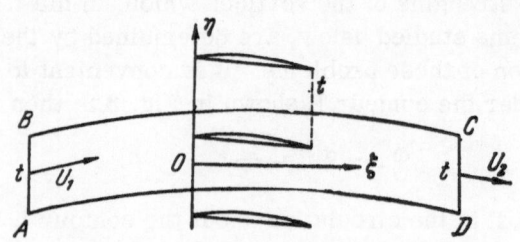

Fig. 3.4. The derivation of Zhukovskii's for a profile cascade.

the motion and denote the velocity of the flow at infinity in front of and behind the cascade by U_1 and U_2, and the pressures by p_1 and p_2, (for an infinite number of profiles, the velocities U_1 and U_2 and also the pressures p_1 and p_2 can be different).

The flow between the profiles in a cascade will be periodic with period t. We draw two streamlines AD and BC at a distance t from one another so that there is a profile between them. These streamlines will of course be congruent.

We calculate the circulation Γ_+ of the velocity about the closed contour ABCD containing the profile, with AB and CD parallel to $O\eta$. Since

$$\Gamma_{BC} + \Gamma_{DA} = 0,$$

if we assume that the velocities and pressures are equalized at infinity we have

$$\Gamma_+ = t(U_{1\eta} - U_{2\eta}), \tag{3.10}$$

where $U_{1\eta}$ and $U_{2\eta}$ are the projections of U_1 and U_2 on the $O\eta$ axis.

The divergence equation for the volume bounded by the cylindrical surface ABCD and two planes parallel to the plane of the diagram at distance unity from one another yields

$$\rho U_{1\xi} t = \rho U_{2\xi} t,$$

where $U_{1\xi}$ and $U_{2\xi}$ are the projections of U_1 and U_2 on the $O\xi$ axis, and so

$$U_{1\xi} = U_{2\xi}. \tag{3.11}$$

From (3.10) and (3.11) it is clear that for irrotational flow about a profile grid with $\Gamma_+ = 0$, the velocities U_1 and U_2 at infinity are equal. When $\Gamma_+ \neq 0$, these velocities are different because of the difference between the components parallel to the grid front.

As we will show below, an important role is played in cascade theory by the quantity equal to one half the geometrical sum of the velocities U_1 and U_2. We will assume that the only sources of disturbance in the medium are the bound vortices of the profile, i.e., the cascade has no free end vortices.

To find the disturbed velocities at large distances from a cascade, we can replace each profile by an attached vortex with circulation Γ_+ (Fig. 3.5).

We can thus write (j' is the unit vector in the direction $O\eta$)

$$U_1 = U_0 + j' \frac{\Gamma_+}{2t}, \qquad U_2 = U_0 - j' \frac{\Gamma_+}{2t}, \tag{3.12}$$

and so

$$\frac{U_1 + U_2}{2} = U_0. \tag{3.13}$$

Let **Y** be the vector describing the lift per unit span of a profile in the cascade; then **Y** will be the force acting from the side of the profile on the fluid. We apply the momentum

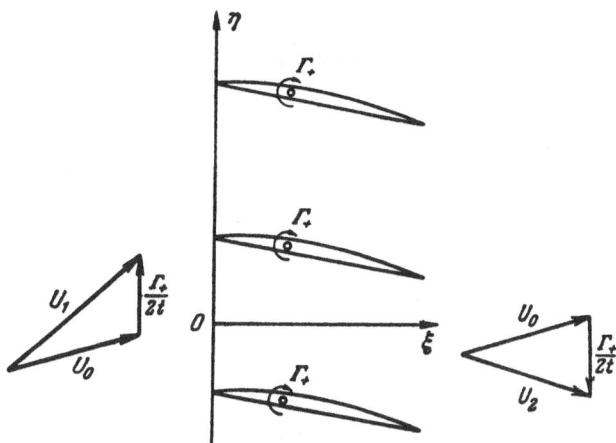

Fig. 3.5. Relative velocities of the medium at infinity in front of and behind a cascade.

theorem to the mass of fluid inside the cylindrical surface ABCD which has generators of unit length. The mass flowing into this volume per second is

$$m = \rho U_{1\xi} t,$$

and the momentum theorem yields

$$m(U_2 - U_1) = -Y + i'(p_1 - p_2) t, \qquad (3.14)$$

where i' is the unit vector in the direction of $O\xi$.

We project Eq. (3.14) on the $O\xi$ and $O\eta$ axes, setting

$$Y = i'X' + j'Y' \qquad (3.14)$$

and use the relation (3.11):

$$X' = (p_1 - p_2) t, \qquad Y' = \rho U_{1\xi} t (U_{1\eta} - U_{2\eta}). \qquad (3.15)$$

By using Bernoulli's equation

$$p_1 + \frac{\rho U_1^2}{2} = p_2 + \frac{\rho U_2^2}{2}$$

we express the pressure difference $p_1 - p_2$ in terms of U_1 and U_2:

$$p_1 - p_2 = \rho (U_{2\eta} - U_{1\eta}) \frac{U_{1\eta} + U_{2\eta}}{2}.$$

Using (3.13), we obtain

$$p_1 - p_2 = \rho U_{0\eta} (U_{2\eta} - U_{1\eta}), \qquad (3.16)$$

where $U_{0\eta}$ is the projection of the translational velocity U_0 of the cascade on the $O\eta$ axis (in other words, the mean velocity of flow in rotational motion). Replacing the difference $U_{1\eta} - U_{2\eta}$ by Γ_+/t according to (3.10) and using (3.16), we obtain instead of (3.15) the relations

$$X' = -\rho U_{0\eta} \Gamma_+, \qquad Y' = \rho U_{0\xi} \Gamma_+, \qquad (3.17)$$

where (3.11) and (3.13) show that $U_{1\xi} = U_{2\xi} = U_{0\xi}$.

The lift is $Y = \sqrt{X'^2 + Y'^2}$ or, from (3.17),

$$Y = \rho U_0 \Gamma_+. \tag{3.18}$$

since

$$U_0 = \sqrt{U_{0\xi}^2 + U_{0\eta}^2}.$$

The direction of the lift Y can be obtained by rotating the mean-velocity vector U_0 (Fig. 3.5) through 90° in the direction opposite to that of the circulation Γ_+. To show this, we multiply the first equation in (3.17) by $U_{0\xi}$ and the second by $U_{0\eta}$ and; this yields

$$X'U_{0\xi} + Y'U_{0\eta} = 0. \tag{3.19}$$

We note that, from (3.17) with $\Gamma_+ > 0$, $U_{0\xi} > 0$ and $U_{0\eta} > 0$, we have $X' < 0$ and $Y' > 0$.

§5. Suction Force

As has already been noted above, in the theoretical analysis of smooth flow over an infinitely thin lifting surface a suction force develops, caused by the flow rounding the thin leading edge of the surface.

At a leading edge, the velocities and rarefactions are theoretically infinite and although the thickness of the wing is infinitesimal the pressure has a component in the direction of the tangent to the wing surface at the leading edge. Calculations carried out for the simplest cases show that the suction force depends weakly on the wing thickness. We may thus assume that its theoretical value obtained for an infinitely thin lifting surface can be used in practice for any thickness.

The suction force explains the paradox obtained when we compare the results of applying the Zhukovskii theorem "in the small" and the Zhukovskii theorem in the form (3.18) for a thin profile. For simplicity, we actually consider an isolated plate of infinite span. From (3.19), its lift Y will be perpendicular to the velocity U_0 of the oncoming flow (Fig. 3.6). On the other hand, if we use Eq. (3.7) for the pressure difference Δp and integrate this difference over the plate, we obtain the normal force N perpendicular to the plate. The difference is caused by the existence of the suction force Q, which is not taken into account in the second case but which is automatically taken into account in the first, based on the momentum theorem.

In real flow over a wing with a very thin leading edge there is of course no infinite rarefaction. For low angles of attack, there is simply a local separation of the flow at the leading edge, leading to a small effect on the lift, but to a very important effect on the drag of the wing. Because of this local separation of the flow, there is practically no suction force and the wing drag is increased. If the leading edge is properly shaped, however, so that it is rounded sufficiently to prevent the existence of flow separation, then a suction force must be taken into account in the determination of the drag. As has already been indicated, the suction force can be approximately determined from data obtained for an infinitely thin lifting surface.

We can thus obtain two limiting theoretical values for the drag of a wing — by taking into account or by not taking into account the suction force. Depending on the way the leading edge is shaped, experimental results, which will be between the two limiting theoretical values, may be close to one or the other.

If we know the magnitude and direction of the normal force N, it is not difficult to find the resistance X_- of an arbitrary lifting surface without suction force. To do this, it is sufficient

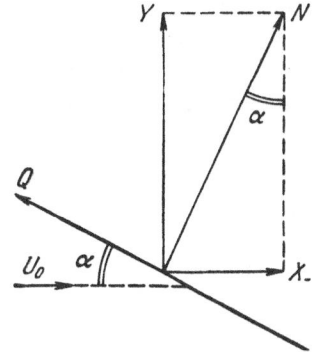

Fig. 3.6. The forces acting on a monoplane wing plate (Q is the suction force, N the normal force, Y the lift, and X_ the induced drag without suction force).

to calculate the projection of N on the direction of the velocity U_0. This is especially simple when the lifting surface is a plate of arbitrary plan form. In this case (Fig. 3.6), for any Mach number M_∞ (and not only for an incompressible medium), we have

$$X_- = N \sin \alpha,$$

or, for small angles of attack α,

$$X_- = Y\alpha. \tag{3.20}$$

It is not difficult to see that this formula will also be valid for annular wings.

For an isolated profile of infinite span or a profile cascade, when there is no resistance in a perfect incompressible medium, the component X_ will be exactly equal to the projection of Q on U_0.

§ 6. General Relations for the Induced Drag of a Lifting Surface

We now study the determination of the drag of a wing of finite span moving in a perfect incompressible medium with a translational motion U_0, constant in magnitude and direction. At subsonic speeds, when there is still no shock wave at the wing surface, the behavior of the drag is related to the presence of a sheet of free vortices extending from behind the wing to infinity. This drag is conventionally called induced drag. The appearance of shock waves can lead to a change in the strength of the bound and free vortices, and this will have an effect on the induced drag. The formulas given below can be used to take this change into account if the strength of the vortices is known. In addition to the fact that there is an irreversible loss at discontinuities, these discontinuities can lead to an extra (wave) drag. The calculation of this drag is a special problem.

We reverse the motion; let the oncoming flow have velocity U_0 (Fig. 3.7). We construct an arbitrary closed surface Σ containing the wing. We consider a coordinate system Oxyz with the Ox axis in the direction of the velocity U_0. In agreement with linear theory we assume that the free vortices are parallel to the undisturbed flow velocity U_0. We denote the induced drag, including the suction force, by X_+, let n be the normal external to Σ (relative to the internal region bounded by this surface). To the gas inside Σ we apply the momentum theorem

$$-X_+ - \iint_\Sigma p \cos(n, x)\, d\sigma = \iint_\Sigma \rho W_{0n} W_{0x}\, d\sigma. \tag{3.21}$$

If Φ is the potential of the perturbed velocities, then the projections of the relative velocities are

$$W_{0x} = U_0 + \frac{\partial \Phi}{\partial x}, \qquad W_{0y} = \frac{\partial \Phi}{\partial y}, \qquad W_{0z} = \frac{\partial \Phi}{\partial z},$$

$$W_{0n} = U_0 \cos(n, x) + \frac{\partial \Phi}{\partial n}. \tag{3.22}$$

For isentropic flow of a compressible gas, the Bernoulli equation is

$$\frac{k}{k-1} \cdot \frac{p}{\rho} + \frac{W_{0x}^2 + W_{0y}^2 + W_{0z}^2}{2} = \frac{k}{k-1} \frac{p_\infty}{\rho_\infty} + \frac{U_0^2}{2},$$

where k is the adiabatic exponent and p_∞ and ρ_∞ are the pressure and the density of the undisturbed gas. We assume that the disturbed velocities $\partial\Phi/\partial x$, $\partial\Phi/\partial y$, and $\partial\Phi/\partial z$ are small quantities of the first order, and then the principal term in the expression for the induced drag will consist of small quantites of the second order. This principal term is determined on the basis of the solution of the linear problem. Hence in all the following formulas we will retain only small quantities of the first and second order and discard higher order terms.

After some transformations using the relation $kp_\infty/\rho_\infty = a_\infty^2$, where a_∞ is the sound velocity in the undisturbed flow, the Bernoulli equation can be written

$$\frac{p}{p_\infty} = \frac{\rho}{\rho_\infty}\left\{ 1 - \frac{k-1}{2a_\infty^2}\left[2U_0\frac{\partial\Phi}{\partial x} + \left(\frac{\partial\Phi}{\partial x}\right)^2 + \left(\frac{\partial\Phi}{\partial y}\right)^2 + \left(\frac{\partial\Phi}{\partial z}\right)^2 \right] \right\}.$$

But

$$\frac{p}{p_\infty} = \left(\frac{\rho}{\rho_\infty}\right)^k, \tag{3.23}$$

and so

$$\frac{p}{p_\infty} = \left\{ 1 - \frac{k-1}{2a_\infty^2}\left[2U_0\frac{\partial\Phi}{\partial x} + \left(\frac{\partial\Phi}{\partial x}\right)^2 + \left(\frac{\partial\Phi}{\partial y}\right)^2 + \left(\frac{\partial\Phi}{\partial z}\right)^2 \right] \right\}^{\frac{k}{k-1}}.$$

If we retain small terms up to the second order inclusive in this equation, we obtain

$$\frac{p}{p_\infty} = 1 - \frac{k}{2a_\infty^2}\left[2U_0\frac{\partial\Phi}{\partial x} + \left(1 - \frac{U_0^2}{a_\infty^2}\right)\left(\frac{\partial\Phi}{\partial x}\right)^2 + \left(\frac{\partial\Phi}{\partial y}\right)^2 + \left(\frac{\partial\Phi}{\partial z}\right)^2 \right]. \tag{3.24}$$

From (3.22) and (3.24), we obtain from (3.21) the formula

$$X_+ = -\int_\Sigma\int p\cos(n,\,x)\,d\sigma - \int_\Sigma\int \rho W_{0n}W_{0x}\,d\sigma =$$

$$= \frac{\rho_\infty}{2}\int_\Sigma\int \cos(n,\,x)\left[\left(1 - \frac{U_0^2}{a_\infty^2}\right)\left(\frac{\partial\Phi}{\partial x}\right)^2 + \left(\frac{\partial\Phi}{\partial y}\right)^2 + \left(\frac{\partial\Phi}{\partial z}\right)^2\right]d\sigma -$$

$$- U_0\int_\Sigma\int (\rho - \rho_\infty)\cos(n,\,x)\frac{\partial\Phi}{\partial x}\,d\sigma - \int_\Sigma\int \rho\frac{\partial\Phi}{\partial x}\frac{\partial\Phi}{\partial n}\,d\sigma, \tag{3.25}$$

since

$$\int_\Sigma\int \rho W_{0n}\,d\sigma = 0, \quad \int_\Sigma\int \cos(n,\,x)\,d\sigma = 0.$$

We now express the difference $\rho - \rho_\infty$ in terms of the disturbed velocities. Since the product $(\rho - \rho_\infty)(\partial\Phi/\partial x)$ occurs in (3.25), we need only determine $(\rho - \rho_\infty)$ correct to the first order. Using the adiabatic relation (3.23), we obtain

$$p = p_\infty\left(\frac{\rho_\infty + \rho - \rho_\infty}{\rho_\infty}\right)^k = p_\infty\left(1 + \frac{\rho - \rho_\infty}{\rho_\infty}\right)^k,$$

or, retaining only small quantities of the first order, we have

$$p - p_\infty = a_\infty^2(\rho - \rho_\infty). \tag{3.26}$$

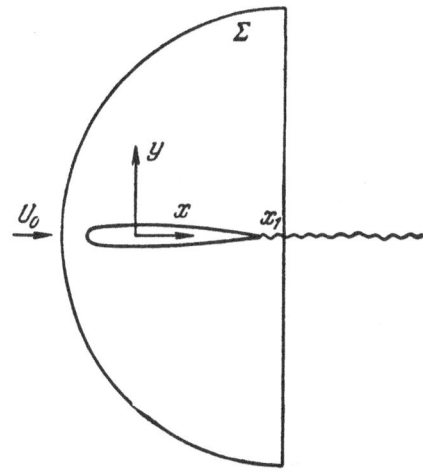

Fig. 3.7. Calculation of induced drag according to the momentum theorem.

Recalling Eq. (3.24) and retaining only small quantities of the first order there, from the linearized adiabatic relation (3.23) we obtain

$$p - p_\infty = -\frac{\rho_\infty}{a_\infty^2} U_0 \frac{\partial \Phi}{\partial x}.$$

The expression for the induced drag, with the suction force taken into account, now becomes

$$X_+ = \frac{\rho_\infty}{2} \int\int_\Sigma \cos(n, x)\left[(1 - \mathbf{M}_\infty^2)\left(\frac{\partial \Phi}{\partial x}\right)^2 + \left(\frac{\partial \Phi}{\partial y}\right)^2 \right.$$
$$\left. + \left(\frac{\partial \Phi}{\partial z}\right)^2\right] d\sigma + \mathbf{M}_\infty^2 \rho_\infty \int\int_\Sigma \cos(n, x)$$
$$\left(\frac{\partial \Phi}{\partial x}\right)^2 d\sigma - \int\int_\Sigma p \frac{\partial \Phi}{\partial x} \frac{\partial \Phi}{\partial n} d\sigma, \tag{3.27}$$

where $\mathbf{M}_\infty^2 = U_0^2/a_\infty^2$ is the Mach number of the undisturbed flow.

For a control surface Σ, we now use the hemisphere with center at the point with coordinates $x = x_1$, $y = 0$, $z = 0$ and the corresponding part of the plane $x_1 = $ const (see Fig. 3.7). We let the radius of the hemisphere increase indefinitely with x_1 fixed. From Theorems 4 and 6 of Chapter II, the integrals in (3.27) taken over the hemisphere are zero. The integrations must be taken over the whole plane $x_1 = $ const, and these integrals will exist; this is shown by the theorems just referred to.

But on the plane $x_1 = $ const we have

$$\cos(n, x) = 1, \qquad \frac{\partial \Phi}{\partial n} = \frac{\partial \Phi}{\partial x},$$

and so instead of (3.27) we have

$$X_+ = \frac{\rho_\infty}{2} \int\int\left[(1 - \mathbf{M}_\infty^2)\left(\frac{\partial \Phi}{\partial x}\right)^2 + \left(\frac{\partial \Phi}{\partial y}\right)^2 + \left(\frac{\partial \Phi}{\partial z}\right)^2\right] d\sigma + \mathbf{M}_\infty^2 \rho_\infty \int\int\left(\frac{\partial \Phi}{\partial x}\right)^2 d\sigma - \int\int p\left(\frac{\partial \Phi}{\partial x}\right)^2 d\sigma,$$

where the integration is over the plane $x_1 = $ const.

When x_1 increases indefinitely (and the plane $x_1 = $ const moves to infinity behind the wing), we obtain

$$X_+ = \frac{\rho_\infty}{2} \int\int\left[\left(\frac{\partial \Phi}{\partial y}\right)^2 + \left(\frac{\partial \Phi}{\partial z}\right)^2\right] d\sigma, \tag{3.28}$$

since here $\partial \Phi/\partial x$ tends to zero like $1/R_0^2$ (see Theorem 4 of Chapter II). Theorem 5 of Chapter II now shows that the integral of (3.28), taken over the whole Oyz plane located at an infinite distance behind the wing, has a meaning.

§7. The Induced Drag of a Monoplane Wing

We consider a monoplane wing of arbitrary plan form. The linearized equation for the potential of the disturbed velocities is

$$(1 - \mathbf{M}_\infty^2)\frac{\partial^2 \Phi}{\partial x^2} + \frac{\partial^2 \Phi}{\partial y^2} + \frac{\partial^2 \Phi}{\partial z^2} = 0. \tag{3.29}$$

When $x_1 \to \infty$ the derivative $\partial\Phi/\partial x$ tends to zero, and so in the infinitely distant plane we have

$$\frac{\partial^2\Phi}{\partial x^2} + \frac{\partial^2\Phi}{\partial y^2} = 0, \tag{3.30}$$

i.e., the velocity potential will be a harmonic function.

The intersection of the trailing vortex sheet of the wing with the plane $x_1 =$ const is along the segment $Oz[-l/2, \; l/2]$, of the Oz axis, where l is the wing span. Using Green's formula and noting that the potential Φ is a harmonic function, we obtain

$$\int\int \left[\left(\frac{\partial\Phi}{\partial x}\right)^2 + \left(\frac{\partial\Phi}{\partial y}\right)^2\right] dy\, dz = \int_{(l)} \Phi\, \frac{\partial\Phi}{\partial n}\, dl.$$

The positive direction of the normal n for the closed contour l is shown in Fig. 3.8. Through the vortex sheet $[-l/2, \; l/2]$ the normal component W_n of the disturbed velocity varies continuously and the normals $\mathbf{n}_+$ and $\mathbf{n}_-$ to the contour l have opposite signs, so

$$W_y = W_{n-} = -W_{n+}.$$

We can now write

$$\int_{(l)} \Phi\, \frac{\partial\Phi}{\partial n}\, dl = \int_{-l/2}^{l/2} (\Phi_+ W_{n+} + \Phi_- W_{n-})\, dz = -\int_{-l/2}^{l/2} (\Phi_+ - \Phi_-)\, W_y\, dz = -\int_{-l/2}^{l/2} \Gamma(z)\, W_y(z)\, dz,$$

where the positive direction of $\Gamma(z)$ in the z section is shown in Fig. 3.8.

Formula (3.28) now becomes

$$X_+ = -\frac{\rho_\infty}{2} \int_{-l/2}^{l/2} \Gamma(z)\, W_y(z)\, dz. \tag{3.31}$$

The strength $\Gamma(z)$ of the free vortices at any section z is equal to the strength of the bound vortices at this section. The velocities $W_y(z)$ must be calculated from the free vortices in the Oyz plane infinitely distant behind the wing, i.e., from the plane vortex strip $[-l/2, \; l/2]$ of the infinite line.

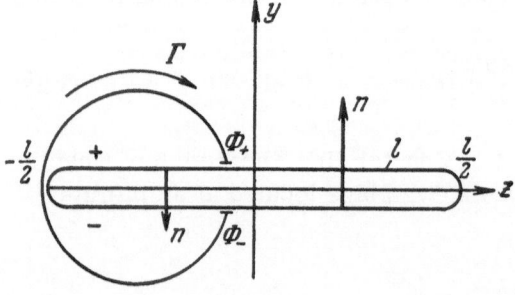

Fig. 3.8. The transformation of the double integral into a line integral for a monoplane wing.

§8. The Induced Drag of an Annular Wing

In this case, the formula (3.28) for the induced drag with suction force, and also Eqs. (3.29) and (3.30) will hold. However, in contrast to the previous case, the intersection of the infinitely distant $\bar{y}z$ plane with the vortex sheet will not be the segment $[-l/2, \; l/2]$. Since the vortex sheet is a semi-infinite circular cylinder for an annular wing in the linear problem, it follows that we can assume that in this scheme the trace from the intersection will be a circle of radius r.

Let Σ_1 and Σ_2 be the internal and external regions determined by this circle (Fig. 3.9). We transform the right-hand side of (3.28); since Φ_- is harmonic, we have

$$\int\!\!\int_{\Sigma_1} \left[\left(\frac{\partial \Phi}{\partial y}\right)^2 + \left(\frac{\partial \Phi}{\partial z}\right)^2 \right] dy\, dz = \int_{(l)} \Phi_- W_{n-}\, dl.$$

Here Φ_- is the limit approached by Φ at a point that approaches l, remaining in Σ_1. The positive direction is chosen so that Σ_1 is on the left, n_- is the normal to l, external to Σ_1, and W_{n-} is the normal component of the disturbed velocity.

We similarly transform the integral through Σ_2:

$$\int\!\!\int_{\Sigma_2} \left[\left(\frac{\partial \Phi}{\partial y}\right)^2 + \left(\frac{\partial \Phi}{\partial z}\right)^2 \right] dy\, dz = \int_{(l)} \Phi_+ W_{n+}\, dl.$$

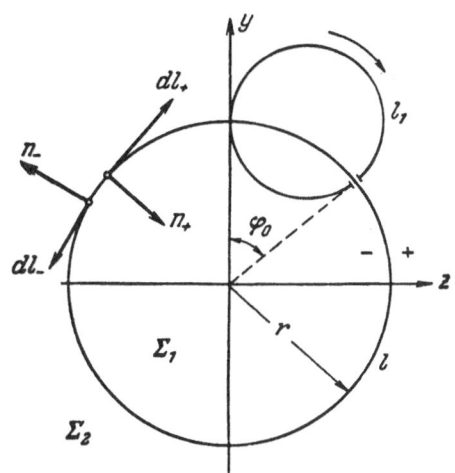

Fig. 3.9. The transformation of the double integral into a line integral for an annular wing.

From (3.28), since $W_{n+} = -W_{n-}$, we have

$$X_+ = \frac{\rho_\infty}{2} \int\!\!\int_{\Sigma_1 + \Sigma_2} \left[\left(\frac{\partial \Phi}{\partial y}\right)^2 + \left(\frac{\partial \Phi}{\partial z}\right)^2 \right] dy\, dz = \frac{\rho_\infty}{2} \int_{(l)} (\Phi_+ - \Phi_-) W_{n+}\, dl. \tag{3.32}$$

We denote by $\Gamma_1(\varphi_0)$ the circulation around the contour l_1 which intersects the contour l for $\varphi = 0$ and $\varphi = \varphi_0$ (Fig. 3.9), and then $\Phi_+ - \Phi_- = \Gamma_1(\varphi_0)$, where the positive direction around l_1 is shown in the diagram. We move the contour l_1 along the free vortices parallel to their axes up to the bound vortices. We do not intersect the vortices, and so $\Gamma_1(\varphi_0)$ will not change. We now consider a closed contour l_2 in the meridian plane for the angle φ_0 surrounding all the bound vortices. We denote the circulation around l_2 by $\Gamma(\varphi_0)$. For the contour l_2 to be converted into l_1, it must intersect all the bound vortices in the section $\varphi = 0$. The circulation changes to $\Gamma(0)$ and so

$$\Gamma(\varphi_0) = \Gamma_1(\varphi_0) + \Gamma(0),$$

hence

$$\Phi_+ - \Phi_- = \Gamma(\varphi_0) - \Gamma(0).$$

Substituting the resulting expression in (3.32) and noting that $dl = r\, d\varphi_0$ and $W_{n+} = -W_r$, we obtain

$$X_+ = -\frac{\rho_\infty}{2} r \int_0^{2\pi} [\Gamma(\varphi_0) - \Gamma(0)]\, W_r(\varphi_0)\, d\varphi_0. \tag{3.33}$$

CHAPTER IV

SOME GENERAL PROPERTIES OF AERODYNAMIC COEFFICIENTS

§ 1. Monoplane Wings

In the linearized problem, the dependence of the basic aerodynamic characteristics on the dimensionless kinematic parameters (1.19) is linear. The only exception is the induced drag, which is a small quantity of the second order, and for which this dependence is quadratic. It is convenient to separate explicitly these dependences of the aerodynamic coefficients on the kinematic parameters.

The aerodynamic load on the wing, i.e., the difference Δp between the pressures on the upper and lower surfaces, can be obtained by using the Zhukovskii theorem "in the small" (3.7). When small quantities of the second and higher orders are discarded, this formula becomes

$$\Delta p = \rho U_0 \gamma_{z+}, \tag{4.1}$$

where γ_{z+} is the strength of the attached layer. The positive direction of rotation of the vortices γ_{z+} with axes parallel to the Oz axis is shown in Fig. 4.1.

In agreement with the above, and recalling the boundary condition (1.30), for a plate of arbitrary shape we can write

$$\gamma_{z+}(x, z) = U_0 \left[\gamma_z^\alpha(x, z) \alpha + \gamma_z^{\omega_z}(x, z) \omega_z + \gamma_z^{\omega_x}(x, z) \omega_x \right], \tag{4.2}$$

where $\gamma_z^\alpha(x, z)$, $\gamma_z^{\omega_z}(x, z)$, $\gamma_z^{\omega_x}(x, z)$ are dimensionless functions.

A rigorous proof that it is possible to express the function γ_{z+} in the form (4.2) can be obtained by an analysis of problems arising in the determination of the dimensionless functions. This proof follows from the existence and uniqueness of the solution of such problems. As we will see below, the determination of $\gamma_z^\alpha(x, z)$, $\gamma_z^{\omega_z}(x, z)$, $\gamma_z^{\omega_x}(x, z)$ reduces to the solution of corresponding integral equations.

We must stress that the representation considered for rotational flow is valid only for stationary flow in which the kinematic parameters (1.19) are independent of the time. In addition to progressive motion of a wing with constant angle of attack α and with $\omega_x = \omega_z = 0$, the following is also possible: steady rotation of the wing about a circle in the Oxy plane, for which α = const, and ω_z = const; steady motion U_0 = const with a steady rotation about the Ox axis ($\omega_z = 0$ and ω_x = const); and naturally a combination of these motions.

43

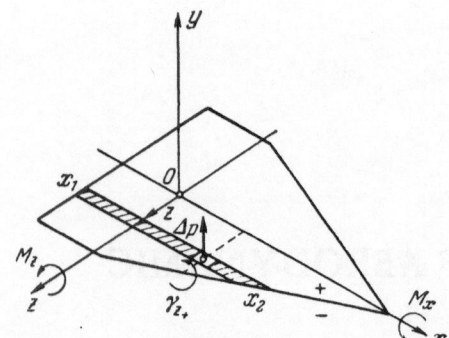

Fig. 4.1. The calculation of aerodynamic forces and moments on a monoplane wing.

From Fig. 4.1, we see that the lift and moments will be

$$Y = \int\int \Delta p \, dx \, dz, \quad M_z = \int\int \Delta p x \, dx \, dz, \\ M_x = -\int\int \Delta p z \, dx \, dz, \tag{4.3}$$

where the integration extends over the whole area of the wing.

We introduce the aerodynamic coefficients, taking the root chord b of the wing as the characteristic linear dimension. From (4.1), (4.2), and (4.3) it follows that the aerodynamic coefficients of a thin wing (a plate), symmetric in plan, can be written

$$c_y = c_y^\alpha \alpha + c_y^{\omega_z}\omega_z, \quad m_x = m_x^{\omega_x}\omega_x, \\ m_z = m_z^\alpha \alpha + m_z^{\omega_z}\omega_z. \tag{4.4}$$

For the aerodynamic load the relations are similar to those in (4.2):

$$\frac{\Delta p}{q} = p_\alpha(x, z)\alpha + p_{\omega_z}(x, z)\omega_z + p_{\omega_x}(x, z)\omega_x, \tag{4.5}$$

where, for a wing symmetrical in plan, the functions $\gamma_z^\alpha(x, z)$, $\gamma_z^{\omega_z}(x, z)$, $p_\alpha(x, z)$, and $p_{\omega_z}(x, z)$ are symmetric, and $\gamma_z^{\omega_x}(x, z)$, $p_{\omega_x}(x, z)$ are antisymmetric functions of z.

The aerodynamic derivatives in (4.2), (4.4), and (4.5) are usually called the rotational-derivative coefficients [32]. The use of the above coefficients in the linear problem leads to the determination of the total and distributed aerodynamic characteristics of a wing for any stationary motion on a curved trajectory.

The rotational derivatives of the total characteristics (4.4) are easily expressed in terms of the corresponding coefficients of the distributed characteristics (4.2) or (4.5).

From (4.2) and (4.3) we obtain

$$c_y^\alpha = \frac{4}{S}\int\int \gamma_z^\alpha \, dx \, dz, \quad c_y^{\omega_z} = \frac{4}{S}\int\int \gamma_z^{\omega_z} \, dx \, dz, \\ m_z^\alpha = \frac{4}{Sb}\int\int \gamma_z^\alpha x \, dx \, dz, \quad m_z^{\omega_z} = \frac{4}{Sb}\int\int \gamma_z^{\omega_z} x \, dx \, dz, \\ m_x^{\omega_x} = -\frac{4}{Sb}\int\int \gamma_z^{\omega_z} x z \, dx \, dz. \tag{4.6}$$

Similarly

$$c_y^\alpha = \frac{2}{S}\int\int p_\alpha \, dx \, dz, \quad c_y^{\omega_z} = \frac{2}{S}\int\int p_{\omega_z} \, dx \, dz, \\ m_z^\alpha = \frac{2}{Sb}\int\int p_\alpha x \, dx \, dz, \quad m_z^{\omega_z} = \frac{2}{Sb}\int\int p_{\omega_z} x \, dx \, dz, \\ m_x^{\omega_x} = -\frac{2}{Sb}\int\int p_{\omega_z} z \, dx \, dz, \tag{4.7}$$

where the integration in (4.6) and (4.7) is taken over one half of the wing (the right half, see Fig. 4.1).

We now describe how to find the aerodynamic characteristics and rotational-derivative coefficients of arbitrary wing sections. For the lift and longitudinal moment of a section relative to the Oz axis, we have

$$dY = \rho U_0\, dz \int_{x_1}^{x_2} \gamma_{z+}\, dx, \quad dM_z = \rho U_0\, dz \int_{x_1}^{x_2} \gamma_{z+}\, x\, dx.$$

The corresponding coefficients are introduced as follows:

$$c'_y = \frac{dY}{qb'\, dz}, \quad m'_z = \frac{dM_z}{qb'^2\, dz}, \tag{4.8}$$

where b' is the chord at the section z = const under consideration; then

$$c'_y = \frac{2}{U_0 b'} \int_{x_1}^{x_2} \gamma_{z+}\, dx, \quad m'_z = \frac{2}{U_0 b'^2} \int_{x_1}^{x_2} \gamma_{z+} x\, dx. \tag{4.9}$$

Here the integration is taken over the relevant section from the trailing to the leading edge (from x_1 to x_2). The characteristics of a section can also be expressed in terms of the rotational-derivative coefficients:

$$c'_y = c_y'^{\alpha}\alpha + c_y'^{\omega_z}\omega_z + c_y'^{\omega_x}\omega_x, \quad m'_z = m_z'^{\alpha}\alpha + m_z'^{\omega_z}\omega_z + m_z'^{\omega_x}\omega_x, \tag{4.10}$$

where, from (4.2), (4.9), and (4.10),

$$\left.\begin{array}{lll} c_y'^{\alpha} = \dfrac{2}{b'} \displaystyle\int_{x_1}^{x_2} \gamma_z^{\alpha}\, dx, & c_y'^{\omega_z} = \dfrac{2}{b'} \displaystyle\int_{x_1}^{x_2} \gamma_z^{\omega_z}\, dx, & c_y'^{\omega_z} = \dfrac{2}{b'} \displaystyle\int_{x_1}^{x_2} \gamma_z^{\omega_x}\, dx, \\[6mm] m_z'^{\alpha} = \dfrac{2}{b'^2} \displaystyle\int_{x_1}^{x_2} \gamma_z^{\alpha} x\, dx, & m_z'^{\omega_z} = \dfrac{2}{b'^2} \displaystyle\int_{x_1}^{x_2} \gamma_z^{\omega_z} x\, dx, & m_z'^{\omega_x} = \dfrac{2}{b'^2} \displaystyle\int_{x_1}^{x_2} \gamma_z^{\omega_x} x\, dx. \end{array}\right\} \tag{4.11}$$

We now consider the calculation of the coefficient of induced drag of a wing. We investigate only the case of steady rectilinear motion, which is of fundamental practical importance. It has already been noted that, depending on the shape of the leading edge of a wing, we can have two limiting theoretical relations for the determination of the induced drag: without suction force X_-, and with suction force X_+. The first case is for a sharp leading edge (for example, a wing designed for supersonic speeds moving at a subsonic speed), and the second for a well rounded smooth leading edge. The respective coefficients will be written in the form

$$\bar{c}_{xi} = \frac{X_-}{qS}, \quad c_{xi} = \frac{X_+}{qS}.$$

Using (3.20) for the induced drag without suction forces, for translational motion without rotation we have

$$\bar{c}_{x_i} = c_y \alpha. \tag{4.12}$$

We use the symbol α_0 for the zero-lift angle of the wing, and we have

$$c_y = c_y^a (\alpha - \alpha_0),$$

and so instead of (4.12) we can write

$$\bar{c}_{xi} = \frac{c_y^2}{c_y^a} + c_y a_0. \tag{4.13}$$

For wings made up of symmetric and symmetrically arranged profiles, in particular for wing-plates of arbitrary shape in plan, we have $\alpha_0 = 0$ and

$$\bar{c}_{xi} = c_y a = \frac{c_y^2}{c_y^a}. \tag{4.14}$$

To find the coefficient of induced drag with suction force c_{xi}, we turn to Eq. (3.31). For simplicity we consider a wing for which $\alpha_0 = 0$; the total circulation of the bound vortices for any section z = const is

$$\Gamma(z) = U_0 l \Gamma_\alpha^*(\bar{z}) \alpha, \qquad \bar{z} = \frac{2z}{l}, \tag{4.15}$$

where l is the span and $\Gamma_\alpha^*(\bar{z})$ is a function of the dimensionless coordinate $\bar{z}$.

The velocity $W_y(z)$ in formula (3.31) will be equal to

$$W_y(z) = -\frac{1}{2\pi} \int_{-l/2}^{l/2} \frac{d\Gamma}{d\zeta} \frac{d\zeta}{z - \zeta},$$

or, using the notation of (4.15), we may write

$$W_y(\bar{z}) = U_0 \alpha w_y(\bar{z}), \qquad w_y(\bar{z}) = -\frac{1}{\pi} \int_{-1}^{1} \frac{d\Gamma_\alpha^*}{d\bar{\zeta}} \frac{d\bar{\zeta}}{\bar{z} - \bar{\zeta}}. \tag{4.16}$$

We express the coefficient of induced drag with suction force in the form

$$c_{xi} = B \frac{c_y^2}{\pi\lambda}. \tag{4.17}$$

In the notation of (4.16) and (4.17), Eq. (3.31) for the coefficient c_{xi} becomes

$$c_{xi} = -\frac{\lambda}{2} \frac{c_y^2}{(c_y^a)^2} \int_{-1}^{1} w_y(\bar{z}) \Gamma_\alpha^*(\bar{z}) \, d\bar{z},$$

and so

$$B = -\frac{\pi\lambda^2}{2(c_y^a)^2} \int_{-1}^{1} w_y(\bar{z}) \Gamma_\alpha^*(\bar{z}) \, d\bar{z}. \tag{4.18}$$

The reason for the introduction of the factor B is as follows. By using (4.16) and (4.18), it is not difficult to show that for an elliptic law of variation of the strength of the bound vortices with span for which

$$\Gamma_\alpha^*(\bar{z}) = \Gamma_\alpha^*(0) \sqrt{1 - \bar{z}^2},$$

we have B = 1. Hence B $\geq$ 1 is the factor by which the induced drag of the wing under consideration exceeds that of a wing with the same coefficient c_y and the same aspect ratio λ, but with an optimum circulation distribution (optimum from the point of view of the value of c_{xi}) [13, 14, 17, 18].

Since the difference between the induced drag X_- without suction force and X_+ with suction force, within the accuracy attained in our calculations, is equal to the suction force Q, it follows that the suction-force coefficient is

$$c_Q = \bar{c}_{xi} - c_{xi},$$

or

$$
c_Q = D\,\frac{c_y^2}{\pi\lambda}\,, \qquad D = \bar{B} - B, \\
B = \frac{\pi\lambda}{c_y^{\alpha}}\,, \qquad \bar{c}_{x_i} = \bar{B}\,\frac{c_y^2}{\pi\lambda}\,.
\qquad (4.19)
$$

§2. Annular Wings

A notable feature of annular wings, which is of considerable theoretical interest, is the fact that independently of the aspect ratio λ of the wing we can obtain an explicit expression for the dependence of the strength of the bound vortices on the angle φ of a cylindrical coordinate system. At the same time, the dependence of this strength for a monoplane wing on the coordinate z is unknown, and this naturally complicates the solution of the problem and makes the results of numerical calculations for a monoplane wing less dependable.

If the boundary conditions (1.31) are used, the strength $\gamma_+(x, \varphi)$ of the bound vortex layer for an annular wing can be given by the formula

$$\gamma_+ (x, \varphi) = U_0 \left[\alpha_0 \gamma_0 (x) + \alpha \gamma_\alpha (x) \cos\varphi + \omega_z \gamma_{\omega_z} (x) \cos\varphi \right]. \qquad (4.20)$$

where $\gamma_0(x)$, $\gamma_\alpha(x)$, and $\gamma_{\omega_z}(x)$ are dimensionless functions of x. An appeal to the boundary condition makes it possible to "guess" the character of the functional dependence $\gamma_+(x, \varphi)$ on the kinematic parameters and, in particular, on the angle φ. A rigorous proof of the validity of (4.20), however, will be obtained from the reasoning in the following paragraph, in which we formulate the problem of determining $\gamma_0(x)$, $\gamma_\alpha(x)$, and $\gamma_{\omega_z}(x)$.

Since Zhukovskii's theorem "in the small" is valid in the problem under consideration, we obtain for the difference between the pressure inside and outside an annular wing the formula

$$\frac{\Delta p}{q} = p_0 (x)\, \alpha_0 + p_\alpha (x)\, \alpha \cos\varphi + p_{\omega_z} (x)\, \omega_z \cos\varphi. \qquad (4.21)$$

For the monoplane wing, we confine our attention in circulation problems to the investigation of steady flow. Rotation of an annular wing about the Ox axis does not disturb the medium. Hence the possible motion reduces to translational motion ($\alpha = \text{const}$, $\omega_z = 0$) or steady rotation about a circle in the Oxy plane ($\alpha = \text{const}$, $\omega_z = \text{const}$).

To find the forces and moments acting on an annular wing without taking the suction forces into account, it is sufficient to integrate the pressure over the surface of the wing. From Fig. 4.2, the projections of the pressure difference Δp on the directions Ox, Or, and Oy are

$$
\Delta p_x = -\Delta p \sin\alpha_0, \\
\Delta p_r = \Delta p \cos\alpha_0, \\
\Delta p_y = \Delta p \cos\varphi \cos\alpha_0.
\qquad (4.22)
$$

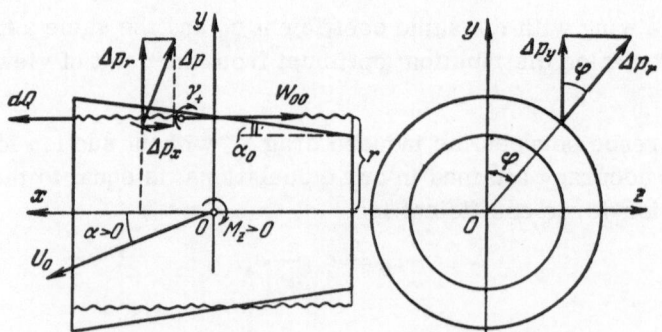

Fig. 4.2. The calculation of the aerodynamic forces
and moment on an annular wing.

Correct to the first order, we now obtain for the tangential force T, the normal force Y, and the longitudinal moment $\overline{M}_z$ the formulas (§1, Chapter I)

$$T = a_0 \int_{-b/2}^{b/2} r \int_0^{2\pi} \Delta p \, d\varphi \, dx,$$

$$Y = \int_{-b/2}^{b/2} r \int_0^{2\pi} \Delta p \cos\varphi \, d\varphi \, dx, \tag{4.23}$$

$$\overline{M}_z = \int_{-b/2}^{b/2} r \int_0^{2\pi} \Delta p \, x \cos\varphi \, d\varphi \, dx + a_0 \int_{-b/2}^{b/2} r^2 \int_0^{2\pi} \Delta p \cos\varphi \, d\varphi \, dx.$$

The aerodynamic coefficients are defined as

$$c_y = \frac{Y}{q2rb}, \qquad \overline{c}_t = \frac{T}{q2rb}, \qquad \overline{m}_z = \frac{\overline{M}_z}{q2rb^2}, \tag{4.24}$$

where r is the mean radius and b is the chord of the annular wing. We write $\overline{x} = x/b$, and assume that r = const in (4.23); this is correct for the linear theory in which the vortex layer is distributed over a cylindrical surface. This yields

$$c_y = \frac{1}{2} \int_{-1/2}^{1/2} \int_0^{2\pi} \frac{\Delta p}{q} \cos\varphi \, d\varphi \, d\overline{x}, \qquad \overline{c}_t = \frac{a_0}{2} \int_{-1/2}^{1/2} \int_0^{2\pi} \frac{\Delta p}{q} \, d\varphi \, d\overline{x},$$

$$\overline{m}_z = \frac{1}{2} \int_{-1/2}^{1/2} \int_0^{2\pi} \frac{\Delta p}{q} \overline{x} \cos\varphi \, d\varphi \, d\overline{x} + a_0 \frac{c_y \lambda}{2}. \tag{4.25}$$

For the monoplane wing, the lift and moment can be expressed in terms of the rotational-derivative coefficient. In the framework of the linear theory, the coefficient c_y does not depend on the suction force. If the wing is cylindrical ($\alpha_0 = 0$), then the moment M_z is also independent of the suction force.

We denote the longitudinal-moment coefficient by m_z, and taking the suction force into account we can write

$$c_y = c_y^\alpha \alpha + c_y^{\omega_z} \omega_z, \qquad m_z = m_z^\alpha \alpha + m_z^{\omega_z} \omega_z,$$

$$\overline{m}_z = \overline{m}_z^\alpha \alpha + \overline{m}_z^{\omega_z} \omega_z, \qquad m_x = 0. \tag{4.26}$$

As in the case of a monoplane wing, it is not difficult to express c_y^α, $c_y^\omega z$, $\overline{m}_z^\alpha$, $\overline{m}_z^\omega$ z in terms of the coefficient in (4.21):

$$c_y^a = \frac{\pi}{2} \int_{-1/2}^{1/2} p_a \, d\overline{x}, \qquad \overline{m}_z^a = \frac{\pi}{2} \int_{-1/2}^{1/2} p_a \overline{x} \, d\overline{x} + \frac{\alpha_0 \lambda}{2} c_y^a,$$
$$c_y^\omega z = \frac{\pi}{2} \int_{-1/2}^{1/2} p_{\omega_z} \, d\overline{x}, \qquad \overline{m}_z^\omega z = \frac{\pi}{2} \int_{-1/2}^{1/2} p_{\omega_z} \overline{x} \, d\overline{x} + \frac{\alpha_0 \lambda}{2} c_y^\omega z. \tag{4.27}$$

To relate the given coefficients to the strength of the bound vortex layer, we must consider the Zhukovskii theorem "in the small." From (3.7) we obtain

$$\Delta p = - \rho \gamma_+ W_{00},$$

where the positive directions for the strength γ_+ of the bound vortices and the relative velocity W_{00} for corresponding points of the vortex layer are shown in Fig. 4.2. Since for small angles of attack we have

$$W_{00} = - U_0 - W_{x0},$$

with the usual accuracy, where W_{x0} is the component of the disturbed velocity at the point of the vortex layer under consideration, it follows that

$$\Delta p = \rho \gamma_+ (U_0 + W_{x0}). \tag{4.28}$$

The disturbed velocities are obtained from the formulas in Chapter II in which γ_+ appears linearly; hence, as in (4.20), we have

$$W_{x0} = U_0 \left[\alpha_0 w_{x0}^{(0)}(x) + \alpha w_{x0}^{(\alpha)}(x) \cos \varphi + \omega_z w_{x0}^{(\omega_z)}(x) \cos \varphi \right]. \tag{4.29}$$

In the determination of the aerodynamic coefficients, the scheme we are using permits us to take terms of order $\alpha_0 \alpha$, $\alpha_0 \omega_z$ and α_0^2 into account, and so (4.20), (4.21), (4.28), and (4.29) yield

$$p_0 = 2\gamma_0 \left(1 + \alpha_0 w_{x0}^{(0)} \right), \quad p_a = 2 \left[\left(1 + \alpha_0 w_{x0}^{(0)} \right) \gamma_a + \alpha_0 \gamma_0 w_{x0}^{(\alpha)} \right],$$
$$p_{\omega_z} = 2 \left[\left(1 + \alpha_0 w_{x0}^{(0)} \right) \gamma_{\omega_z} + \alpha_0 \gamma_0 w_{x0}^{(\omega_z)} \right] \tag{4.30}$$

We now turn to the determination of the aerodynamic characteristics of an annular wing, taking the suction force into account. For simplicity we confine ourselves to the most important case — that of translational motion. From (4.20), the total circulation of the bound vortices in this case is

$$\Gamma(\varphi) = U_0 r \left(\alpha_0 \Gamma_0 + \alpha \Gamma_a \cos \varphi \right). \tag{4.31}$$

We obtain an expression for the velocity $W_r(\varphi_0)$ in (3.33) from an infinite cylindrical system of free vortices. For Fig. 4.3, we see that

$$W_r(\varphi_0) = \frac{1}{4\pi r} \int_0^{2\pi} \frac{d\Gamma}{d\varphi} \operatorname{ctg} \frac{\varphi - \varphi_0}{2} \, d\varphi,$$

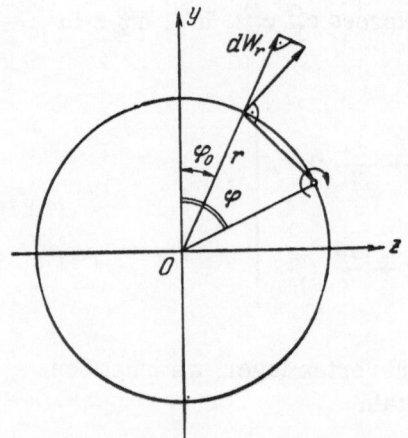

Fig. 4.3. Calculation of the radial velocity at infinity behind an annular wing.

while from (4.31) we have

$$\frac{d\Gamma}{d\varphi} = -U_0 r \Gamma_a \alpha \sin\varphi;$$

using the relation

$$\int_0^{2\pi} \sin\varphi \, \text{ctg} \, \frac{\varphi - \varphi_0}{2} \, d\varphi = 2\pi \cos\varphi_0.$$

we now obtain

$$W_r(\varphi_0) = -\frac{U_0}{2}\Gamma_a \alpha \cos\varphi_0. \tag{4.32}$$

Equations (4.31), (4.32), and (3.33) yield

$$X_+ = q \, \frac{\pi r^2}{2} \Gamma_a^2 \alpha^2.$$

and using this relation we obtain the induced drag coefficient taking into account the suction force, similarly introduced in (4.24):

$$c_{xi} = \frac{X_{i+}}{q2rb}, \qquad c_{xi} = \frac{\pi\lambda}{8} \Gamma_a^2 \alpha^2. \tag{4.33}$$

For a cylindrical wing, for which $\alpha_0 = 0$, the formula (4.33) can be written

$$c_{xi} = \frac{c_y^2}{2\pi\lambda}. \tag{4.34}$$

In agreement with formula (1.2), the induced-drag coefficient of an annular wing without suction force is

$$\bar{c}_{xi} = c_t + c_y \alpha = c_t + \frac{c_y^2}{c_y^a} = c_t + c_y^\alpha \alpha^2, \tag{4.35}$$

since here $\alpha = 0$, we have $c_y = 0$.

We now easily find the coefficient c_Q of the suction force Q to be

$$c_Q = \bar{c}_{xi} - c_{xi}, \qquad c_Q = \frac{Q}{q2rb}.$$

Using (4.33), we obtain

$$c_Q = c_t + \left(c_y^\alpha - \frac{\pi\lambda}{8} \Gamma_a^2 \right) \alpha^2. \tag{4.36}$$

Since the suction force influences not only the drag but also the longitudinal moment of an annular wing, we must find the law governing the variation of this force for varying φ. To do this we must first establish the structure of this law.

The suction force at the leading edge is obtained from an expression involving γ_+^2 [24]. Hence, recalling (4.20), for the coefficient c_Q' of the suction force dQ of a section, for translation motion we have

$$c'_Q = c_0 \alpha_0^2 + c_1 \alpha_0 \alpha \cos \varphi + c_2 \alpha^2 \cos^2 \varphi, \tag{4.37}$$

where c_0, c_1 and c_2 are constants and

$$c'_Q = \frac{dQ}{qbr \, d\varphi}.$$

By integrating (4.37), we obtain the suction-force coefficient of the whole wing

$$c_Q = \pi c_0 \alpha_0^2 + \frac{\pi}{2} c_2 \alpha^2. \tag{4.38}$$

From (4.36) and (4.38) we obtain

$$c_0 = \frac{c_t}{\pi \alpha_0^2}, \qquad c_2 = \frac{2}{\pi} \left(c_y^\alpha - \frac{\pi \lambda}{8} \Gamma_\alpha^2 \right). \tag{4.39}$$

To find the constant c_1 we use the fact that $c'_Q \geq 0$ for any values of $\alpha \cos \varphi$, and so the quadratic equation

$$c_2 \alpha^2 \cos^2 \varphi + c_1 \alpha_0 \alpha \cos \varphi + c_0 \alpha_0^2 = 0$$

has multiple roots. The condition

$$(c_1 \alpha_0)^2 = 4 c_0 \alpha_0^2 c_2$$

yields

$$c_1^2 = 4 c_0 c_2. \tag{4.40}$$

We now establish the sign of c_1, remembering that for $\alpha_0 > 0$ and $\alpha > 0$ (Fig. 4.2) the suction force is greater on the upper than on the lower section of the wing. From (4.37) this leads to the inequality

$$c_0 \alpha_0^2 + c_1 \alpha_0 \alpha + c_2 \alpha^2 > c_0 \alpha_0^2 - c_1 \alpha_0 \alpha + c_2 \alpha^2,$$

and so

$$c_1 > 0. \tag{4.41}$$

Similar reasoning shows that (4.41) also holds for $\alpha_0 < 0$.

The extra moment due to the suction force is (see Fig. 4.2)

$$\Delta M_z = - qbr^2 \int_0^{2\pi} c'_Q \cos \varphi \, d\varphi = - qbr^2 c_1 \alpha_0 \alpha \pi,$$

and so, writing

$$m_z = \overline{m}_z + \Delta m_z, \qquad m_z^\alpha = \overline{m}_z^\alpha + \Delta m_z^\alpha, \tag{4.42}$$

we obtain

$$\Delta m_z = \Delta m_z^\alpha \alpha, \qquad \Delta m_z^\alpha = - \frac{\pi \lambda}{4} c_1 \alpha_0. \tag{4.43}$$

§3. Profile Cascades

In the boundary condition (1.32), the parameters β_Γ and ε are constants for a given cascade and are independent of the angle of attack α, while ϑ is related to α. From Fig. 1.7. we see that

$$\alpha_\Gamma = \vartheta - \beta_\Gamma = \alpha + \alpha_0,$$

and so the condition (1.32) becomes

$$-\frac{W_n}{U_0} = \sin(\alpha + \alpha_0)\cos\varepsilon + \cos(\alpha + \alpha_0)\sin\varepsilon. \tag{4.44}$$

We can thus establish a structural formula for the circulation of the bound vortices:

$$\gamma = U_0[\gamma_1(s)\sin(\alpha + \alpha_0) + \gamma_2(s)\cos(\alpha + \alpha_0)], \tag{4.45}$$

where s is the arc length of a profile.

The total circulation in the Zhukovskii theorem (3.18) will be

$$\Gamma_+ = U_0\sin(\alpha + \alpha_0)\int \gamma_1(s)\,ds + U_0\cos(\alpha + \alpha_0)\int \gamma_2(s)\,ds,$$

where the integration extends over the whole length of the profile. Since the angle of attack α is taken for circulation-free flow, it follows that $\Gamma_+ = 0$ for $\alpha = 0$. This condition permits us to transform the expression for Γ_+ into

$$\Gamma_+ = \frac{U_0 b}{\cos\alpha_0}\sin\alpha\int \gamma_1\,d\bar{s}, \qquad \bar{s} = \frac{s}{b}. \tag{4.46}$$

From the Zhukovskii theorem (3.18), the lift (1.17) can now be expressed in the form

$$c_y = c_y^\alpha \sin\alpha, \qquad c_y^\alpha = \frac{2}{\cos\alpha_0}\int \gamma_1\,d\bar{s}. \tag{4.47}$$

Since we can consider the nonlinear problem in cascade theory, the lift will depend on whether or not a suction force acts on the profile. Eqs. (4.47) are obtained by taking suction force into account. We can also find c_y^α without taking this force into account; we use the Zhukovskii theorem "in the small," but we will not consider this here.

In the calculation of the moment M_Z from this theorem, the suction forces can be taken into account as follows. If M_Z is calculated relative to the leading edge of the plane the inflow force yields no moment. To find M_Z for another axis parallel to the original axis, we must know the lift Y. We will find values of the moment M_Z both with the suction force taken into account and not taken into account.

The Zhukovskii theorem "in the small" for a profile grid yields

$$\Delta p = \rho(U_{0t} + W_t)\gamma,$$

where $U_{0t} + W_t$ is the tangential component of the relative velocity at the profile. From Fig. 1.11 we obtain

$$U_{0t} = U_0\cos(\vartheta - \beta_\Gamma + \varepsilon),$$

$$W_t = W_\xi\cos(\beta_\Gamma - \varepsilon) + W_\eta\sin(\beta_\Gamma - \varepsilon).$$

and so

$$U_{0t} + W_t = U_0 \cos(\alpha + \alpha_0 + \varepsilon) + W_\xi \cos(\beta_\Gamma - \varepsilon) + W_\eta \sin(\beta_\Gamma - \varepsilon). \tag{4.48}$$

For disturbed velocities, as in (4.45), we can write

$$\left. \begin{aligned} W_\xi &= U_0 \left[w_\xi^{(1)} \sin(\alpha + \alpha_0) + w_\xi^{(2)} \cos(\alpha + \alpha_0) \right], \\ W_\eta &= U_0 \left[w_\eta^{(1)} \sin(\alpha + \alpha_0) + w_\eta^{(2)} \cos(\alpha + \alpha_0) \right]. \end{aligned} \right\} \tag{4.49}$$

After some transformations we obtain

$$\Delta p = \rho U_0^2 \left[\gamma_2 U_2 + (\gamma_1 U_1 - \gamma_2 U_2) \sin^2(\alpha + \alpha_0) + (\gamma_1 U_2 + \gamma_2 U_1) \frac{\sin 2(\alpha + \alpha_0)}{2} \right], \tag{4.50}$$

where

$$\left. \begin{aligned} U_1 &= -\sin\varepsilon + w_\xi^{(1)} \cos(\beta_\Gamma - \varepsilon) + w_\eta^{(1)} \sin(\beta_\Gamma - \varepsilon), \\ U_2 &= \cos\varepsilon + w_\xi^{(2)} \cos(\beta_\Gamma - \varepsilon) + w_\eta^{(2)} \sin(\beta_\Gamma - \varepsilon). \end{aligned} \right\} \tag{4.51}$$

From Fig. 4.4 we obtain the following expression for the moment M_z relative to the leading edge of the profile:

$$M_z = -\int \Delta p \, (x \cos\varepsilon - y \sin\varepsilon) \, ds.$$

The moment coefficient

$$m_z = \frac{M_z}{qb^2}$$

can be expressed in the form

$$m_z = -B_1 - B_2 \sin 2(\alpha + \alpha_0) - B_3 \sin^2(\alpha + \alpha_0),$$

where

$$\left. \begin{aligned} B_1 &= 2\int \gamma_2 U_2 \bar{h} \, d\bar{s}, & B_2 &= \int (\gamma_1 U_2 + \gamma_2 U_1) \bar{h} \, d\bar{s}, \\ B_3 &= 2\int (\gamma_1 U_1 - \gamma_2 U_2) \bar{h} \, d\bar{s}, & \bar{h} &= \frac{x \cos\varepsilon - y \sin\varepsilon}{b}, \quad \bar{s} = \frac{s}{b}. \end{aligned} \right\} \tag{4.52}$$

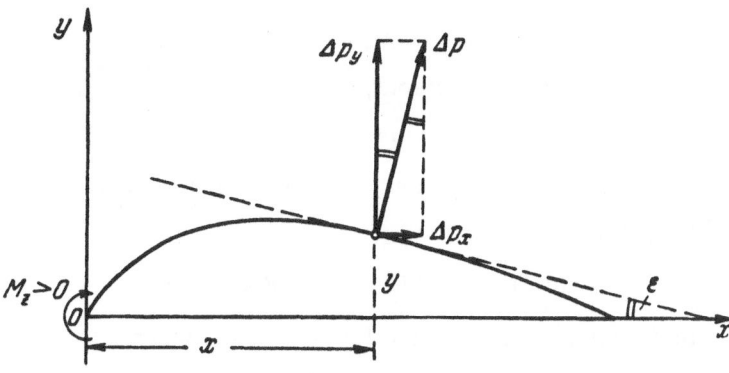

Fig. 4.4. The calculation of the aerodynamic moment on a profile in a grid.

This coefficient can also be written

$$\left.\begin{aligned}
m_z &= -m_{z0} - \frac{m_z^\alpha}{2} \sin 2\alpha - \frac{m_z^{\alpha\alpha}}{2} \sin^2 \alpha, \\
m_{z0} &= B_1 + B_2 \sin 2\alpha_0 + B_3 \sin^2 \alpha_0, \\
\frac{m_z^\alpha}{2} &= B_2 \cos 2\alpha_0 + \frac{B_3}{2} \sin 2\alpha_0, \\
\frac{m_z^{\alpha\alpha}}{2} &= -2B_2 \sin 2\alpha_0 + B_3 \cos 2\alpha_0.
\end{aligned}\right\}
\tag{4.53}$$

§4. The Center of Pressure, the Focus, and the Lift-Drag Ratio

In the investigation of uniform and rectilinear motion of a wing in various practical problems, the ideas of "center of pressure" and "lifting-surface focus" are very important.

The center of pressure (c. p.) is a certain point on the line of action of the resultant of the aerodynamic forces acting on the lifting surface. For a monoplane wing-plate the center of pressure will be the name given to the intersection of this line with the plane of the wing; for an annular wing its intersection with the axis of symmetry; and for a profile in a cascade its intersection with the geometrical chord of the profile.

In general the pitching-moment coefficient depends on the angle of attack. However, as will be shown below, for monoplane and annular wings in the linear theory and for profile cascades in the exact theory, there is a position of the Oz axis relative to which the moment is independent of α. The point F characterizing this position is called the focus or aerodynamic center of the lifting surface.

We start with a consideration of monoplane and annular wings. Let the aerodynamic coefficients c_y and m_z relative to any coordinate system $Oxyz$ be known for an annular or monoplane wing. Since the resultant force Y through the center of pressure (Fig. 4.5) is equivalent to a combination of a resultant force Y and a moment M_z corresponding to a center of pressure O, we have

$$M_z = Y x_{c.p.}$$

and so

$$\bar{x}_{c.p.} = \frac{m_z}{c_y}, \qquad \bar{x}_{c.p.} = \frac{x_{c.p.}}{b}. \tag{4.54}$$

We write M_F for the moment of the wing about an axis parallel to Oz and passing through the focus F (Fig. 4.5); then if x_α is the coordinate of F we have

$$M_F = M_z - Y x_\alpha$$

or, converting to the corresponding coefficients,

$$m_F b = m_z b - c_y x_\alpha.$$

For small angles of attack, c_y and m_z for monoplane and annular wings depend linearly on α, and so in the most general case we have

$$c_y = c_y^\alpha (\alpha - \alpha_0), \qquad m_z = m_{z0}^* + m_z^\alpha \alpha.$$

For a monoplane flat-plate wing and for an arbitrary annular wing

$$\alpha_0 = 0, \qquad m_{z0}^* = 0.$$

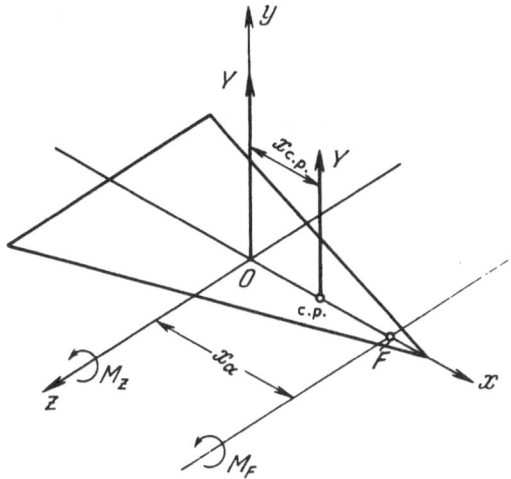

Fig. 4.5. The center of pressure (c.p.) and the focus (F) of a monoplane wing.

In the general case

$$m_F = m_{z0}^* + m_z^a \alpha - c_y^a (\alpha - \alpha_0) \frac{x_a}{b},$$

and hence the moment coefficient m_F is independent of the angle of attack if

$$m_z^a \alpha - c_y^a \alpha \frac{x_a}{b} = 0.$$

Thus the dimensionless coordinates of the focus F are

$$\bar{x}_a = \frac{m_z^a}{c_y^a}, \qquad \bar{x}_a = \frac{x_a}{b}. \qquad (4.55)$$

A comparison of (4.54) and (4.55) easily shows that, although in general $\bar{x}_{c.p.} \neq \bar{x}_a$, for any annular wing and for a monoplane flat-plate wing the center of pressure is at the focus.

We now turn to the consideration of an arbitrary thin profile of a cascade. Let the moment M_z of the profile relative to an origin O located at the nose and the lift Y be known. From Fig. 4.6 we see that

$$M_z = - Y x_{c.p.} \cos (\vartheta - \beta_r),$$

and so

$$(\vartheta - \beta_r = \alpha + \alpha_0),$$

$$\bar{x}_{c.p.} = - \frac{m_z}{c_y \cos (\alpha + \alpha_0)}, \qquad \bar{x}_{c.p.} = \frac{x_{c.p.}}{b}. \qquad (4.56)$$

For the point F we can write

$$M_F = M_z + Y' x_F - X' y_F.$$

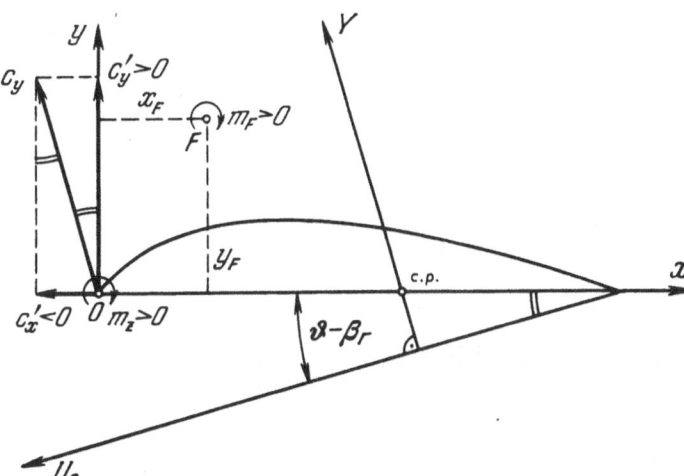

Fig. 4.6. The center of pressure (c.p.) and the focus (F) of a profile in cascade.

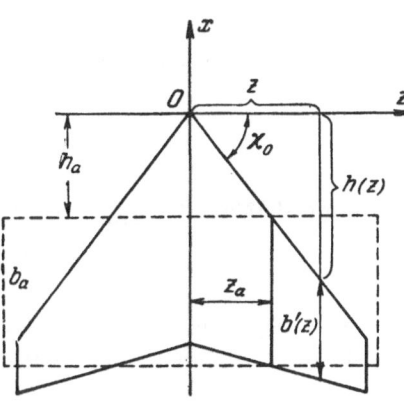

Fig. 4.7. The mean aerodynamic chord b_a.

or

$$m_F b = m_z b + c_y' x_F - c_x' y_F,$$

where x_F and y_F are the coordinates of F and c_x' and c_y' are the projections of the coefficient c_y on the Ox and Oy axes. From Fig. 4.6 we conclude that

$$c_y' = c_y \cos(a + a_0), \qquad c_x' = -c_y \sin(a + a_0);$$

recalling also the relations (4.47) and (4.53), we obtain

$$m_F = - m_{z0} - \frac{m_z^a}{2} \sin 2a - \frac{m_z^{aa}}{2} \sin^2 a + \bar{x}_F c_y^a \sin a \cos(a + a_0) + \bar{y}_F c_y^a \sin a \sin(a + a_0).$$

It is not difficult now to establish that the coefficient m_F will be independent of the angle of attack α for the following coordinates of F:

$$\bar{x}_F = \frac{1}{c_y^a} \left(m_z^a \cos a_0 - \frac{m_z^{aa}}{2} \sin a_0 \right),$$

$$\bar{y}_F = \frac{1}{c_y^a} \left(m_z^a \sin a_0 + \frac{m_z^{aa}}{2} \cos a_0 \right), \qquad (4.57)$$

$$\bar{x}_F = \frac{x_F}{b}, \qquad \bar{y}_F = \frac{y_F}{b}.$$

Important characteristics of wings are the lift-drag ratio

$$k = \frac{c_y}{c_x}$$

and the maximum lift-drag ratio k_{max}; the latter is attained by a given wing for a fixed $\mathbf{M}_\infty$ for a certain small angle of attack.

In the calculation k and k_{max}, we must take into account all forms of resistance acting on the wing, including the frictional resistance. For small angles of attack the drag coefficient can be given by the formula

$$c_x = c_{x0} + A c_y^2, \qquad (4.58)$$

where c_{x0} is the drag for zero lift and A is a factor characterizing the change in drag due to a change in angle of attack, and is independent of c_y. Since

$$k = \frac{c_y}{c_{x0} + A c_y^2},$$

the condition that for $k = k_{max}$ we must have $dk/dc_y = 0$ yields

$$A c_{y*}^2 = c_{x0}, \qquad (4.59)$$

where c_{y*} is the value of c_y for $k = k_{max}$. It is thus plain that for $k = k_{max}$, the induced drag is equal to the resistance for zero lift. Substituting c_{y*} from (4.59), in the expression for k, we obtain

$$k_{max} = \frac{1}{2\sqrt{A c_{x0}}}. \qquad (4.60)$$

Equation (4.60) holds in particular for monoplane and annular wings.

§5. The Mean Aerodynamic Chord of a Monoplane Wing

If the longitudinal moment m_z of a wing and the coordinate $\overline{x}_\alpha$ of the focus are related to the magnitude and position of the root chord b of the wing, then m_z and $\overline{x}_\alpha$ will depend strongly on the geometrical parameters of the wing, and particularly strongly on the sweepback χ_0. The coefficients m_z and $\overline{x}_\alpha$ become considerably more stable when they are related to the so-called mean aerodynamic chord of the wing (MAC).

This is the name given to the chord of a rectangular wing with the same area S as the given wing, and also the same lift and longitudinal moment under the same conditions of flight.

To find the precise chord b_a of such a wing and its position relative to the original wing, we must know the aerodynamic characteristics of both wings. In this case, however, there is no practical sense in this idea. It is more expedient to introduce an equivalent wing whose parameters can be determined simply and do not depend on $\mathbf{M}_\infty$, although in this case c_y and m_z will only be approximately equal.

Coefficients of wing sections will be indicated by primes and all characteristics of the equivalent rectangular wing will have the subscript "a". For the determination of the parameters of the equivalent wing, we have the conditions

$$Y_a = Y, \qquad M_{za} = M_z, \qquad S_a = S. \tag{4.61}$$

Let h_a be the distance from the Oz axis to the leading edge b_a, and let $h(z)$ be the distance from Oz to the leading edge $b'(z)$ (Fig. 4.7); we denote the coordinates of the centers of pressure of the wing sections relative to the corresponding leading edges by $x'_{c.p.}$ and $x'_{c.p.a}$. From the first two conditions (4.61) we have

$$b_a \int_{-l_a/2}^{l_a/2} c'_{ya}\, dz = \int_{-l/2}^{l/2} c'_y b'\, dz; \qquad b_a \int_{-l_a/2}^{l_a/2} c'_{ya}\left(h_a + x'_{c.p.a}\right) dz = \int_{-l/2}^{l/2} b' c'_y \left(h + x'_{c.p.}\right) dz,$$

where l_a and l are the wing spans. Setting c'_{ya} = const, c'_y = const, $\overline{x}_{c.p.a} = x_{c.p.a}/b_a$ = const, and $\overline{x}_{c.p.} = x'_{c.p.}/b'$ = const, and using the relation $b_a l_a = S_a = S$, we obtain

$$c'_{ya} = c'_y,$$

$$h_a + b_a \overline{x}'_{c.p.a} = \frac{1}{S} \int_{-l/2}^{l/2} h b'\, dz + \frac{\overline{x}'_{c.p.}}{S} \int_{-l/2}^{l/2} b'^2\, dz.$$

The geometrical parameters of the equivalent wing will be independent of $\mathbf{M}_\infty$ if we also have $\overline{x}'_{c.p.a} = x'_{c.p.}$ and set

$$h_a = \frac{1}{S} \int_{-l/2}^{l/2} h b'\, dz, \qquad b_a = \frac{1}{S} \int_{-l/2}^{l/2} b'^2\, dz. \tag{4.62}$$

We now consider a wing symmetric in plan with a constant sweepback relative to the leading and trailing edge. Then b' will be a linear function of $\overline{z}$ [see formula (1.7)]; moreover Fig. 4.7 shows that

$$h = z\, \text{tg}\, \chi_0.$$

After some transformation and using the relations in §2, Chapter, I, we obtain from (4.62) the formulas

$$\frac{b_a}{b} = \frac{2}{3}\left(1 + \frac{1}{\eta(\eta+1)}\right), \qquad \frac{h_a}{b} = \frac{\lambda \, \mathrm{tg}\,\chi_0}{12}\cdot\frac{\eta+2}{\eta}. \tag{4.63}$$

If z_a is the section at which the leading edges of the two wings intersect, then from Fig. 4.7 we obtain

$$z_a = \frac{h_a}{\mathrm{tg}\,\chi_0}.$$

Hence the expression for z_a can be written

$$\bar{z}_a = \frac{1}{3}\cdot\frac{\eta+2}{\eta+1}, \qquad \bar{z}_a = \frac{z_a}{l/2}. \tag{4.64}$$

It is not difficult to see that when $\bar{z} = \bar{z}_a$ the chord of the wing under consideration will be b_a.

It follows from (4.63) that b_a/b is independent of geometrical wing parameters such as λ and χ_0. The mean aerodynamic chord depends only on the taper ratio η and varies in the range

$$\frac{2}{3} \leqslant \frac{b_a}{b} \leqslant 1.$$

The position of b_a, determined by h_a, depends on all geometrical parameters of the wing.

§6. The Calculation of Aerodynamic Coefficients

It is sometimes necessary for a given wing to convert the rotational-derivative coefficients from certain characteristic dimensions to others. As an example we consider the case in which the characteristic linear dimensions in Eqs. (1.4) and (1.19) are identical.

In one case let

$$\left.\begin{array}{l} c_y = \dfrac{Y}{qS}, \qquad m_x = \dfrac{M_x}{qSl_1}, \qquad m_z = \dfrac{M_z}{qSl_2}, \\[2mm] \omega_x = \dfrac{\Omega_x l_1}{U_0}, \qquad \omega_z = \dfrac{\Omega_z l_2}{U_0}, \qquad m_x = m_x^{\omega_x}\omega_x, \\[2mm] c_y = c_y^{\alpha}\alpha + c_y^{\omega_z}\omega_z, \qquad m_z = m_z^{\alpha}\alpha + m_z^{\omega_z}\omega_z \end{array}\right\} \tag{4.65}$$

and in the other let

$$\left.\begin{array}{l} c_y^* = \dfrac{Y}{qS^*}, \qquad m_x^* = \dfrac{M_x}{qS^*l_1^*}, \qquad m_z^* = \dfrac{M_z}{qS^*l_2^*}, \\[2mm] \omega_x^* = \dfrac{\Omega_x l_1^*}{U_0}, \qquad \omega_z^* = \dfrac{\Omega_z l_2^*}{U_0}, \qquad m_x^* = m_{x*}^{\omega_x}\omega_x^*, \\[2mm] c_y^* = c_{y*}^{\alpha}\alpha + c_{y*}^{\omega_z}\omega_z^*, \qquad m_z^* = m_{z*}^{\alpha}\alpha + m_{z*}^{\omega_z}\omega_z^*. \end{array}\right\} \tag{4.66}$$

Using the fact that (4.65) and (4.66) yield identical representations for Y, M_x and M_z obtain the following relations between the rotational derivative coefficients:

$$c_{y*}^\alpha = \frac{S}{S^*}\, c_y^\alpha, \quad m_{z*}^\alpha = \frac{Sl_2}{S^* l_2^*}\, m_z^\alpha, \quad c_{y*}^{\omega_z} = \frac{Sl_2}{S^* l_2^*}\, c_y^{\omega_z},$$

$$m_{z*}^{\omega_z} = \frac{Sl_2^2}{S^* l_2^{*2}}\, m_z^{\omega_z}, \quad m_{x*}^{\omega_x} = \frac{Sl_1^2}{S^* l_1^{*2}}\, m_x^{\omega_x}. \qquad (4.67)$$

In applications it is important to know how to convert rotational-derivative coefficients from one position of the coordinate system to another [32]. Let there be two coordinate systems with parallel axes related to the body, so that its motion can be described in two ways. One approach is to consider the motion as a translation with velocity U_0 of the point O together with rotation with angular velocity Ω about an axis passing through O; the other is to consider the motion a translation of O_1 with the vector Ω passing through O_1. For these two different methods of considering the same motion the aerodynamic forces and moments will be identical. If we use the same characteristic linear dimension $l_1 = l_2 = b$ and area S in the two representations of the motion, then the aerodynamic coefficients will also be identical. The kinematic parameters and rotational-derivative coefficients, however, will differ.

Let the position of the new origin O_1 relative to the old origin O be given by the vector

$$r_0 = ix_0 + jy_0 + kz_0.$$

Since the velocity of O_1 relative to O is given by the vector $\Omega \times r_0$, the angle of attack α in the new coordinate system is

$$\alpha_1 = \alpha + \omega_x \zeta_0 - \omega_z \xi_0, \quad \xi_0 = \frac{x_0}{b}, \quad \zeta_0 = \frac{z_0}{b},$$

$$\eta_0 = \frac{y_0}{b}, \quad \omega_x = \frac{\Omega_x b}{U_0}, \quad \omega_z = \frac{\Omega_z b}{U_0}. \qquad (4.68)$$

We first consider the case when the kinematic parameters are determined for the new origin O_1 while the center of reduction of the forces is the original origin O. Such relations can be of interest, for example, in the analysis of experimental results obtained from models in rotational motion relative to different centers but with all forces determined for the same position.

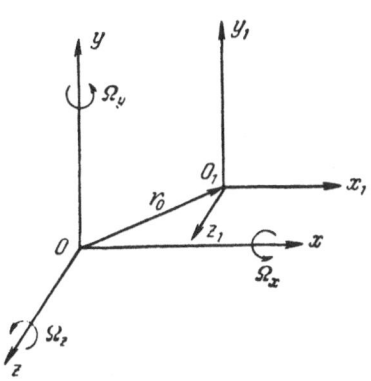

Fig. 4.8. The conversion from one coordinate system Oxyz to another $O_1 x_1 y_1 z_1$.

We consider a thin, annular or monoplane wing-plate of arbitrary shape in plan. The Oxz plane is parallel to the plane of the monoplane wing, the effect of the angle of slip β will not be considered and we set $\beta = 0$. Then the rotation Ω_y of the wing will not disturb the surrounding medium. For an annular wing the axes of the Oxyz system will be located so as to eliminate the need to consider the rotation Ω_y; this clearly does not cause any loss of generality. For the original system Oxyz we now have, with linear accuracy,

$$c_y = c_y^\alpha \alpha + c_y^{\omega_x}\omega_x + c_y^{\omega_z}\omega_z, \quad c_z = 0,$$

$$m_x = m_x^\alpha \alpha + m_x^{\omega_x}\omega_x + m_x^{\omega_z}\omega_z, \quad m_y = 0,$$

$$m_z = m_z^\alpha \alpha + m_z^{\omega_x}\omega_x + m_z^{\omega_z}\omega_z, \qquad (4.69)$$

and when the kinematic parameters are referred to the $O_1 x_1 y_1 z_1$ system we have

$$c_y = c_{y0}^{\alpha}\alpha_1 + c_{y0}^{\omega x}\omega_x + c_{y0}^{\omega z}\omega_z, \quad m_x = m_{x0}^{\alpha}\alpha_1 + m_{x0}^{\omega x}\omega_x + m_{x0}^{\omega z}\omega_z,$$
$$m_z = m_{z0}^{\alpha}\alpha_1 + m_{z0}^{\omega x}\omega_x + m_{z0}^{\omega z}\omega_z. \quad \text{(4.70)}$$

We substitute α_1 from (4.68) into (4.70) and obtain new expressions for the coefficients c_y, m_x and m_z identical to (4.69). Comparing the terms in these expressions representing the same kinematic parameters, we obtain

$$c_{y0}^{\alpha} = c_y^{\alpha}, \qquad\qquad m_{x0}^{\alpha} = m_x^{\alpha}, \quad m_{z0}^{\alpha} = m_z^{\alpha},$$
$$c_{y0}^{\omega x} = c_y^{\omega x} - c_y^{\alpha}\zeta_0, \qquad c_{y0}^{\omega z} = c_y^{\omega z} + c_y^{\alpha}\xi_0,$$
$$m_{x0}^{\omega x} = m_x^{\omega x} - m_x^{\alpha}\zeta_0, \quad m_{x0}^{\omega z} = m_x^{\omega z} + m_x^{\alpha}\xi_0, \quad \text{(4.71)}$$
$$m_{z0}^{\omega x} = m_z^{\omega x} - m_z^{\alpha}\zeta_0, \quad m_{z0}^{\omega z} = m_z^{\omega z} + m_z^{\alpha}\xi_0.$$

We now investigate the case when the kinematic parameters are determined for the new origin O_1 which is the center of action of the forces. All quantities referred to the new center O_1 will have the subscript "1". A change in the center of action does not change the resultant force $\mathbf{R}$, while the resultant moment changes by an amount $\mathbf{r}_0 \times \mathbf{R}$. Discarding c_X as a small quantity of the second order, we obtain

$$m_{x1} = m_x - c_z\eta_0 + c_y\zeta_0,$$
$$m_{y1} = m_y + c_z\xi_0, \quad \text{(4.72)}$$
$$m_{z1} = m_z - c_y\xi_0.$$

It is thus clear that we have $m_{y1} = 0$ under the above limitations.

We express the left- and right-hand sides of (4.72) in terms of the rotational-derivative coefficients with kinematic parameters for the center O_1. Since we obtain identitites in α_1, ω_X, and ω_Z, it follows that we can equate corresponding terms on the left and right in (4.72), and using (4.71) we obtain

$$m_{x1}^{\alpha} = m_x^{\alpha} + c_y^{\alpha}\zeta_0, \quad m_{z1}^{\alpha} = m_z^{\alpha} - c_y^{\alpha}\xi_0,$$
$$m_{x1}^{\omega x} = m_x^{\omega x} - m_x^{\alpha}\zeta_0 - c_y^{\alpha}\zeta_0^2 - c_y^{\omega x}\zeta_0,$$
$$m_{z1}^{\omega x} = m_z^{\omega x} - m_z^{\alpha}\zeta_0 + c_y^{\alpha}\xi_0\zeta_0 - c_y^{\omega x}\xi_0, \quad \text{(4.73)}$$
$$m_{x1}^{\omega z} = m_x^{\omega z} + m_x^{\alpha}\xi_0 + c_y^{\alpha}\xi_0\zeta_0 + c_y^{\omega z}\zeta_0,$$
$$m_{z1}^{\omega z} = m_z^{\omega z} + m_z^{\alpha}\xi_0 - c_y^{\alpha}\xi_0^2 - c_y^{\omega z}\xi_0.$$

The coefficients of the rotational derivatives of the forces do not change with a change in the center of action of the forces, and so

$$c_{y1}^{\alpha} = c_{y0}^{\alpha}, \quad c_{y1}^{\omega x} = c_{y0}^{\omega x}, \quad c_{y1}^{\omega z} = c_{y0}^{\omega z}. \quad \text{(4.74)}$$

The rotational-derivative coefficients for the pressure and circulation are calculated in the same way as the coefficients for the forces, i.e., from relations of the type (4.74).

The conversion formulas for a monoplane wing can be further simplified if we consider wings of symmetric shape in plan and refer all the original data to an origin O at some point of the wing surface in the plane of symmetry xOy (Fig. 4.5); we obtain

$$m_x^a = 0, \quad c_y^{\omega x} = 0, \quad m_x^{\omega z} = 0, \quad m_z^{\omega x} = 0.$$

For this most important case, the conversion formulas are

$$\left.\begin{array}{l}
c_{y1}^a = c_{y0}^a = c_y^a, \quad c_{y1}^{\omega x} = c_{y0}^{\omega x} = -c_y^a \zeta_0, \quad c_{y1}^{\omega z} = c_{y0}^{\omega z} = c_y^{\omega z} + c_y^a \xi_0, \\[4pt]
m_{x0}^{\omega x} = m_x^{\omega x}, \quad m_{x0}^{\omega z} = 0, \quad m_{z0}^a = m_z^a, \quad m_{z0}^{\omega x} = -m_z^a \zeta_0, \\[4pt]
m_{z0}^{\omega z} = m_z^{\omega z} + m_z^a \xi_0, \quad m_{x1}^a = c_y^a \zeta_0, \quad m_{x1}^{\omega x} = m_x^{\omega x} - c_y^a \zeta_0^2, \\[4pt]
m_{x1}^{\omega z} = c_y^a \xi_0 \zeta_0 + c_y^{\omega z} \zeta_0, \quad m_{z1}^a = m_z^a - c_y^a \xi_0, \\[4pt]
m_{z1}^{\omega x} = c_y^a \xi_0 \zeta_0 - m_z^a \zeta_0, \quad m_{z1}^{\omega z} = m_z^{\omega z} + (m_z^a - c_y^{\omega z}) \xi_0 - c_y^a \xi_0^2.
\end{array}\right\} \quad (4.75)$$

In the case of annular wings, we will assume that all the original data are obtained for an origin O on the axis of symmetry of the wing (Fig. 4.2). Because of the axial symmetry of the wing, we need only consider changes in the position of O in any meridian plane, for example $\zeta = 0$. Using the symmetry of the wing and the fact that a rotation about Ox will not disturb the flow, we obtain

$$c_y^{\omega x} = m_x^a = m_x^{\omega x} = m_x^{\omega z} = 0.$$

For an annular wing the conversion formulas are thus

$$\left.\begin{array}{l}
c_{y1}^a = c_{y0}^a = c_y^a, \quad c_{y1}^{\omega z} = c_{y0}^{\omega z} = c_y^{\omega z} + c_y^a \xi_0, \quad m_{x0}^{\omega x} = m_{x0}^{\omega z} = 0, \\[4pt]
m_{z0}^a = m_z^a, \quad m_{z0}^{\omega x} = 0, \quad m_{z0}^{\omega z} = m_z^{\omega z} + m_z^a \xi_0, \quad m_{x1}^a = 0, \\[4pt]
m_{x1}^{\omega x} = 0, \quad m_{x1}^{\omega z} = 0, \quad m_{z1}^a = m_z^a - c_y^a \xi_0, \quad m_{z1}^{\omega x} = 0, \\[4pt]
m_{z1}^{\omega z} = m_z^{\omega z} + (m_z^a - c_y^{\omega z}) \xi_0 - c_y^a \xi_0^2.
\end{array}\right\} \quad (4.76)$$

We finally consider the conversion from the coordinate $\bar{x}_\alpha$ of the focus F in the Oxyz system to the coordinate $\bar{x}_F$ relative to the nose of the mean aerodynamic chord.

Let x_F be the distance from F to the leading edge of the mean aerodynamic chord ($x_F > 0$ if F is behind the leading edge — in Fig. 4.9 we have $x_F > 0$) and let x_0 be the distance from O to the leading edge. Then from Fig. 4.9 we have

$$x_F = x_0 - h_a - x_a,$$

or

$$\bar{x}_F = \frac{b}{b_a}\left(\bar{x}_0 - \bar{x}_a - \frac{h_a}{b}\right), \quad \bar{x}_F = \frac{x_F}{b_a},$$

$$\bar{x}_0 = \frac{x_0}{b}, \quad \bar{x}_a = \frac{x_a}{b}. \quad (4.77)$$

For an annular wing, as for a rectangular wing, $h_a = 0$, $b_a = b$, and so

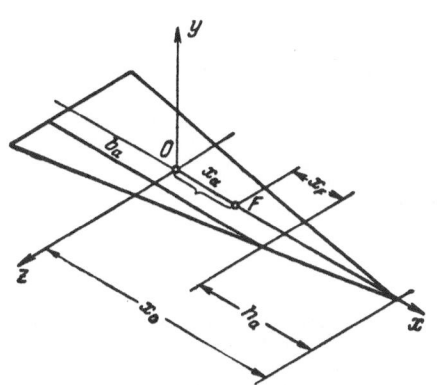

Fig. 4.9. Relation between the dimensionless coordinates $\bar{x}_F$ and $\bar{x}_\alpha$.

$$\bar{x}_F = \bar{x}_0 - \bar{x}_a, \quad \bar{x}_0 = \frac{x_0}{b}, \quad \bar{x}_a = \frac{x_a}{b}. \quad (4.78)$$

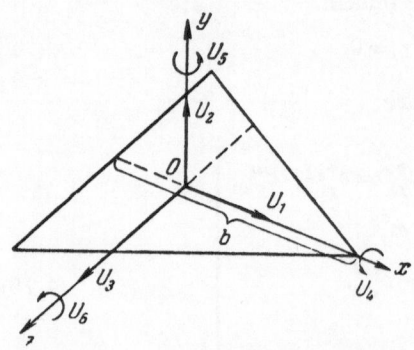

Fig. 4.10. The calculation of the apparent masses of a monoplane wing.

§7. The Apparent Masses of a Monoplane Wing

In the calculation of apparent masses of thin monoplane wings, we will use the lifting-vortex-sheet method. We will derive relations relating the apparent masses to the strength of the bound vortex layer.

To use the Zhukovskii theorem "in the small" (3.9), we must find the relative velocities at points on the wing surface. For a plate of arbitrary plan form, the disturbed velocities W_{x0} and W_{z0} at any point s_0 of the plate are zero. Moreover, at any point of the plate the normal component of the relative velocity is zero because of the boundary condition:

$$W_{00y} = W_{00n} = 0.$$

On the surface of the plate we have y = 0, and the velocity (1.27) will be zero (Fig. 4.10)

$$\mathbf{W}_0^* = \mathbf{i}(U_1 + U_5 z) + \mathbf{j}(U_2 + U_6 x - U_4 z) + \mathbf{k}(U_3 - U_5 x).$$

Hence

$$\mathbf{W}_{00} = \mathbf{W} - \mathbf{W}^* = -\mathbf{i}(U_1 + U_5 z) - \mathbf{k}(U_3 - U_5 x). \tag{4.79}$$

For l it is most convenient to use the Oz direction, and for τ the Ox direction. Then the Zhukovskii theorem can be written

$$\frac{\Delta p}{\rho} = W_{00z}\gamma_z - W_{00x}\gamma_x + \frac{\partial \Gamma}{\partial t}. \tag{4.80}$$

Using the fact that motion of the wing corresponding to the velocities U_1, U_3, and U_5 does not disturb the medium, we express γ_x, γ_z, and Γ as linear combinations of the remaining kinematic parameters:

$$\left.\begin{aligned}
\gamma_x &= U_2 \gamma_{x2} + U_4 b \gamma_{x4} + U_6 b \gamma_{x6}, \\
\gamma_z &= U_2 \gamma_{z2} + U_4 b \gamma_{z4} + U_6 b \gamma_{z6}, \\
\Gamma &= U_2 b \Gamma_2 + U_4 b^2 \Gamma_4 + U_6 b^2 \Gamma_6.
\end{aligned}\right\} \tag{4.81}$$

Here γ_{x_i}, γ_{z_i}, and Γ_i are dimensionless; for the projection of the vector $\mathbf{W}_{00}$ on the Oy and Oz axes we obtain from (4.79) the formulas

$$W_{00x} = -U_1 - U_5 z, \qquad W_{00z} = -U_3 + U_5 x. \tag{4.82}$$

Only the kinematic parameters U_i depend everywhere on the time t.

Equation (4.80) for the pressure difference can now be written

$$\frac{\Delta p}{\rho} = (-U_3 + U_5 x)(U_2 \gamma_{x2} + U_4 b \gamma_{x4} + U_6 b \gamma_{x6}) +$$
$$+ (U_1 + U_5 z)(U_2 \gamma_{z2} + U_4 b \gamma_{z4} + U_6 b \gamma_{z6}) + \frac{dU_2}{dt} b \Gamma_2 + \frac{dU_4}{dt} b^2 \Gamma_4 + \frac{dU_6}{dt} b^2 \Gamma_6. \tag{4.83}$$

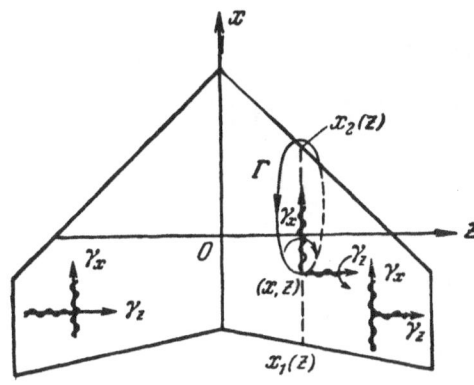

Fig. 4.11. The determination of $\Gamma(x, z)$.

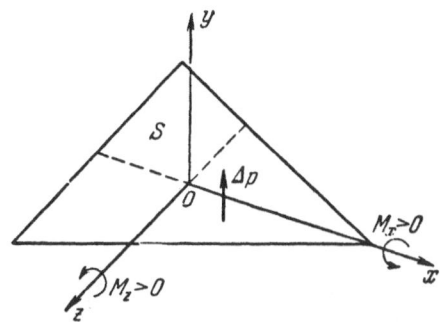

Fig. 4.12. The determination of the aerodynamic moments and forces for irrotational flow.

We note that if $x_1(z)$ and $x_2(z)$ are the coordinates of the trailing and leading edges of the section z = const, then (Fig. 4.11)

$$\left.\begin{aligned}
\Gamma_2(x, z) &= \int_x^{x_2} \gamma_{z2}(x, z)\,dx, \\
\Gamma_4(x, z) &= \int_x^{x_2} \gamma_{z4}(x, z)\,dx, \\
\Gamma_6(x, z) &= \int_x^{x_2} \gamma_{z6}(x, z)\,dx.
\end{aligned}\right\} \quad (4.84)$$

Since we are considering irrotational flow about the wing, it follows that for each section z = const the total circulation of all the vortices must be zero:

$$\int_{x_1}^{x_2} \gamma_{z2}\,dx = 0, \quad \int_{x_1}^{x_2} \gamma_{z4}\,dx = 0, \quad \int_{x_1}^{x_2} \gamma_{z6}\,dx = 0. \quad (4.85)$$

Similar relations will hold for sections x = const:

$$\int_{z_1}^{z_2} \gamma_{x2}\,dz = 0, \quad \int_{z_1}^{z_2} \gamma_{x4}\,dz = 0, \quad \int_{z_1}^{z_2} \gamma_{x6}\,dz = 0. \quad (4.86)$$

Let the wing be symmetric in plan and let the plane of symmetry be Oxy. Then it is clear that for motion with velocities U_2 and U_6 the flow will be symmetric, and so

$$\left.\begin{aligned}
\gamma_{z2}(x, -z) &= \gamma_{z2}(x, z), & \gamma_{z6}(x, -z) &= \gamma_{z6}(x, z), \\
\gamma_{x2}(x, -z) &= -\gamma_{x2}(x, z), & \gamma_{x6}(x, -z) &= -\gamma_{x6}(x, z), \\
\Gamma_2(x, -z) &= \Gamma_2(x, z), & \Gamma_6(x, -z) &= \Gamma_6(x, z).
\end{aligned}\right\} \quad (4.87)$$

For motion with velocity U_4, the flow will be antisymmetric and

$$\begin{aligned}
\gamma_{z4}(x, -z) &= -\gamma_{z4}(x, z), \\
\gamma_{x4}(x, -z) &= \gamma_{x4}(x, z), \\
\Gamma_4(x, -z) &= -\Gamma_4(x, z).
\end{aligned} \quad (4.88)$$

We now find the forces and moments acting on the wing, expressing them relative to the axes of a bound coordinate system (Fig. 4.12):

$$Y = \iint_S \Delta p\,ds, \quad M_x = -\iint_S \Delta p z\,ds, \quad M_z = \iint_S \Delta p x\,ds. \quad (4.89)$$

We substitute (4.83) in the right-hand sides of the relations in (4.89). Assuming that the wing is symmetric in plan and using (4.87) and (4.88), we obtain

$$\frac{1}{\rho} Y = - U_3 U_4 b \int \int \gamma_{x4}\, ds + U_4 U_5 b \int \int \gamma_{x4} x\, ds +$$

$$+ U_1 U_2 \int \int \gamma_{22}\, ds + U_1 U_6 b \int \int \gamma_{26}\, ds + U_4 U_5 b \int \int \gamma_{24} z\, ds +$$

$$+ \frac{dU_2}{dt} b \int \int \Gamma_2\, ds + \frac{dU_6}{dt} b^2 \int \int \Gamma_6\, ds,$$

$$\frac{1}{\rho} M_z = - U_3 U_4 b \int \int \gamma_{x4} x\, ds + U_4 U_5 b \int \int \gamma_{x4} x^2\, ds +$$

$$+ U_1 U_2 \int \int \gamma_{22} x\, ds + U_1 U_6 b \int \int \gamma_{26} x\, ds +$$

$$+ U_4 U_5 b \int \int \gamma_{24} z x\, ds + \frac{dU_2}{dt} b \int \int \Gamma_2 x\, ds + \frac{dU_6}{dt} b^2 \int \int \Gamma_6 x\, ds,$$

$$\frac{1}{\rho} M_x = U_2 U_3 \int \int \gamma_{x2} z\, ds - U_2 U_5 \int \int \gamma_{x2} z x\, ds +$$

$$+ U_3 U_6 b \int \int \gamma_{x6} z\, ds - U_5 U_6 b \int \int \gamma_{x6} x z\, ds - U_1 U_4 b \int \int \gamma_{24} z\, ds -$$

$$- U_2 U_5 \int \int \gamma_{22} z^2\, ds - U_5 U_6 b \int \int \gamma_{26} z^2\, ds + \frac{dU_4}{dt} b^2 \int \int \Gamma_4 z\, ds.$$

$$(4.90)$$

Comparing these expressions with those obtained from the general dynamical theory [see (1.46)], we obtain expressions for the apparent masses in terms of the circulation.

In the calculation of the apparent masses, it is convenient to use the formulas

$$\lambda_{22} = - \rho b \int \int \Gamma_2\, ds, \qquad \lambda_{44} = - \rho b^2 \int \int \Gamma_4 z\, ds,$$
$$\lambda_{66} = - \rho b^2 \int \int \Gamma_6 x\, ds, \qquad \lambda_{26} = - \rho b^2 \int \int \Gamma_6\, ds. \qquad (4.91)$$

The apparent-mass coefficients of a wing of arbitrary shape in plan will be introduced as follows:

$$k_{22} = \frac{\lambda_{22}}{\rho S b}, \qquad k_{44} = \frac{\lambda_{44}}{\rho S b^3}, \qquad k_{66} = \frac{\lambda_{66}}{\rho S b^3}, \qquad k_{26} = \frac{\lambda_{26}}{\rho S b^2}. \qquad (4.92)$$

We then have

$$k_{22} = - \frac{1}{S} \int_S \int \Gamma_2\, ds, \qquad k_{44} = - \frac{1}{Sb} \int_S \int \Gamma_4 z\, ds,$$
$$k_{66} = - \frac{1}{Sb} \int_S \int \Gamma_6 x\, ds, \qquad k_{26} = - \frac{1}{S} \int_S \int \Gamma_6\, ds. \qquad (4.93)$$

As for the rotational-derivative coefficients, we can use the reasoning of the preceding paragraph to obtain formulas for the transition of the apparent masses from one coordinate system to another.

In the present case, instead of (4.75) we obtain

$$(k_{22})_1 = k_{22}, \qquad (k_{44})_1 = k_{44}, \qquad (k_{26})_1 = k_{26} - k_{22} \xi_0,$$
$$(k_{66})_1 = k_{66} - 2 k_{26} \xi_0 + k_{22} \xi_0^2. \qquad (4.94)$$

§ 8. The Apparent Masses of an Annular Wing

As in the case of monoplane wings, we obtain expressions for the apparent masses of a cylindrical annular wing in terms of the strength of the bound vortices. The axes of these vortices are in planes parallel to the axes of symmetry of the annular wing, and so the Zhukovskii theorem "in the small" has the form (Fig. 4.2)

$$\frac{\Delta p}{\rho} = -W_{00}\gamma + \frac{\partial \Gamma}{\partial t}. \tag{4.95}$$

Motion with velocities U_1 and U_4 does not disturb the medium and moreover, for an annular wing, it is sufficient to investigate motion characterized by the parameters U_2 and U_6; hence

$$\left. \begin{array}{l} \gamma = (U_2\gamma_2 + U_6 r\gamma_6)\cos\varphi, \\ \Gamma = (U_2 r\Gamma_2 + U_6 r^2\Gamma_6)\cos\varphi. \end{array} \right\} \tag{4.96}$$

where γ_2, γ_6, Γ_2 and Γ_6 are dimensionless functions independent of the time and the kinematic parameters U_2 and U_6 are functions of t.

Since

$$W_{00} = -U_1 + U_6 y,$$

it follows that (4.95) can be written

$$\frac{\Delta p}{\rho} = \left[(U_1 - U_6 y)(U_2\gamma_2 + U_6 r\gamma_6) + \frac{dU_2}{dt} r\Gamma_2 + \frac{dU_6}{dt} r^2\Gamma_6 \right] \cos\varphi. \tag{4.97}$$

The lift Y and the moment M_Z can be obtained in two different ways. One method is to use (4.23), in which we must set $\alpha_0 = 0$, and (4.97); the other is to use (1.49). By comparing the resulting expressions, it is not difficult to obtain the desired relation. The apparent-mass coefficients for annular wings are introduced in the form

$$k_{22} = \frac{\lambda_{22}}{\rho S b}, \qquad k_{66} = \frac{\lambda_{66}}{\rho S b^3}, \qquad k_{26} = \frac{\lambda_{26}}{\rho S b^2}, \tag{4.98}$$

where

$$S = 2rb.$$

To calculate these coefficients it is then most convenient to use the formulas

$$\left. \begin{array}{l} k_{22} = -\dfrac{\pi\lambda}{4b} \displaystyle\int_{-b/2}^{b/2} \Gamma_2\, dx, \qquad k_{66} = -\dfrac{\pi\lambda^2}{8b^2} \displaystyle\int_{-b/2}^{b/2} \Gamma_6 x\, dx, \\[4mm] k_{26} = -\dfrac{\pi\lambda^2}{8b} \displaystyle\int_{-b/2}^{b/2} \Gamma_6\, dx. \end{array} \right\} \tag{4.99}$$

The formulas for the conversion of these coefficients to the other axes are similar to those in (4.75):

$$\left. \begin{array}{l} (k_{22})_1 = k_{22}, \qquad (k_{26})_1 = k_{26} - k_{22}\xi_0, \\ (k_{66})_1 = k_{66} - 2k_{26}\xi_0 + k_{22}\xi_0^2. \end{array} \right\} \tag{4.100}$$

CHAPTER V

THE CONVERSION OF AERODYNAMIC CHARACTERISTICS TO VALUES FOR HIGH SUBSONIC VELOCITIES

§ 1. The Basic Transformation

If the velocity of flight U_0 is low compared to the velocity of sound in the medium, then the medium in which the body moves can be assumed to be incompressible. This is because compressibility has practically no effect when the disturbance p' of the pressure generated by the motion of the body is very small compared to the pressure p_∞ in the undisturbed medium:

$$\frac{p'}{p_\infty} \ll 1.$$

Denoting the dynamic pressure, the density, and the Mach number in the undisturbed flow by the subscript ∞, we can write

$$\frac{q_\infty}{p_\infty} = \frac{\rho_\infty U_0^2}{2 p_\infty} = \frac{k}{2} \mathbf{M}_\infty^2,$$

and so

$$\frac{p'}{p_\infty} = \frac{k}{2} \bar{p} \mathbf{M}_\infty^2, \tag{5.1}$$

where $\bar{p} = p'/q_\infty$ is the pressure coefficient. This coefficient basically depends on the coordinates of the point where $\bar{p}$ is determined, on the shape of the body, and on its position in the flow. The dependence of $\bar{p}$ on $\mathbf{M}_\infty$ is relatively slight for subsonic velocities, since an increase in $\mathbf{M}_\infty$ will always lead to an increase in p'/p_∞. Moreover for increasing body thickness and increasing angle of attack with fixed $\mathbf{M}_\infty$, the maximum magnitude of $\bar{p}$ will increase, the value of $|p'/p_\infty|_{max}$ will increase, and the effects of compressibility of the medium will be felt for lower $\mathbf{M}_\infty$.

Hence the effect of compressibility of the medium will be stronger for higher $\mathbf{M}_\infty$ in the oncoming flow and for stronger disturbances caused by the body for small $\mathbf{M}_\infty$.

In the linear theory, the effect of compressibility can be taken into account approximately by reducing the problem to that of the flow of an incompressible fluid about a deformed body. If systematic data for incompressible media are available, then the problem can be reduced to a simple conversion of the available characteristics. We will show how to perform this conversion, and will explain its basis.

The application of certain affine transformation, called the Prandtl-Glauert transformation [9, 41], reduces the continuity equation (1.18) to the form (1.23) for an incompressible medium. We replace the coordinates x, y, and z of the physical space by new coordinates x_M,

67

y_M, and z_M, related to the old coordinates by the equations

$$x = x_M \sqrt{1 - M_\infty^2}, \qquad y = y_M, \qquad z = z_M. \tag{5.2}$$

Since

$$\frac{dx_M}{dx} = \frac{1}{\sqrt{1 - M_\infty^2}}, \qquad \frac{dy_M}{dy} = 1, \qquad \frac{dz_M}{dz} = 1,$$

we have

$$\left.\begin{array}{l} \dfrac{\partial \Phi}{\partial x} = \dfrac{\partial \Phi}{\partial x_M} \dfrac{dx_M}{dx} = \dfrac{\partial \Phi}{\partial x_M} \dfrac{1}{\sqrt{1 - M_\infty^2}}, \\[4mm] \dfrac{\partial \Phi}{\partial y} = \dfrac{\partial \Phi}{\partial y_M}, \quad \dfrac{\partial \Phi}{\partial z} = \dfrac{\partial \Phi}{\partial z_M}, \end{array}\right\} \tag{5.3}$$

and the continuity equation (1.18) becomes

$$\frac{\partial^2 \Phi}{\partial x_M^2} + \frac{\partial^2 \Phi}{\partial y_M^2} + \frac{\partial^2 \Phi}{\partial z_M^2} = 0. \tag{5.4}$$

Hence in the (x_M, y_M, z_M) space, the potential of the disturbed velocities $\Phi(x_M, y_M, z_M)$ satisfies an equation identical to the equation of continuity for an incompressible medium. Hence the (x_M, y_M, z_M) space can be considered as a fictitious (auxiliary) space filled with an incompressible fluid.

In addition to the disturbed velocities of the gas (1.20), we introduce the disturbed velocities of the fictitious incompressible medium

$$W_{xM} = \frac{\partial \Phi}{\partial x_M}, \qquad W_{yM} = \frac{\partial \Phi}{\partial y_M}, \qquad W_{zM} = \frac{\partial \Phi}{\partial z_M}. \tag{5.5}$$

Then from (5.3) we can find the relation between the disturbed velocities in the compressible medium (the gas) and the velocities (5.5):

$$W_{xcom} = \frac{W_{xM}}{\sqrt{1 - M_\infty^2}}, \qquad W_{ycom} = W_{yM}, \qquad W_{zcom} = W_{zM}. \tag{5.6}$$

The velocities in (5.6) are taken at the points (x, y, z) and (x_M, y_M, z_M) respectively, the relation between these points being given by (5.2).

The above considerations, however, do not completely solve the problem of converting aerodynamic characteristics to values that can be used for high subsonic velocities. This is because the equation of continuity is not the only condition that must be satisfied in the flow about a body. We must also show that in the incompressible-medium space we can find a lifting surface such that if we know the solution for this surface in an incompressible fluid we can satisfy all the conditions of the problem in the space filled with gas.

§2. The Geometrical Parameters of the Transformed Wing

One of the conditions of the problem is the boundary condition related to the smooth flow past a lifting surface. In this connection we must answer two questions. First, how will we choose the shape of the lifting surface, and second, what must be the form of the boundary conditions on this surface in the fictitious space of incompressible gas so that the above-mentioned conditions will be satisfied on the original lifting surface in a gas flow?

To answer the first question, we consider how the original surface will change under the transformation (5.2). In the (x_M, y_M, z_M) space all the linear dimensions in the direction of the Ox axis are increased by a factor $1/\sqrt{1-M_\infty^2}$, while the linear dimensions in the directions of the Oy and Oz axes remained unchanged. Hence the shape of the lifting surface in the fictitious space of incompressible fluid will be different from the shape of the original surface in the gas space. All the geometrical parameters will be indicated by the subscript "M".

We consider a monoplane wing symmetric in plan, with a constant sweepback from the leading to the trailing edge (Fig. 5.1). Its shape in plan, as we have already noted, will be determined by three dimensionless parameters — the aspect ratio λ, the taper ratio η, and the sweepback from the leading edge χ_0. The root and tip chords of the corresponding transformed wing and the span will be

$$b_M = \frac{b}{\sqrt{1-M_\infty^2}}, \quad b_{\kappa M} = \frac{b_\kappa}{\sqrt{1-M_\infty^2}}, \quad l_M = l.$$

Hence from Fig. 5.1. we find that the area and the values of the parameter Δ_M are given by the formulas

$$S_M = \frac{b_M + b_{\kappa M}}{2} l_M = \frac{S}{\sqrt{1-M_\infty^2}}, \quad \Delta_M = \frac{\Delta}{\sqrt{1-M_\infty^2}},$$

and since

$$\lambda_M = \frac{l_M^2}{S_M}, \quad \eta_M = \frac{b_M}{b_{\kappa M}}, \quad \operatorname{tg}\chi_{0M} = \frac{2\Delta_M}{l_M},$$

the dimensionless geometrical parameters of the transformed wing will be given in terms of the parameters of the original wing by the formulas

$$\lambda_M = \lambda\sqrt{1-M_\infty^2}, \quad \eta_M = \eta, \quad \operatorname{tg}\chi_{0M} = \frac{\operatorname{tg}\chi_0}{\sqrt{1-M_\infty^2}}. \tag{5.7}$$

For an annular wing we can write (Fig. 5.2):

$$b_M = \frac{b}{\sqrt{1-M_\infty^2}}, \quad D_M = D.$$

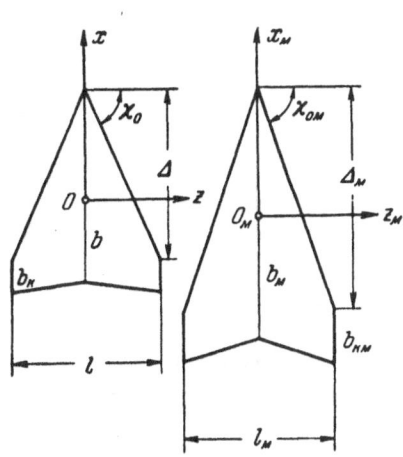

and, since the aspect ratio of the transformed wing is

$$\lambda_M = \frac{D_M}{b_M},$$

we find that it is related to the aspect ratio of the original wing by the equation

$$\lambda_M = \lambda\sqrt{1-M_\infty^2}. \tag{5.8}$$

Fig. 5.1. Original and transformed monoplane wings.

We must now find all the parameters in the boundary conditions. These parameters are the angular velocities ω_{xM} and ω_{zM}, the angle of attack α_M, and the taper angle α_{0M} of an annular wing, and they must be selected so that the solution of the

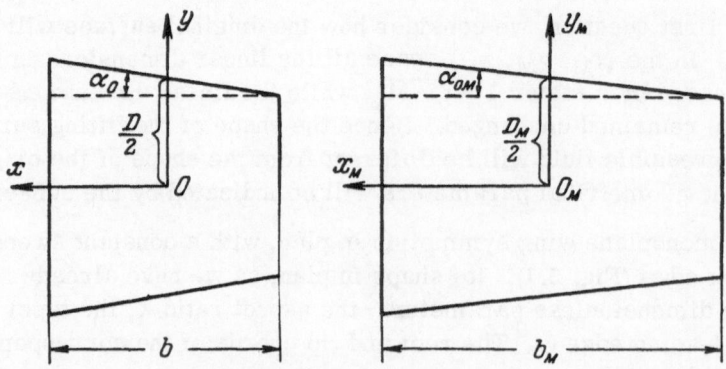

Fig. 5.2. The basic and the transformed annular wing.

problem for the transformed wing will satisfy in particular the boundary conditions imposed on the original wing.

The boundary condition (1.30) for the original monoplane wing can be written

$$\frac{1}{U_0}\frac{\partial\Phi}{\partial y} = -a + \omega_z\frac{x_0}{b} - \omega_{x1}\frac{2z_0}{l}, \quad \omega_z = \frac{\Omega_z b}{U_0}, \quad \omega_{x1} = \frac{\Omega_x l}{2U_0} \tag{5.9}$$

and for the transformed wing

$$\frac{1}{U_{0M}}\frac{\partial\Phi}{\partial y_M} = -a_M + \omega_{zM}\frac{x_{0M}}{b_M} - \omega_{x1M}\frac{2z_{0M}}{l_M}. \tag{5.10}$$

If we set

$$U_{0M} = U_0, \quad a_M = a, \quad \omega_{zM} = \omega_z, \quad \omega_{x1M} = \omega_{x1}, \tag{5.11}$$

then for the corresponding base points and transformed wing

$$\frac{x_M}{b_M} = \frac{x}{b}, \quad \frac{2z_M}{l_M} = \frac{2z}{l}$$

the boundary conditions (5.10) and (5.11) will be identical.

We proceed similarly for the annular wing, for which the original boundary conditions are (see (1.31)

$$\frac{1}{U_0}\frac{\partial\Phi}{\partial r} = -\alpha_0 - a\cos\varphi_0 + \omega_z\frac{x_0}{b}\cos\varphi_0, \tag{5.12}$$

and the conditions for the transformed wing are

$$\frac{1}{U_{0M}}\frac{\partial\Phi}{\partial r_M} = -\alpha_{0M} - a_M\cos\varphi_0 + \omega_{zM}\frac{x_{0M}}{b_M}\cos\varphi_0, \tag{5.13}$$

here it is sufficient to set

$$U_{0M} = U_0, \quad a_{0M} = \alpha_0, \quad a_M = a, \quad \omega_{zM} = \omega_z. \tag{5.14}$$

Thus by finding $\Phi(x_M, y_M, z_M)$ in the fictitious space of the incompressible fluid for the usual boundary conditions and the Chaplygin-Zhukovskii conditions, we find a velocity potential

$$\Phi(x, y, z) = \Phi(x_M, y_M, z_M),$$

which will satisfy all the conditions of the problem in gas flow about the original wing.

§3. The Conversion of Aerodynamic Characteristics of a Wing to Values for High Subsonic Velocities

In the solution of the problem for a wing in an incompressible medium, the velocity potential Φ is not usually calculated, and only the circulation, the pressure, and the aerodynamic coefficients are determined. It is thus expedient to obtain formulas that will permit us to carry out the direct transition from these results to the corresponding characteristics of a wing in a gas flow.

The disturbed velocities in the gas are obtained from the Eq. (5.6). By using the linearized Bernoulli equation (1.21) and Eqs. (5.6), it is not difficult to establish the relation between the pressure coefficient $\overline{p}_M = p'_M/q_\infty$ in the transformed space of incompressible fluid and the same coefficient $\overline{p}_{com} = p'/q_\infty$ in the gas flow:

$$\overline{p}_{com} = \frac{\overline{p}_M}{\sqrt{1 - M_\infty^2}}. \tag{5.15}$$

From this result and from expressions of type (4.5), we obtain

$$p_{a\,com} = \frac{p_{aM}}{\sqrt{1 - M_\infty^2}}, \quad p_{\omega_z com} = \frac{p_{\omega_z M}}{\sqrt{1 - M_\infty^2}}, \quad p_{\omega_x com} = \frac{p_{\omega_x M}}{\sqrt{1 - M_\infty^2}}. \tag{5.16}$$

We now consider the rotational-derivative coefficients for the wing as a whole. Let the aerodynamic coefficients $c_{y\,com}$, $m_{x1\,com}$, and $m_{z\,com}$ of a monoplane wing for some $M_\infty < 1$ be referred respectively to S, Sl, and Sb, where S is the area, l the span, and b the root chord of the wing, and let the coefficients of the transformed wing be referred to S_M, $S_M l_M$, $S_M b_M$. In the first case we can write

$$\left.\begin{aligned}
c_{y\,com} &= c_{y\,com}^\alpha \alpha + c_{y\,com}^{\omega_z} \omega_z, \quad & m_{x1\,com} &= m_{x1\,com}^{\omega_{x1}} \omega_{x1}, \\
m_{z\,com} &= m_{z\,com}^\alpha \alpha + m_{z\,com}^{\omega_z} \omega_z, \quad & \omega_{x1} &= \frac{\Omega_x l}{2U_0}, \quad \omega_z = \frac{\Omega_z b}{U_0}
\end{aligned}\right\} \tag{5.17}$$

and in the second case, using (5.11), we write

$$\left.\begin{aligned}
c_{yM} &= c_{yM}^\alpha \alpha + c_{yM}^{\omega_z} \omega_z, \quad & m_{x1\,M} &= m_{x1\,M}^{\omega_{x1}} \omega_{x1}, \\
m_{zM} &= m_{zM}^\alpha \alpha + m_{zM}^{\omega_z} \omega_z.
\end{aligned}\right\} \tag{5.18}$$

From Eqs. (5.16) and (4.7), we obtain the relation between the rotational-derivative coefficients (5.18) of the transformed wing in an incompressible medium and the required rotational-derivative coefficients (5.17):

$$c_{y\,com}^\alpha = \frac{c_{yM}^\alpha}{\sqrt{1 - M_\infty^2}}, \quad c_{y\,com}^{\omega_z} = \frac{c_{yM}^{\omega_z}}{\sqrt{1 - M_\infty^2}},$$

$$m_{x1\,com}^{\omega_{x1}} = \frac{m_{x1\,M}^{\omega_{x1}}}{\sqrt{1 - M_\infty^2}}, \quad m_{z\,com}^\alpha = \frac{m_{zM}^\alpha}{\sqrt{1 - M_\infty^2}}, \quad m_{z\,com}^{\omega_z} = \frac{m_{zM}^{\omega_z}}{\sqrt{1 - M_\infty^2}}. \tag{5.19}$$

From (4.55) and (4.77) it is now easily shown that the relative coordinate of the focus of the wing under consideration in the gas flow will be equal to the corresponding coordinate of the transformed wing in the incompressible flow:

$$\bar{x}_{F\,com} = \bar{x}_{FM}. \tag{5.20}$$

The formulas for the conversion of the corresponding characteristics of wing sections are similar.

The resulting formulas reduce the problem of the calculation of flow about a wing in a subsonic gas flow to a similar problem for a transformed wing in an incompressible fluid. This is particularly convenient in the determination of various aerodynamic characteristics of a wing for high subsonic velocities if systematic results are available for a series of wings in a compressible medium.

We will indicate the arrangement of the calculations needed to obtain aerodynamic characteristics of a monoplane wing at a high subsonic velocity. For a wing with given geometrical parameters λ, χ_0, and η and for the Mach number M_∞, we will obtain the parameters (5.7) of the transformed wing. Then from λ_M, χ_{0M}, and η_M, we calculate the characteristics of the transformed wing in an incompressible medium (for example C_{yM}^α and M_{zM}^α). The conversion of these characteristics to the required values is performed by using (5.6), (5.16), (5.19), and (5.20).

It is clear from (5.8) that when M_∞ tends to 1, the aspect ratio λ_M of the transformed wing decreases and tends to zero, independently of the elongation of the original wing (for finite λ). Below (see Chapters IX, XI, and XII), we will show that the rotational-derivative coefficients of a monoplane (or annular) wing for very small aspect ratio λ in an incompressible medium are

$$c_i = A_i\lambda. \tag{5.21}$$

where the c_i are certain of the coefficients and the A_i are constants.

Using (5.21) and the rule for the conversion of aerodynamic characteristics to their values for high subsonic velocities, we obtain the limiting values of the c_i for an annular or monoplane wing of arbitrary shape in plan for $M_\infty \to 1$:

$$c_{i\,com} = A_i\lambda.$$

where the original aspect ratio λ is not necessarily small.

Among other conclusions, it follows from this result that all the coefficients of wings of very small aspect ratio, with the form (5.21) for small Mach numbers, are independent of M_∞ for values up to $M_\infty = 1$.

Hence, for any finite λ, the formally linear theory leads to a method for converting aerodynamic characteristics to their values for any $M_\infty \leq 1$. In other words, we may use this method as long as it is sufficiently accurate. The limits of applicability of the linear theory must be established by a comparison of results obtained from its use with experimental or calculated results that are known to be more accurate. We can only say here that the linear theory will be more accurate when the disturbances generated by the body are smaller. In particular a decrease in the angle of attack, a lengthening of the wing, or an increase in the sweepback angle will all tend to decrease the strength of the disturbances caused by the presence of the wing in the flow, and so will increase the reliability of results obtained by the linear theory and increase the range of M_∞ in which it can be used.

The linear theory can be used to take into account the effect of compressibility both on the induced drag without suction and with suction forces.

In the first case, the problem is completely solved by using (5.7) and (5.19). In fact it follows from (4.19) that

$$\bar{B}_{com} = \frac{\pi\lambda}{c^a_{y\,com}} \qquad \bar{B}_M = \frac{\pi\lambda_M}{c^a_{yM}},$$

and so

$$\bar{B}_{com} = \bar{B}_M. \tag{5.22}$$

In the second case, (4.17) shows that the problem reduces to that of finding the relation between B_{com} and B_M. Both these factors are obtained by using (4.18):

$$B_{com} = -\frac{\pi\lambda^2}{2\left(c^a_{ycom}\right)^2} \int_{-1}^{1} w_{y\,com}\Gamma^*_{a\,com}\,d\bar{z},$$

$$B_M = -\frac{\pi\lambda^2_M}{2\left(c^a_{yM}\right)^2} \int_{-1}^{1} w_{yM}\Gamma^*_{aM}\,d\bar{z}.$$

We now obtain formulas for the conversion of the circulation. From (4.1), (4.5), (5.2), and (5.19) we have

$$\gamma_{z+com} = \frac{\gamma_{zM}}{\sqrt{1-M^2_\infty}} \qquad \Gamma_{+com} = \Gamma_{+M}. \tag{5.23}$$

since $\Gamma_{+com} = \gamma_{z+com}\,\Delta x$, $\Gamma_{+M} = \gamma_{z+M}\Delta x_M$.

Noting that from (5.5) we have

$$w_{y\,com} = w_{yM},$$

we easily establish a dependence for B similar to (5.22):

$$B_{com} = B_M. \tag{5.24}$$

Since from (4.19) we have

$$D_{com} = \bar{B}_{com} - B_{com}, \quad D_M = \bar{B}_M - B_M,$$

we finally obtain

$$D_{com} = D_M. \tag{5.25}$$

Hence the factors $\bar{B}$ for induced drag without suction force, B for induced drag with suction force, and the suction force D in a gas flow are equal respectively to the factors for the transformed wing in an incompressible medium.

4. The Conversion of Aerodynamic Characteristics of a Profile Cascade and Multiplane to Values for High Subsonic Velocities

We now consider a cascade of thin profiles (Fig. 5.3). We direct the Ox axis along a chord b of any plan; using the fact that the flow is plane−parallel, we write the transformation (5.2)

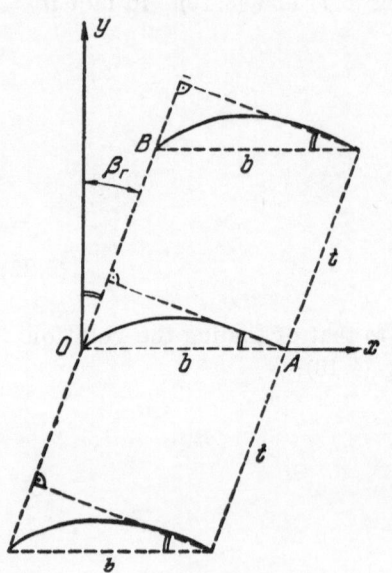

Fig. 5.3. The coordinate system for a profile cascade.

in the form

$$x = x_M \sqrt{1 - \mathbf{M}_\infty^2}, \qquad y = y_M. \tag{5.26}$$

The velocity in the gas flow will be calculated as it was for (5.6):

$$W_{x\,com} = \frac{W_{xM}}{\sqrt{1 - \mathbf{M}_\infty^2}}, \qquad W_{y\,com} = W_{yM}. \tag{5.27}$$

We now find the geometrical parameters of the transformed profile cascade. After the variable change (5.26), all the linear dimensions parallel to the Ox axis increase by the factor $1/\sqrt{1 - \mathbf{M}_\infty^2}$, while in the Oy direction there is no change. Using the fact that from Fig. 5.3 we have

$$OB = t, \qquad OA = b,$$

and changing the variables, we obtain

$$\left.\begin{aligned}
t_M \cos\beta_{\Gamma M} &= t \cos\beta_\Gamma, \\
t_M \sin\beta_{\Gamma M} &= \frac{t \sin\beta_\Gamma}{\sqrt{1 - \mathbf{M}_\infty^2}}, \\
b_M &= \frac{b}{\sqrt{1 - \mathbf{M}_\infty^2}}.
\end{aligned}\right\} \tag{5.28}$$

We have now a formula for the calculation of the geometrical angle of stagger of the transformed profile cascade:

$$\operatorname{tg}\beta_{\Gamma M} = \frac{\operatorname{tg}\beta_\Gamma}{\sqrt{1 - \mathbf{M}_\infty^2}}. \tag{5.29}$$

To find the relative gap $\bar{t}_M$ of the transformed cascade, we obtain from (5.29) the relation

$$\sin\beta_{\Gamma M} = \frac{\sin\beta_\Gamma}{\sqrt{1 - \mathbf{M}_\infty^2 \cos^2\beta_\Gamma}}$$

and substituting this expression in (5.28) we obtain

$$t_M = t\, \frac{\sqrt{1 - \mathbf{M}_\infty^2 \cos^2\beta_\Gamma}}{\sqrt{1 - \mathbf{M}_\infty^2}}.$$

From this relation between t_M and t and from the relation between b_M and b (5.28), we have

$$\bar{t}_M = \bar{t}\,\sqrt{1 - \mathbf{M}_\infty^2 \cos^2\beta_\Gamma}, \qquad \bar{t}_M = \frac{t_M}{b_M}, \qquad \bar{t} = \frac{t}{b}. \tag{5.30}$$

For the original profile cascade in a gas flow, the boundary condition (1.32), after we use (1.15), becomes

$$-\frac{W_{n\,com}}{U_0} = \sin\alpha_\Gamma\cos\varepsilon + \cos\alpha_\Gamma\sin\varepsilon,\tag{5.31}$$

and for the transformed cascade in an incompressible medium

$$-\frac{W_{nM}}{U_{0M}} = \sin\alpha_{\Gamma M}\cos\varepsilon_M + \cos\alpha_{\Gamma M}\sin\varepsilon_M.\tag{5.32}$$

We set

$$U_{0M} = U_0, \quad \alpha_{\Gamma M} = \alpha_\Gamma, \quad \bar{f}_M = \bar{f}.\tag{5.33}$$

and then at corresponding points of the transformed and original profiles, i.e., for $x_M/b_M = x/b$, we will have $\varepsilon_M = \varepsilon$ and the boundary conditions (5.31) and (5.32) will be identical. Hence, carrying out the calculation of the flow about the profile cascade in the (x_M, y_M) plane of the fictitious incompressible flow with the boundary conditions (5.32), we obtain the disturbed-velocity potential

$$\Phi(x_M, y_M) = \Phi(x, y).$$

This potential will satisfy the continuity equation (1.18) for the gas with the boundary condition (5.31) and the other conditions of our problem.

To derive conversion formulas so that we can avoid calculating the potential, we use the linearized Bernoulli equation. Using (5.27), we find as in §3 that

$$\bar{p}_{com} = \frac{\bar{p}_M}{\sqrt{1-M_\infty^2}}, \quad c^\alpha_{y\,com} = \frac{c^\alpha_{y,M}}{\sqrt{1-M_\infty^2}} \quad m^\alpha_{z\,com} = \frac{m^\alpha_{zM}}{\sqrt{1-M_\infty^2}}.\tag{5.34}$$

When M_∞ increases, the relative gap $\bar{t}_M$ of the transformed profile cascade decreases, as can be seen from (5.30). If $\beta_\Gamma = 0$, then when $M_\infty \to 1$, we have $\bar{t}_M \to 0$. We will also show that, for very small t,

$$c^\alpha_y = B_1\bar{t},\tag{5.35}$$

where B_1 is a constant. Hence in the case we are considering, if $M_\infty \to 1$ we have the following limiting value of $c^\alpha_{y\,com}$:

$$c^\alpha_{y\,com} = B_1\bar{t},\tag{5.36}$$

where the original $\bar{t}$ is not assumed to be small.

Hence the linear theory for finite $\bar{t}$ also permits the calculation of c^α_y for large M_∞ up to $M_\infty = 1$. We can judge the accuracy of the results of such a calculation, as was mentioned above, only by a comparison of results obtained by the linear theory with results of more accurate calculations or of experiment. The fact that we are here considering a lifting surface of infinite span is disadvantageous for the linear theory.

We should stress that for a cascade of thin profiles in an incompressible medium we will derive a linear theory, while at the same time the conversion of these characteristics to high subsonic velocities has been carried out by linearizing the fundamental equations. Hence in the present case the accuracy of the conversion to high subsonic numbers M_∞ will be lower than that of the original data. Strictly speaking this conversion cannot be extended to nonlinear terms in (4.47) and (4.53).

All the results obtained in this section apply equally well not only to profile cascades but also to multiplanes of infinite span made up of a finite number of identical and similarly oriented profiles.

CHAPTER VI

THE FLOW REVERSAL THEOREM AND ITS CONSEQUENCES

§ 1. The Theorem

The flow reversal theorem gives the relation between the aerodynamic characteristics of a wing in a direct flow and in a reversed flow, defined as a flow with the direction of U_0 reversed. It is based on the fundamental linearized equations of gas dynamics. In the framework of linear theory both the theorem and its consequences are exact, holding not only for subsonic but also for supersonic velocities. This theorem can be proved both for stationary motion and for harmonic vibrations of a wing [34].

We consider stationary motion of a thin monoplane or annular wing in a flow of compressible gas. This motion can be characterized in general not only by the velocity U_0, but also by angular velocities independent of the time. As before, we introduce a moving coordinate system Oxyz fixed to the body (Fig. 6.1). The motion of the fluid outside the body and its trailing vortex sheet are assumed to be potential flows, and we denote the potential of the disturbed motion by Φ. Without changing the coordinate system or the angular velocities Ω_x, Ω_y, or Ω_z, we reverse the direction of the translational motion (we replace U_0 by $-U_0$). The first motion, corresponding to U_0 will be called direction motion, while the second corresponding to $-U_0$ will be called reversed. All functions for the first case will have á subscript "+", all functions for the second a subscript "−".

In agreement with (1.18), for the potentials Φ_+ and Φ_- of the disturbed velocities we have the equations

$$\left.\begin{array}{l} (1 - \mathbf{M}_\infty^2)\,\dfrac{\partial^2\Phi_+}{\partial x^2} + \dfrac{\partial^2\Phi_+}{\partial y^2} + \dfrac{\partial^2\Phi_+}{\partial z^2} = 0, \\[2mm] (1 - \mathbf{M}_\infty^2)\,\dfrac{\partial^2\Phi_-}{\partial x^2} + \dfrac{\partial^2\Phi_-}{\partial y^2} + \dfrac{\partial^2\Phi_-}{\partial z^2} = 0. \end{array}\right\} \tag{6.1}$$

The disturbed pressures, given by the Bernoulli equation (1.21), are

$$p'_+ = \rho_\infty U_0\,\frac{\partial\Phi_+}{\partial x}, \qquad p'_- = -\rho_\infty U_0\,\frac{\partial\Phi_-}{\partial x}. \tag{6.2}$$

We will obtain the basic relation for the flow reversal theorem for a monoplane wing moving in an unbounded space filled with a perfect gas [34]. To do this, we consider the following volume integral which in view of (6.1) is equal to zero:

$$I_0 = \int\int\int \left[p'_+ \left(\frac{\partial^2\Phi_-}{\partial x^2} + \frac{\partial^2\Phi_-}{\partial y^2} + \frac{\partial^2\Phi_-}{\partial z^2} - \mathbf{M}_\infty^2\,\frac{\partial^2\Phi_-}{\partial x^2} \right) - \right.$$
$$\left. - \Phi_- \left(\frac{\partial^2 p'_+}{\partial x^2} + \frac{\partial^2 p'_+}{\partial y^2} + \frac{\partial^2 p'_+}{\partial z^2} - \mathbf{M}_\infty^2\,\frac{\partial^2 p'_+}{\partial x^2} \right) \right] dv = 0, \tag{6.3}$$

where the integration is taken through the region outside the wing. It should be noted that the potentials Φ_- and Φ_+ satisfy Eq. (6.1) everywhere outside the lifting surface and outside the

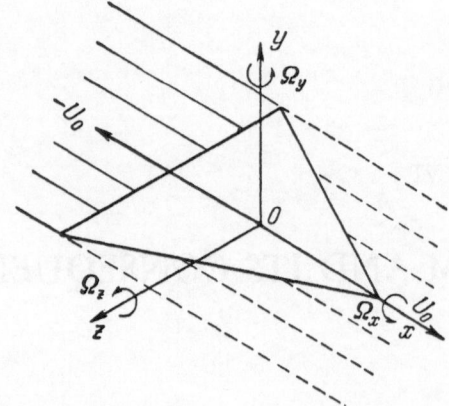

Fig. 6.1. Direct (U_0) and reverse ($-U_0$) wing motion.

trailing vortex sheet behind it. The potential Φ_+ corresponds to a vortex sheet along the negative part of the Ox axis, the potential Φ_- to a vortex sheet along the positive part. The disturbed pressure p'_+ tends to zero for $x \to +\infty$ and, from the Chaplygin-Zhukovskii condition, is zero at the trailing edge of a direct wing. The pressure p' tends to zero when $x \to -\infty$ and is equal to zero on the trailing edge of the wing. We take the section by the plane $y = 0$ and apply Green's formula to the integral I_0. Noting that

$$I_0 = \int\int\int \left[\frac{\partial}{\partial x}\left(p'_+ \frac{\partial \Phi_-}{\partial x} \right) + \frac{\partial}{\partial y}\left(p'_+ \frac{\partial \Phi_-}{\partial y} \right) + \frac{\partial}{\partial z}\left(p'_+ \frac{\partial \Phi_-}{\partial z} \right) \right] dv -$$

$$- \int\int\int \left[\frac{\partial}{\partial x}\left(\Phi_- \frac{\partial p'_+}{\partial x} \right) + \frac{\partial}{\partial y}\left(\Phi_- \frac{\partial p'_+}{\partial y} \right) + \frac{\partial}{\partial z}\left(\Phi_- \frac{\partial p'_+}{\partial z} \right) \right] dv -$$

$$- M_\infty^2 \int\int\int \left[\frac{\partial}{\partial x}\left(p'_+ \frac{\partial \Phi_-}{\partial x} \right) - \frac{\partial}{\partial x}\left(\Phi_- \frac{\partial p'_+}{\partial x} \right) \right] dv,$$

we write

$$I_0 = \int\int \left(p'_+ \frac{\partial \Phi_-}{\partial n} - \Phi_- \frac{\partial p'_+}{\partial n} \right) ds - M_\infty^2 \int\int \left(p'_+ \frac{\partial \Phi_-}{\partial x} - \Phi_- \frac{\partial p'_+}{\partial x} \right) \cos(nx)\, ds = 0. \qquad (6.4)$$

where n is the external normal relative to the region of integration. In (6.4) the integration is taken over both sides of a section close to $y = 0$: over the upper where $y = 0 + \varepsilon$ and over the lower where $y = 0 - \varepsilon$. On these surfaces $\cos(n, x) = 0$, and thus (6.4) yields

$$\int\int \left(p'_+ \frac{\partial \Phi_-}{\partial n} - \Phi_- \frac{\partial p'_+}{\partial n} \right) ds = 0. \qquad (6.5)$$

The normal velocities vary continuously through the vortex sheet (see Chapter II). The external normals to the upper and lower surfaces of the section, however, have opposite directions. Hence

$$\left(\frac{\partial \Phi_+}{\partial n} \right)_{y=0+\varepsilon} = -\left(\frac{\partial \Phi_+}{\partial n} \right)_{y=0-\varepsilon} \qquad \left(\frac{\partial \Phi_-}{\partial n} \right)_{y=0+\varepsilon} = -\left(\frac{\partial \Phi_-}{\partial n} \right)_{y=0-\varepsilon} \qquad (6.6)$$

For the vortex scheme for a wing corresponding to the linear theory and in the plane $y = 0$, the disturbed velocities W_x on the upper and lower sides of the section differ in sign, and from (6.2) we have

$$\left(p'_+ \right)_{y=0+\varepsilon} = -\left(p'_+ \right)_{y=0-\varepsilon}, \qquad \left(p'_- \right)_{y=0+\varepsilon} = -\left(p'_- \right)_{y=0-\varepsilon}. \qquad (6.7)$$

Using the relations (6.6) and (6.7) and the corresponding relations for $\partial p'/\partial n$ and Φ we can convert (6.5) into an equation containing a single integral taken over the upper side of the section:

$$\int\int \left(\Phi_- \frac{\partial p'_+}{\partial y} - p'_+ \frac{\partial \Phi_-}{\partial y} \right) ds = 0.$$

We transform the first term of our relation so that it takes the same form as the second. Using (6.2) we can write

$$\frac{\partial p'_+}{\partial y} = \rho_\infty U_0 \frac{\partial W_{x+}}{\partial y} = \rho_\infty U_0 \frac{\partial W_{y+}}{\partial x},$$

since the condition that there be no vortices means that

$$\frac{\partial W_{x+}}{\partial y} = \frac{\partial W_{y+}}{\partial x}.$$

In the integration with respect to x we use integration by parts and obtain

$$\int\int \Phi_- \frac{\partial p'_+}{\partial y} ds = \rho_\infty U_0 \int\int \Phi_- \frac{\partial W_{y+}}{\partial x} dx\, dy = \rho_\infty U_0 \int\int W_{y+} \frac{\partial \Phi_-}{\partial x} dx\, dy,$$

since for $x = -\infty$ we have $\Phi_- = 0$, and for $x = +\infty$ we have $W_{y+} = 0$. Substituting

$$\int\int \Phi_- \frac{\partial p'_+}{\partial y} ds = \int\int p'_- W_{y+} ds,$$

in (6.5), we obtain the basic relation in the flow reversal theorem

$$\int_S\int p'_+ W_{y-}\, ds = \int_S\int p'_- W_{y+}\, ds. \qquad (6.8)$$

The integration in this relation is taken only over the surface of the wing, and so everywhere outside the wing (as is also clear from (6.6) for example) the disturbed pressures p'_+ and p'_- are zero, since on a free surface the pressure difference is zero.

Hence the flow reversal theorem yields an integral relation between the disturbed pressures p'_+ and p'_- on the upper surface of the wing in the direct and reverse flows, and also between the downwash W_{y+} and W_{y-} on a wing in these two flows. These last two quantities can of course be replaced by their values as determined by the boundary conditions.

Since the pressures on the lower and upper surfaces of a wing differ only in sign, it follows that p'_+ and p'_- in (6.8) can be replaced by the pressure differences (the load) at the corresponding points. This yields

$$\int_S\int \Delta p_+ W_{y-}\, ds = \int_S\int \Delta p_- W_{y+}\, ds. \qquad (6.9)$$

§2. Some Generalizations of the Theorem

The results of the preceding section are easily extended to the case of a wing moving past an interface or extended to apply to annular wings, etc. [35].

We now consider an annular wing, and replace its bound and free vortices by a layer on the wing's cylindrical surface. The equations (6.1) and (6.2) will hold in this case, and so reasoning similar to that used above leads to a relation of the form (6.4). Now, however, the integration is taken over the two sides of the circular cylindrical surface $r' = r + \varepsilon$ and $r' = r - \varepsilon$, where $r' = r$ is the equation of the surface. On this surface $\cos(n, x) = 0$, and so

$$I_0 = \int\int_{r'=r+\varepsilon}\left(p'_+ \frac{\partial \Phi_-}{\partial n} - \Phi_- \frac{\partial p'_+}{\partial n}\right) ds + \int\int_{r'=r-\varepsilon}\left(p'_+ \frac{\partial \Phi_-}{\partial n} - \Phi_- \frac{\partial p'_+}{\partial n}\right) ds = 0. \qquad (6.10)$$

Here again the subscript " + " refers to the direct and " − " to the reversed annular wing.

The normal derivative $\partial \Phi_- / \partial n$, as in the previous case, changes sign in the first and second integrals only, due to the change in direction of n from the first to the second surface. Hence the first and third terms in (6.10) can be written in the form

$$I'_0 = \int \int \Delta p_+ W_{r-} \, ds. \tag{6.11}$$

where Δp_+ is the pressure difference (the load) on the annular wing.

The second and fourth terms can be transformed into a similar form by the same method that was used for a monoplane wing:

$$\int \int \Phi_- \frac{\partial p'_+}{\partial n} ds = \int \int \Phi_- \frac{\partial p'_+}{\partial r} ds = \rho_\infty U_0 \int \int \Phi_- \frac{\partial W_x}{\partial r} ds =$$
$$= \rho_\infty U_0 \int \int \Phi_- \frac{\partial W_{r+}}{\partial x} ds = \rho_\infty U_0 \int \int W_{r+} \frac{\partial \Phi_-}{\partial x} ds = \int \int W_{r+} p' \, ds.$$

Since this is true for both the outer and inner boundaries, we can write

$$I''_0 = - \int \int \Delta p_- W_{r+} \, ds. \tag{6.12}$$

Using the relation

$$I_0 = I'_0 + I''_0 = 0.$$

we obtain for an annular wing the equation

$$\int\limits_S \int \Delta p_+ W_{r-} \, ds = \int\limits_S \int \Delta p_- W_r \, ds. \tag{6.13}$$

where the integration is taken over one surface of the wing (external or internal).

It can be shown that the flow reversal theorem is also valid more generally. For example, let a wing with an infinite cylindrical body of arbitrary cross section move parallel to the generators of the interface which is also represented by an arbitrary cylindrical surface. This surface can be liquid, solid, or a combination of the two; it is only important that on it we have $\cos(n, x) = 0$. The last case corresponds, for example, to a model in a wind tunnel with an open, closed, or combined operating section.

For the problems under consideration, the integration with respect to I_0 must be taken not only over the surface of the wing, but also over the interface S_2 and the surface S_3 of the body. To generalize the theorem it is sufficient to show that the integration over S_2 and S_3 does not yield any new terms.

Since $\cos(n, x) = 0$ on S_2 and S_3 by hypothesis, it follows from (6.4) that it is sufficient to show that

$$\int\limits_{S_2} \int \left(p'_+ \frac{\partial \Phi_-}{\partial n} - \Phi_- \frac{\partial p'_+}{\partial n} \right) ds + \int\limits_{S_3} \int \left(p'_+ \frac{\partial \Phi_-}{\partial n} - \Phi_- \frac{\partial p'_+}{\partial n} \right) ds = 0. \tag{6.14}$$

A solid wall is a stream surface, and so the normal component of the disturbed velocity on such a surface is zero; we also have

$$\frac{\partial p'}{\partial n} = \rho_\infty U_0 \frac{\partial}{\partial n}\left(\frac{\partial \Phi}{\partial x}\right) = \rho_\infty U_0 \frac{\partial}{\partial x}\left(\frac{\partial \Phi}{\partial n}\right) = 0.$$

Hence, on a solid boundary,

$$\frac{\partial \Phi_+}{\partial n} = \frac{\partial \Phi_-}{\partial n} = 0, \qquad \frac{\partial p'_+}{\partial n} = \frac{\partial p'_-}{\partial n} = 0. \tag{6.15}$$

On a free (fluid) surface, the condition that there be no pressure difference is satisfied. The boundary in this case is deformed and will no longer be a stream surface. Within the accuracy of the linearized scheme, we can assume that this condition is, however, satisfied on the original (undeformed) boundary. The requirement that $\Delta p = 0$ when $\partial \Phi / \partial n \neq 0$ leads to the condition that the disturbed pressure close to a free surface be zero. Since the relation

$$p' = \rho_\infty U_0 \frac{\partial \Phi}{\partial x} = 0$$

is satisfied identically at all points of a free boundary, it follows that here

$$p'_+ = p'_- = 0, \qquad \Phi_+ = \Phi_- = 0. \tag{6.16}$$

Using (6.15) and (6.16), which hold at all points of corresponding surfaces, we can easily prove (6.14).

From the above results, it follows in particular that the flow reversal theorem holds for motion of a wing close to an interface.

§3. Some Consequences of the Theorem

The flow reversal theorem was proved under the assumption that the coordinate system remains invariant both for the direct and reversed flow. We will prove a relation between the rotational-derivative coefficients for a direct and reversed monoplane wing, using for each case the standard coordinate system. An original (direct) wing and the coordinate system related to it are shown in Fig. 6.2, and a reversed wing with the corresponding axes is shown in Fig. 6.3. The origin O cannot be taken at the midpoint of the root chord, since then the centering will be different for the direct and reversed wings.

We can say that the reversed wing is obtained by rotating the direct wing through an angle of 180° about the Oy axis, the coordinate system Oxyz and the translational velocity U_0 remaining unchanged. Then the leading edge for the reversed wing will be the same as the trailing edge for the direct wing. The geometrical parameters of the direct ("+") and reversed ("−") wings will be related by the equations

$$\lambda_+ = \lambda_-, \qquad \eta_+ = \eta_-, \qquad \chi_{0+} = -\chi_{1-}, \tag{6.17}$$

where χ_1 is the sweepback angle relative to the trailing edge.

Between the coordinates for the direct and reversed wings we have the relations

$$x_+ = -x_-, \qquad y_+ = y_-, \qquad z_+ = -z_-$$

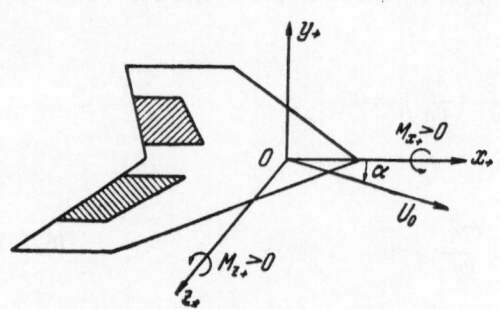

Fig. 6.2. The original (direct) wing.

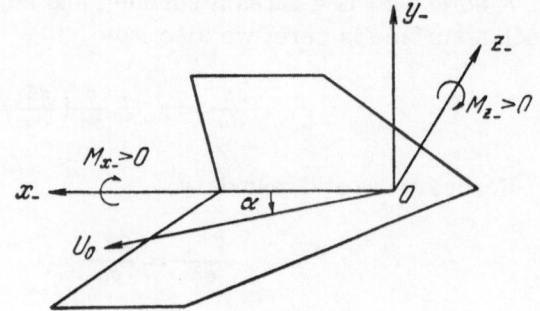

Fig. 6.3. The reversed wing.

or in dimensionless form

$$\xi_+ = -\xi_-, \quad \eta_+ = \eta_-, \quad \zeta_+ = -\zeta_-,$$
$$\xi = \frac{x}{b}, \quad \eta = \frac{y}{b}, \quad \zeta = \frac{z}{b}, \qquad (6.18)$$

where b is the root chord.

We denote the distance from the leading edge of the nose of the wing to the origin O by χ_T; then we have the following relations between the centerline of the direct and reversed wings:

$$\bar{x}_{T+} = 1 - \bar{x}_{T-}, \quad \bar{x}_T = \frac{x_T}{b} \qquad (6.19)$$

By considering various combinations of motions of direct and reversed wings, we can establish the relation between all their rotational-derivative coefficients. We note here that if we divide (6.9) by q_∞ and U_0 we obtain

$$\int_S \int \frac{\Delta p_+}{q} \frac{W_{y-}}{U_0} \, ds = \int_S \int \frac{\Delta p_-}{q} \frac{W_{y+}}{U_0} \, ds. \qquad (6.20)$$

Let the direct and reversed wings perform translational motion, i.e., from (1.30) and (4.5) let

$$\frac{W_{y+}}{U_0} = -\alpha, \quad \frac{\Delta p_+}{q} - p_{\alpha+}\alpha, \quad \frac{W_{y-}}{U_0} = -\alpha, \quad \frac{\Delta p_-}{q} = p_{\alpha-}\alpha. \qquad (6.21)$$

Then (6.20) becomes

$$\int_S \int p_{\alpha+} \, ds = \int_S \int p_{\alpha-} \, ds,$$

or from (4.7) we have

$$c_{y+}^{\alpha} = c_{y-}^{\alpha}. \qquad (6.22)$$

We also consider the case when the direct and reversed wings perform steady rotations about longitudinal axes. In this case

$$\frac{W_{y+}}{U_0} = -\omega_x \zeta_{0+}, \qquad \frac{\Delta p_+}{q} = p_{\omega_{x+}} \omega_x,$$
$$\frac{W_{y-}}{U_0} = -\omega_x \zeta_{0-}, \qquad \frac{\Delta p_-}{q} = p_{\omega_{x-}} \omega_x \qquad\qquad (6.23)$$

and (6.20) yields

$$\int_S \int p_{\omega_{x+}} \zeta_{0-} \, ds = \int_S \int p_{\omega_{x-}} \zeta_{0+} \, ds,$$

since

$$\zeta_{0+} = -\zeta_{0-},$$

using (4.7) we now obtain

$$m^{\omega_x}_{x+} = m^{\omega_x}_{x-}, \qquad\qquad (6.24)$$

We now write

$$\frac{W_{y+}}{U_0} = \omega_z \xi_{0+}, \qquad \frac{\Delta p_+}{q} = p_{\omega_{z+}} \omega_z,$$
$$\frac{W_{y-}}{U_0} = \omega_z \xi_{0-}, \qquad \frac{\Delta p_-}{q} = p_{\omega_{z-}} \omega_z, \qquad\qquad (6.25)$$

and obtain from (6.20) the relation

$$\int_S \int p_{\omega_{z+}} \xi_{0-} \, ds = \int_S \int p_{\omega_{z-}} \xi_{0+} \, ds,$$

but $\xi_{0-} = -\xi_{0+}$, and so

$$m^{\omega_z}_{z+} = m^{\omega_z}_{z-}. \qquad\qquad (6.26)$$

Finally if we write

$$\frac{W_{y+}}{U_0} = \omega_z \xi_{0+}, \qquad \frac{\Delta p_+}{q} = p_{\omega_{z+}} \omega_z, \qquad \frac{W_{y-}}{U_0} = -\alpha, \qquad \frac{\Delta p_-}{q} = p_{\alpha-} \alpha, \qquad (6.27)$$

or

$$\frac{W_{y+}}{U_0} = -\alpha, \qquad \frac{\Delta p_+}{q} = p_{\alpha+} \alpha,$$
$$\frac{W_{y-}}{U_0} = \omega_z \xi_{0-}, \qquad \frac{\Delta p_-}{q} = p_{\omega_{z-}} \omega_z, \qquad\qquad (6.28)$$

we have

$$m^{\alpha}_{z+} = c^{\omega_z}_{y-}, \qquad m^{\alpha}_{z-} = c^{\omega_z}_{y+}. \qquad\qquad (6.29)$$

Hence for the problems under consideration all the rotational-derivative coefficients of the reversed wing can be expressed in terms of these same coefficients for the direct wing. The resulting equations are exact in the linear theory for any Mach number. If for the direct wing we take $\bar{x}_{T+} = 0.5$, then $\bar{x}_{T-} = 0.5$ for the reversed wing, while for other cases the coefficients (6.26) and (6.29) must be taken for different centerlines for the direct and reversed wings. This does not refer to c_y^α or $m_x^{\omega x}$, which do not change when the coordinate x changes [see (4.76)].

The reader will easily see that the relations (6.22), (6.26), and (6.29) will also hold for annular wings [(6.24) no longer holds, since $m_x^{\omega x} = 0$ for an annular wing].

§ 4. The Calculation of Overall Aerodynamic Effects of the Deformation of a Monoplane Wing

There is a large class of problems, important in practice, in which it is necessary to find the aerodynamic effect of a deformation of the wing. Such questions arise, for example, in problems of aeroelasticity. We must sometimes choose the deflection of a mean wing surface that will produce a desired aerodynamic property of the wing or of the apparatus as a whole, etc.

It is often not necessary to find the local aerodynamic loads, and it is sufficient to find the total effect due to the deformation. Under these conditions, the whole calculation can be greatly simplified if it is carried out by using certain consequences of the flow reversal theorem.

Let the equation of a deformed monoplane wing be written in the form

$$\eta = \Delta f (\xi, \zeta), \tag{6.30}$$

where ξ, η, and ζ are the dimensionless Cartesian coordinates (6.18) and Δ is a parameter determining the scale of the deformation. As we will see below, it is most convenient to take some characteristic value, perhaps the largest value, of the part of the local angle of attack generated by the deflection of the surface.

For this surface the general form of the boundary condition (1.28) holds; we linearize this condition, assuming that Δ is small. It should be noted that in the linear theory it is essential to assume that the derivative $\partial \eta / \partial \xi$ be small, although $\partial \eta / \partial \zeta$ need not in general be small compared to unity. Examples of such lifting surfaces are annular, V-shaped, cruciform, and other wings. The extension to them of the theory we have described leads to no difficulties in principle.

Let n be the normal to the wing surface passing from the lower to the upper surface (forming an acute angle with Oy). Then

$$\left.\begin{aligned}
\cos(n,\ x) &= -\frac{\frac{\partial f}{\partial \xi}\Delta}{\sqrt{1+\left(\frac{\partial f}{\partial \xi}\Delta\right)^2+\left(\frac{\partial f}{\partial \zeta}\Delta\right)^2}}, \\[2mm]
\cos(n,\ y) &= \frac{1}{\sqrt{1+\left(\frac{\partial f}{\partial \xi}\Delta\right)^2+\left(\frac{\partial f}{\partial \zeta}\Delta\right)^2}}, \\[2mm]
\cos(n,\ z) &= -\frac{\frac{\partial f}{\partial \zeta}\Delta}{\sqrt{1+\left(\frac{\partial f}{\partial \xi}\Delta\right)^2+\left(\frac{\partial f}{\partial \zeta}\Delta\right)^2}}.
\end{aligned}\right\} \tag{6.31}$$

Assuming that Δ is small and linearizing the expressions in (6.31) with respect to this parameter, for the point $\xi_0 = x_0/b$, $\zeta_0 = z_0/b$ considered in (1.29), we can write

$$\cos(n, x) = -\frac{\partial f}{\partial \xi_0}\Delta, \quad \cos(n, y) = 1, \quad \cos(n, z) = -\frac{\partial f}{\partial \zeta_0}\Delta.$$

Retaining the same accuracy in the boundary condition (1.28), we have

$$\frac{W_y}{U_0} = -\frac{\partial f}{\partial \xi_0}\Delta - \alpha + \omega_z \xi_0 - \omega_x \zeta_0. \tag{6.32}$$

The coefficients of the aerodynamic forces and moments for the deformed wing can be written in a form similar to (4.4):

$$\left. \begin{aligned} c_y &= c_y^\alpha \alpha + c_y^\omega z \omega_z + c_y^\Delta \Delta, \quad m_x = m_x^{\omega x} \omega_x + m_x^\Delta \Delta, \\ m_z &= m_z^\alpha \alpha + m_z^{\omega z} \omega_z + m_z^\Delta \Delta. \end{aligned} \right\} \tag{6.33}$$

For the aerodynamic loads, instead of (4.5) we have

$$\frac{\Delta p}{q} = p_\alpha(x, z)\alpha + p_{\omega_z}(x, z)\omega_z + p_{\omega_x}(x, z)\omega_x + p_\Delta(x, z)\Delta, \tag{6.34}$$

where in addition to the relations (4.7) we have

$$c_y^\Delta = \frac{1}{S}\int\!\!\int_S p_\Delta \, ds, \quad m_z^\Delta = \frac{1}{Sb}\int\!\!\int_S p_\Delta x \, ds, \quad m_x^\Delta = -\frac{1}{Sb}\int\!\!\int_S p_\Delta z \, ds. \tag{6.35}$$

If the wing is symmetric in plan, then the formulas in (6.35) can be written with the integrations over half the wing if we double the result.

From the boundary condition (6.32) and the basic relation of the flow reversal theorem (6.20), we easily obtain convenient formulas for calculating the total characteristics of deformed wings. With this object in view, we consider (6.20) for the combinations of conditions for which direct wings become deformed wings and the reversed wings become plain surfaces.

Let the boundary conditions and the aerodynamic loads of the direct and reversed wings be

$$\left. \begin{aligned} \frac{W_{y+}}{U_0} &= -\left(\frac{\partial f}{\partial \xi_0}\right)_+ \Delta, \quad \frac{\Delta p_+}{q} = p_{\Delta_+}\Delta, \\ \frac{W_{y-}}{U_0} &= \omega_z \xi_{0_-}, \quad \frac{\Delta p_-}{q} = p_{\omega_{z_-}}\omega_z. \end{aligned} \right\} \tag{6.36}$$

Then (6.20) becomes

$$\int\!\!\int_S p_{\Delta_+}\xi_{0_-} \, ds = -\int\!\!\int_S p_{\omega_{z_-}}\left(\frac{\partial f}{\partial \xi_0}\right)_+ ds$$

or, from (6.18),

$$\int\!\!\int_S p_{\Delta_+}\xi_{0+} \, ds = \int\!\!\int_S p_{\omega_{z_-}}\left(\frac{\partial f}{\partial \xi_0}\right)_+ ds,$$

and so using (6.35) we obtain

$$m_{z+}^{\Delta} = \frac{1}{S} \int_S \int p_{\omega z_-} \left(\frac{\partial f}{\partial \xi_0} \right)_+ ds. \tag{6.37}$$

For the direct and reversed wings we also set

$$\frac{W_{y+}}{U_0} = -\left(\frac{\partial f}{\partial \xi_0} \right)_+ \Delta, \quad \frac{p_+}{q} = p_{\Delta+}\Delta, \quad \left. \right\}$$
$$\frac{W_{y-}}{U_0} = -\alpha, \quad \frac{p_-}{q} = p_{\alpha-}\alpha, \quad \left. \right\} \tag{6.38}$$

and then (6.20) yields

$$\int_S \int p_{\Delta+}\, ds = \int_S \int p_{\alpha-} \left(\frac{\partial f}{\partial \xi_0} \right)_+ ds,$$

hence from (6.35) we obtain

$$c_{y+}^{\Delta} = \frac{1}{S} \int_S \int p_{\alpha-} \left(\frac{\partial f}{\partial \xi_0} \right)_+ ds. \tag{6.39}$$

We finally write

$$\frac{W_{y+}}{U_0} = -\left(\frac{\partial f}{\partial \xi_0} \right)_+ \Delta, \quad \frac{p_+}{q} = p_{\Delta+}\Delta, \quad \left. \right\}$$
$$\frac{W_{y-}}{U_0} = -\omega_x \zeta_{0-}, \quad \frac{p_-}{q} = p_{\omega x_-}\omega_x, \quad \left. \right\} \tag{6.40}$$

and in this case

$$\int_S \int p_{\Delta+}\zeta_{0-}\, ds = \int_S \int p_{\omega x_-} \left(\frac{\partial f}{\partial \xi_0} \right)_+ ds,$$

or, using (6.18) and (6.35), we obtain

$$m_{x+}^{\Delta} = \frac{1}{S} \int_S \int p_{\omega x_-} \left(\frac{\partial f}{\partial \xi_0} \right)_+ ds. \tag{6.41}$$

From (6.37), (6.39), and (6.41) it follows that the over-all characteristics of the deformed wing, for any law of deformation $f(\xi, \zeta)$ can be obtained by a simple integration if the aerodynamic loads on the reversed, fixed wing are known. Hence, instead of investigating the concrete deformation under consideration each time we wish to find the over-all effect, we need only solve three special problems and find the functions p_α, $p_{\omega z}$, and $p_{\omega x}$ for the reversed, fixed wing.

If the direct and reversed deformed wings are considered in this way, then we obtain the following relation between their characteristics:

$$\int_S \int p_{\Delta-} \left(\frac{\partial f}{\partial \xi_0} \right)_+ ds = \int_S \int p_{\Delta+} \left(\frac{\partial f}{\partial \xi_0} \right)\, ds. \tag{6.42}$$

§5. The Calculation of the Aerodynamic Moments and Lift Due to the Deflection of Flaps and Ailerons

The deflection of a flap or an aileron on a wing can be considered as a special case of a wing deformation. The basic relations of the previous section can therefore be used to develop a rather simple method of calculation of the extra lift and extra pitching moment caused by a deflection of flaps, or the banking moment caused by ailerons.

Let flaps or ailerons (Fig. 6.2) be attached to a wing of arbitrary plan form, and assume that this is a direct wing. The angle of inclination δ_f of a flap will be taken to be positive when the flap is deflected downwards. Ailerons on the two halves of a wing are deflected in different directions, and we will assume that the angles of inclination of the right and left ailerons have identical absolute values. To be definite, we assume that the flaps and ailerons are located at the trailing edge of the wing, although it is not difficult to investigate more general cases.

Here it is natural to take for Δ the angle of deflection of a flap δ_f or an aileron δ_a, where δ_a will be understood to be the angle of deflection of the right aileron ($\delta_a > 0$ when the aileron is lowered).

The derivative $\partial f/\partial \xi_0$ occurring in all the numerical formulas obtained in the previous section will have the following values in our reasoning:

$$\frac{\partial f}{\partial \xi_0} = \begin{cases} 0 \text{ on the wing outside the flap or aileron} \\ 1 \text{ on a lowered flap or aileron} \end{cases}$$

We assume that the monoplane wing is symmetric in plan, so that Eqs. (6.33) can be written

$$c_y = c_y^\alpha \alpha + c_y^{\omega_z} \omega_z + c_y^{\delta_f} \delta_f, \qquad m_x = m_x^{\omega_x} \omega_x + m_x^{\delta_a} \delta_a,$$
$$m_z = m_z^\alpha \alpha + m_z^{\omega_z} \omega_z + m_z^{\delta_f} \delta_f. \tag{6.43}$$

The derivatives $c_y^{\delta_f}$, $m_x^{\delta_a}$, $m_z^{\delta_f}$, characterize the effectiveness of the controls of the wing for a given centerline, and they obviously depend on the wing geometry, the wing controls, and also on the Mach number. It was noted in connection with Eqs. (6.43) that a deflection of the flaps leads to symmetric aerodynamic loads, and a deflection of the ailerons leads to anti-symmetric loads.

To find the relevant coefficients, we need only use (6.37), (6.39), and (6.41), substituting in them the appropriate values of $\partial f/\partial \xi_0$. It is easily seen that the regions of integration in these expressions will be the shaded regions of the wings in Fig. 6.2, i.e., the integration will be extended over the part of the wing occupied by a flap or an aileron. Hence

$$\left. \begin{array}{c} c_y^{\delta_f} = \dfrac{1}{S} \int\!\!\int\limits_{S_f} p_{a_-} \, ds, \qquad m_z^{\delta_f} = \dfrac{1}{S} \int\!\!\int\limits_{S_f} p_{\omega z_-} \, ds, \\[2em] m_x^{\delta_a} = \dfrac{1}{S} \int\!\!\int\limits_{S_a} p_{\omega x_-} \, ds, \end{array} \right\} \tag{6.44}$$

that is, for the calculation of the over-all characteristics of a wing with deflected flaps or ailerons with any geometry, we need not solve the whole problem anew for every flap or aileron of every concrete form. It is enough to solve the three basic problems for the reversed wing and to determine p_{α_-}, $p_{\omega z_-}$, and $p_{\omega x_-}$. We then use (6.44) and perform a simple integration to obtain the coefficients $c_y^{\delta_f}$, $m_z^{\delta_f}$ and $m_x^{\delta_a}$ for a flap or aileron of arbitrary shape for a wing with the given shape in plan.

It is not difficult to obtain equations of the type (4.75) for the conversion of control characteristics to another wing centerline. For a parallel displacement along the Ox axis (Fig. 4.8) we have

$$c_{y1}^{\delta_f} = c_y^{\delta_f}, \quad m_{x1}^{\delta_a} = m_x^{\delta_a}, \quad m_{z1}^{\delta_f} = m_z^{\delta_f} - c_y^{\delta_f}\xi_0. \tag{6.45}$$

§ 6. Experimental Verification of Some Consequences of the Flow Reversal Theorem

The flow reversal theorem and all its consequences are exact in the framework of linear theory for all Mach numbers. However the fundamental assumptions forming the basis of the linear theory are verified to a certain extent by experiment. Hence the appearance of regions where these consequences are not verified by experiment yields a method of determining the range of applicability of linear theory.

When low subsonic velocities are being considered, one of the fundamental conditions is the requirement that the flow about the wing be smooth. Every disturbance of this smooth flow will lead to some deviation from theoretical values. From this point of view, the results of the following experiments of B. I. Ul'yanov are very typical and well understood.

These experiments were carried out at low velocities for two series of triangular and swept-back wings with a value of λ close to 3, and 8% symmetric profiles of the NACA type. In the first series, the reversed wings were obtained by rotating the direct wings through 180° in the flow, and so they had reversed profiles (Fig. 6.4). In the second series, the reversed wings had the same profiles as the direct wings (Fig. 6.5). In both cases, the geometrical parameters of the wings characterizing their shape in plan satisfied the conditions (6.17).

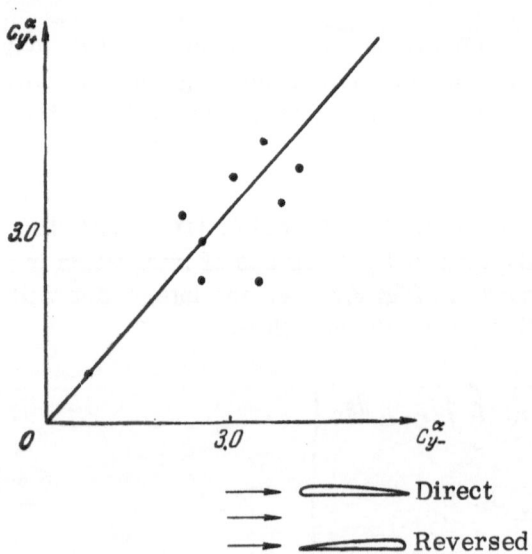

Hence the direct–wing profiles in the first case had rounded noses and sharp tails, while the reversed-wing profiles on the other hand had sharp noses and rounded tails. In the second case, all profiles had rounded noses and sharp tails.

The experimental results are shown in a coordinate system in which the derivatives c_{y+}^{α} for the direct wings are the ordinates and the derivatives c_{y-}^{α} of the reversed wings are the abscissas. The straight line inclined at an angle of 45° to these axes gives the relation for the reversibility theorem.

It can be seen from Fig. 6.4 that for the first series, in which conditions ensuring smooth flow about the reversed wings did not hold, the relation

$$c_{y+}^{\alpha} = c_{y-}^{\alpha}$$

Fig. 6.4. A comparison of the characteristics of direct and reversed monoplane wings with various forms of leading edges.

is not at all well satisfied. On the other hand if smooth flow is ensured about both the direct and

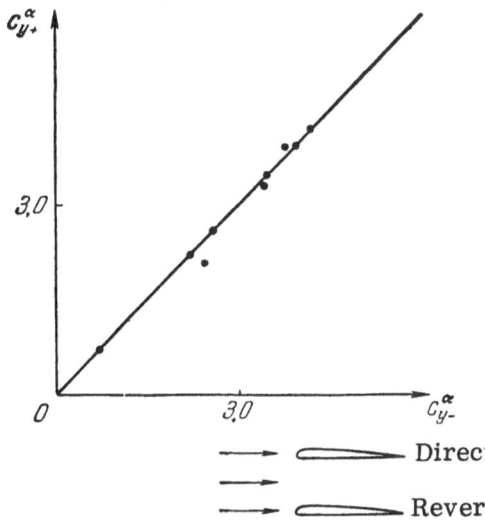

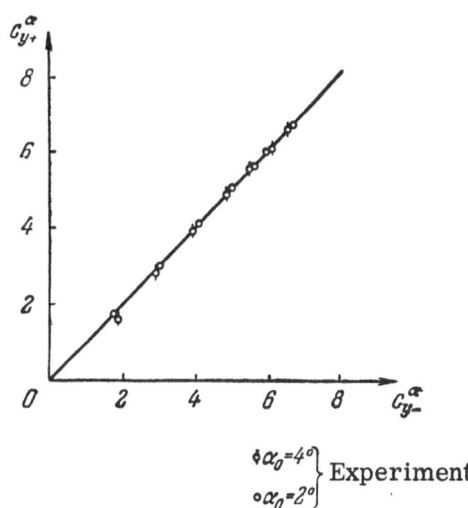

Fig. 6.5. A comparison of the characteristics of direct and reversed monoplane wings with rounded leading edges.

Fig. 6.6. A comparison of direct and reversed conical annular wings with slightly rounded leading edges.

reversed wings, then this relation is well confirmed by the experimental results (Fig. 6.5). Satisfactory agreement between experiment and theory was also obtained for small angles of attack with thin unrounded wings with sharp noses and tails.

Similar results were also obtained for another wing characteristic which can be determined rather accurately by experiment — the damping coefficient in bank $m_X^{\omega x}$. Theory yields the value 1 for the ratio $m_{X+}^{\omega x}/m_{X-}^{\omega x}$, while in experiments with a trianglar wing and $\chi_0 = 45°$, the values of this ratio were found to be 1.41 in the first case, 1.03 in the second, and 1.04 in the third (for a thin plate with sharp ends).

Experimental studies of annular wings at low velocities carried out by L. N. Kravchenko and the author also confirmed the above conclusions. Figure 6.6 shows a comparison of the lifting properties of direct and reversed conical annular wings with conical angles $\alpha_0 = \pm 2°$ and $\alpha_0 = \pm 4°$. The elongation of the wings was varied through wide limits and the sections were those of thin plates with a slight rounding of the nose and tail.

It is interesting to note that conversion to high Mach numbers yields qualitatively the same results as long as there is no compression discontinuity or break in the flow or any combination of these phenomena. The stronger these phenomena become, the worse will be the agreement between experiment and linear theory. For swept-back wings and wings of low aspect ratio, for which irregularities are smoothed out, the results of linear theory may be in good agreement with experimental results up to $M_\infty = 1$.

CHAPTER VII

FUNDAMENTAL VORTEX SYSTEMS

§ 1. Horseshoe Vortices

One of the fundamental vortex systems that will be used in the calculation of flow past monoplane wings is the usual horseshoe vortex. It consists of a segment of a bound vortex of span l_0 and two semi-infinite vortex filaments arising at the ends of the bound vortex and extending in the direction of the undisturbed flow U_0 (Fig. 7.1). All vortices of this system have a constant strength Γ_+ which can be expressed by the relation

$$\Gamma_+ = U_0 l_0 \Gamma. \qquad (7.1)$$

where Γ is a dimensionless constant [36].

The original basic formula for the calculation of the velocity field of this vortex system, also considered in §2, is as follows. Let AB be an arbitrary vortex segment of constant strength Γ_+ and let M be the point where we wish to find the disturbed velocity W. We pass a plane through AB and M; we join A and B to M; we denote the angles at the vertices A and B by α_1 and α_2 and the distance from M to AB by r (Fig. 7.2). Then from the Biot-Savart formula (2.1) we easily obtain

$$W = \frac{\Gamma_+}{4\pi r} (\cos \alpha_1 + \cos \alpha_2). \qquad (7.2)$$

where the velocity W is perpendicular to the plane AMB and has the same direction as Γ_+.

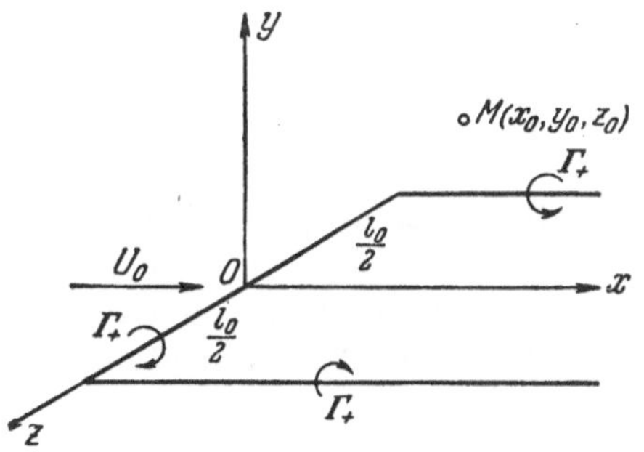

We introduce the coordinate system shown in Fig. 7.1, with the Ox axis parallel to the free vortices.

The projections of the total velocity of the horseshoe vortices W_x, W_y, and W_z will be calculated from the velocities U_x and U_y due to the bound vortex, and from V_y and V_z due to the free vortices, where clearly

$$U_z = 0, \quad V_x = 0.$$

We introduce the dimensionless velocities w_x, w_y, and w_z by writing

$$W_x = \frac{U_0 \Gamma}{2\pi} w_x, \quad W_y = \frac{U_0 \Gamma}{2\pi} w_y, \quad W_z = \frac{U_0 \Gamma}{2\pi} w_z. \quad (7.3)$$

We similarly introduce the dimensionless velocities u_x, u_y and v_y, v_z, corresponding to the attached and free vortices.

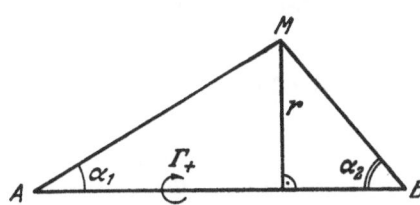

Fig. 7.1. An ordinary (straight) horseshoe vortex.

Fig. 7.2. A vortex filament AB.

91

Let $M(x_0, y_0, z_0)$ be the point at which velocities of the vortex system are to be calculated. As a characteristic linear dimension we use half the span of the attached vortex $l_0/2$. The dimensionless coordinates of M will be denoted by ξ_0, η_0, and ζ_0, where

$$\xi_0 = \frac{2x_0}{l_0}, \qquad \eta_0 = \frac{2y_0}{l_0}, \qquad \zeta_0 = \frac{2z_0}{l_0}. \tag{7.4}$$

Using the fact that (7.2) is linear, we write

$$\left. \begin{aligned} w_x(\xi_0, \eta_0, \zeta_0) &= u_x(\eta_0, \xi_0, \zeta_0), \\ w_y(\xi_0, \eta_0, \zeta_0) &= u_y(\xi_0, \eta_0, \zeta_0) + v_y(\xi_0, \eta_0, \zeta_0), \\ w_z(\xi_0, \eta_0, \zeta_0) &= v_z(\xi_0, \eta_0, \zeta_0). \end{aligned} \right\} \tag{7.5}$$

After some calculation, we easily obtain the following expressions for the dimensionless velocities:

$$\left. \begin{aligned} u_x(\xi_0, \eta_0, \zeta_0) &= \frac{\eta_0}{\xi_0^2 + \eta_0^2} \left(\frac{1-\zeta_0}{r(\xi_0, \eta_0, -\zeta_0)} + \frac{1+\zeta_0}{r(\xi_0, \eta_0, \zeta_0)} \right) \\ u_y(\xi_0, \eta_0, \zeta_0) &= -\frac{\xi_0}{\xi_0^2 + \eta_0^2} \left(\frac{1-\zeta_0}{r(\xi_0, \eta_0, -\zeta_0)} + \frac{1+\zeta_0}{r(\xi_0, \eta_0, \zeta_0)} \right) \\ v_y(\xi_0, \eta_0, \zeta_0) &= -\frac{1-\zeta_0}{\eta_0^2 + (1-\zeta_0)^2} \left(1 + \frac{\xi_0}{r(\xi_0, \eta_0, -\zeta_0)} \right) - \\ &\qquad - \frac{1+\zeta_0}{\eta_0^2 + (1+\zeta_0)^2} \left(1 + \frac{\xi_0}{r(\xi_0, \eta_0, \zeta_0)} \right) \\ v_z(\xi_0, \eta_0, \zeta_0) &= -\frac{\eta_0}{\eta_0^2 + (1-\zeta_0)^2} \left(1 + \frac{\xi_0}{r(\xi_0, \eta_0, -\zeta_0)} \right) + \\ &\qquad + \frac{\eta_0}{\eta_0^2 + (1+\zeta_0)^2} \left(1 + \frac{\xi_0}{r(\xi_0, \eta_0, \zeta_0)} \right) \\ r(\xi_0, \eta_0, \zeta_0) &= \sqrt{\xi_0^2 + \eta_0^2 + (1+\zeta_0)^2} \end{aligned} \right\} \tag{7.6}$$

§2. Oblique Horseshoe Vortices

We consider the vortex system to consist of a bound vortex with constant strength Γ_+ along its span and axis, inclined at an angle χ to the perpendicular to the velocity U_0 of the oncoming flow, and two free vortex filaments with axes parallel to U_0 (Fig. 7.3). It is convenient to use this system in the calculation of characteristics of swept-back wings. The strength of the free vortices will also be constant and equal to Γ_+ along its length.

We introduce the coordinate system shown in Fig. 7.3, where the Ox axis is directed downstream and parallel to U_0, the free and bound vortices are in the Oxz plane, and the origin is in the middle of the bound vortex. The free vortices are parallel to the Ox axis; the angle between the Oz axis and the bound vortex will be denoted by χ, and the distance between the free vortices by l_0.

To find the flow about the wing, we must know the velocities produced by the indicated system in the Oxz plane. Let $M(x_0, 0, z_0)$ be a point in this plane (Fig. 7.4). We join M to the point A with coordinates $[-(l_0/2)\tan\chi, 0, l_0/2]$, and to the point B with coordinates $[(l_0/2)\tan\chi, 0, l_0/2]$. The distance MB will be denoted by a, the distance MA by b, and the length of the perpendicular from M to AB by h. From the cosine formula we have

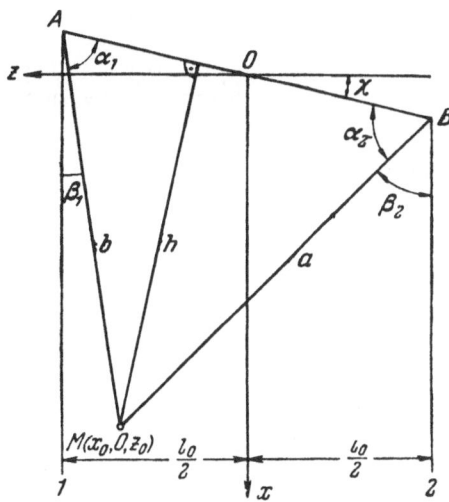

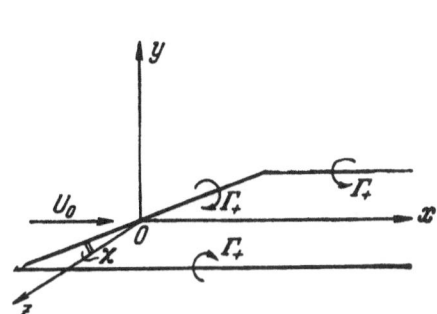

Fig. 7.3. An oblique horseshoe vortex.

Fig. 7.4. The calculation of velocities due to an oblique horseshoe vortex.

$$\cos\alpha_1 = \frac{b^2 + \dfrac{l_0^2}{\cos^2\chi} - a^2}{2b\dfrac{l_0}{\cos\chi}}, \qquad \cos\alpha_2 = \frac{a^2 + \dfrac{l_0^2}{\cos^2\chi} - b^2}{2a\dfrac{l_0}{\cos\chi}},$$

and

$$a = \frac{l_0}{2}\sqrt{(\xi_0 - \operatorname{tg}\chi)^2 + (\zeta_0 + 1)^2}, \qquad b = \frac{l_0}{2}\sqrt{(\xi_0 + \operatorname{tg}\chi)^2 + (\zeta_0 - 1)^2},$$

$$h = b\sin\alpha_1, \qquad \cos\beta_1 = \frac{x_0 + \dfrac{l_0}{2}\operatorname{tg}\chi}{b}, \qquad \cos\beta_2 = \frac{x_0 - \dfrac{l_0}{2}\operatorname{tg}\chi}{a},$$

where ξ_0 and ζ_0 have the same meaning as in (7.4).

After some transformations we obtain

$$\left.\begin{aligned}
\cos\alpha_1 &= \frac{\xi_0\sin\chi - \zeta_0\cos\chi + \dfrac{1}{\cos\chi}}{\sqrt{(\xi_0 + \operatorname{tg}\chi)^2 + (1-\zeta_0)^2}}, & \cos\alpha_2 &= \frac{\dfrac{1}{\cos\chi} - \xi_0\sin\chi + \zeta_0\cos\chi}{\sqrt{(\xi_0 - \operatorname{tg}\chi)^2 + (1+\zeta_0)^2}}, \\[2mm]
\cos\beta_1 &= \frac{\xi_0 + \operatorname{tg}\chi}{\sqrt{(\xi_0 + \operatorname{tg}\chi)^2 + (1-\zeta_0)^2}}, & \cos\beta_2 &= \frac{\xi_0 - \operatorname{tg}\chi}{\sqrt{(\xi_0 - \operatorname{tg}\chi)^2 + (1+\zeta_0)^2}}, \\[2mm]
& \multicolumn{3}{c}{h = \frac{l_0}{2}(\xi_0\cos\chi + \zeta_0\sin\chi).}
\end{aligned}\right\} \tag{7.7}$$

The strength of the vortex will again be written as in (7.1) and the downwash from the bound vortex in the case under consideration will be U_y; from the free vortices it will be V_y. From (7.2) we obtain

$$U_y = -\frac{\Gamma_+}{4\pi h}(\cos\alpha_1 + \cos\alpha_2), \qquad V_y = -\frac{\Gamma_+}{4\pi}\left(\frac{1+\cos\beta_1}{\dfrac{l_0}{2} - z_0} + \frac{1+\cos\beta_2}{\dfrac{l_0}{2} + z_0}\right). \tag{7.8}$$

These velocities can be expressed in the form

$$U_y = \frac{U_0\Gamma}{2\pi}\, u_y\,(\xi_0,\ \zeta_0,\ \chi), \qquad V_y = \frac{U_0\Gamma}{2\pi}\, v_y\,(\xi_0,\ \zeta_0,\ \chi), \tag{7.9}$$

and the use of (7.7) and (7.8) easily yields the relations

$$\left.\begin{aligned}
u_y\,(\xi_0,\ \zeta_0,\ \chi) &= -\frac{1}{\xi_0\cos\chi+\zeta_0\sin\chi}\left[\frac{\xi_0\sin\chi-\zeta_0\cos\chi+\dfrac{1}{\cos\chi}}{\sqrt{(\xi_0+\operatorname{tg}\chi)^2+(1-\zeta_0)^2}}+\right.\\
&\qquad\left.+\frac{\dfrac{1}{\cos\chi}-\xi_0\sin\chi+\zeta_0\cos\chi}{\sqrt{(\xi_0-\operatorname{tg}\chi)^2+(1+\zeta_0)^2}}\right],\\
v_y\,(\xi_0,\ \zeta_0,\ \chi) &= -\frac{1}{1-\zeta_0}\left[1+\frac{\xi_0+\operatorname{tg}\chi}{\sqrt{(\xi_0+\operatorname{tg}\chi)^2+(1-\zeta_0)^2}}\right]-\\
&\qquad-\frac{1}{1+\zeta_0}\left[1+\frac{\xi_0-\operatorname{tg}\chi}{\sqrt{(\xi_0-\operatorname{tg}\chi)^2+(1+\zeta_0)^2}}\right].
\end{aligned}\right\} \tag{7.10}$$

The total downwash will be denoted by W_y, and

$$W_y = \frac{U_0\Gamma}{2\pi}\, w_y\,(\xi_0,\ \zeta_0,\ \chi), \quad w_y\,(\xi_0,\ \zeta_0,\ \chi) = u_y\,(\xi_0,\ \zeta_0,\ \chi) + v_y\,(\xi_0,\ \zeta_0,\ \chi). \tag{7.11}$$

For $\chi = 0$ the relations (7.10) and (7.11) yield the same result as (7.6) with $\eta_0 = 0$.

§ 3. General Relations for Annular Vortices

We use the rectangular coordinate system shown in Fig. 7.5, the annular vortex being located in the Oyz plane with the x axis directed along the axis of the cylindrical coordinate system (x, r, φ), in which the angle φ is measured from the y axis and the radius of the annular vortex is assumed to be r.

We first consider the velocities generated by a bound annular vortex. From the Biot-Savart law, the velocity at the point $M(x_0,\ r_0,\ \varphi_0)$ of an element of the annular vortex $rd\varphi$ corresponding to the point $A(0,\ r\ \varphi)$ will be

$$dU = \frac{\Gamma_+}{4\pi R^{*2}}\sin\beta r\, d\varphi, \tag{7.12}$$

where Γ_+ is the strength of the vortex element, R^* is the distance from the point A to M, and β is the angle between the vortex axis and the radius vector $R*$. The velocity dU is perpendicular to the plane passing through the tangent to the circle at the point A and through the point M.

The equation of this plane is

$$\frac{r-r_0\cos(\varphi-\varphi_0)}{x_0 r}\,x+\frac{\cos\varphi}{r}\,y+\frac{\sin\varphi}{r}\,z-1=0. \tag{7.13}$$

Moreover

$$\left.\begin{aligned}
\sin\beta &= \frac{\sqrt{x_0^2+[r-r_0\cos(\varphi-\varphi_0)]^2}}{R^*},\\
R^* &= \sqrt{x_0^2+r^2+r_0^2-2rr_0\cos(\varphi-\varphi_0)}.
\end{aligned}\right\} \tag{7.14}$$

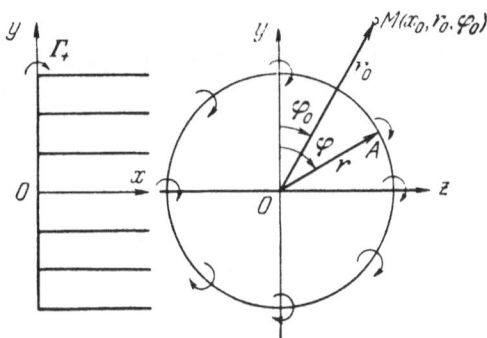

Fig. 7.5. Annular vortex.

The direction cosines of the normal to the plane (7.13) are

$$\left.\begin{aligned}
\cos(n,\ x) &= -\frac{r - r_0 \cos(\varphi - \varphi_0)}{\sqrt{x_0^2 + [r - r_0 \cos(\varphi - \varphi_0)]^2}}, \\[4pt]
\cos(n,\ y) &= -\frac{x_0 \cos\varphi}{\sqrt{x_0^2 + [r - r_0 \cos(\varphi - \varphi_0)]^2}}, \\[4pt]
\cos(n,\ z) &= -\frac{x_0 \sin\varphi}{\sqrt{x_0^2 + [r - r_0 \cos(\varphi - \varphi_0)]^2}},
\end{aligned}\right\} \quad (7.15)$$

and from these formulas we easily obtain expressions for the projections on the axes of a rectangular coordinate system of the velocities generated by an annular vortex. Using the formulas for the conversion of these projections to projections on the axes of the cylindrical coordinate system

$$U_r = U_y \cos\varphi_0 + U_z \sin\varphi_0, \qquad U_\varphi = U_z \cos\varphi_0 - U_y \sin\varphi_0,$$

we obtain

$$\left.\begin{aligned}
U_x &= -\frac{r}{4\pi} \int_0^{2\pi} \frac{\Gamma_+ \, [r - r_0 \cos(\varphi - \varphi_0)] \, d\varphi}{R^{*3}}, \\[4pt]
U_r &= -\frac{x_0 r}{4\pi} \int_0^{2\pi} \frac{\Gamma_+ \cos(\varphi - \varphi_0) \, d\varphi}{R^{*3}}, \\[4pt]
U_\varphi &= -\frac{x_0 r}{4\pi} \int_0^{2\pi} \frac{\Gamma_+ \sin(\varphi - \varphi_0) \, d\varphi}{R^{*3}}.
\end{aligned}\right\} \quad (7.16)$$

We turn now to the derivation of formulas for the velocities generated by the cylindrical vortex layer modeling the free vortices. Since the strength Γ_+ of the bound vortex varies along the length of the vortex, free vortices will be shed from the annular vortex parallel to the x axis with linear strength $\gamma = (d\Gamma_+ / d\varphi) \cdot 1/r$. From Eqs. (7.2) as applied to a semi-infinite vortex filament, we find at the point $M(x_0, r_0, \varphi_0)$ the velocities due to an element $r \, d\varphi$ of the cylindrical layer. Taking into account the fact that this velocity will be perpendicular to the plane through the vortex filament and the point M, we obtain

$$\left.\begin{aligned}
V_x &= 0, \quad V_r = \frac{r}{4\pi} \int_0^{2\pi} \frac{d\Gamma_+}{d\varphi} \frac{\sin(\varphi - \varphi_0)}{r^2 + r_0^2 - 2rr_0 \cos(\varphi - \varphi_0)} \left(1 + \frac{x_0}{R^*}\right) d\varphi, \\[4pt]
V_\varphi &= \frac{1}{4\pi} \int_0^{2\pi} \frac{d\Gamma_+}{d\varphi} \frac{r_0 - r \cos(\varphi - \varphi_0)}{r^2 + r_0^2 - 2rr_0 \cos(\varphi - \varphi_0)} \left(1 + \frac{x_0}{R^*}\right) d\varphi.
\end{aligned}\right\} \quad (7.17)$$

The total velocities generated by the vortex system will be

$$W_x = U_x + V_x, \quad W_r = U_r + V_r, \quad W_\varphi = U_\varphi + V_\varphi. \tag{7.18}$$

§ 14. Annular Vortices of Constant Strength

We consider an annular vortex with a strength Γ_+ which does not vary with the angle φ [11]. We introduce the dimensionless strength Γ of the vortex and use as a characteristic linear dimension with radius r of the annular vortex:

$$\Gamma_+ = U_0 r \Gamma. \tag{7.19}$$

In the present case there will be no free vortices; we will denote the velocities due to the annular vortex by U_{x1} and U_{r1}, and we obtain

$$W_{x1} = U_{x1}, \quad W_{r1} = U_{r1}, \quad W_{\varphi 1} = U_{\varphi 1} = V_{x1} = V_{r1} = V_{\varphi 1} = 0.$$

From (7.16) we now have

$$\left. \begin{aligned} W_{x1} &= -\frac{U_0 r^2 \Gamma}{4\pi} \int_0^{2\pi} \frac{r - r_0 \cos\psi}{\left[x_0^2 + r^2 + r_0^2 - 2rr_0 \cos\psi\right]^{3/2}} \, d\psi, \\ W_{r1} &= -\frac{U_0 x_0 r^2 \Gamma}{4\pi} \int_0^{2\pi} \frac{\cos\psi \, d\psi}{\left[x_0^2 + r^2 + r_0^2 - 2rr_0 \cos\psi\right]^{3/2}}. \end{aligned} \right\} \tag{7.20}$$

In our investigation of the velocity field due to annular vortices, we use the following notation for the dimensionless coordinates:

$$\xi = \frac{x}{r}, \quad \xi_0 = \frac{x_0}{r}, \quad \rho_0 = \frac{r_0}{r}. \tag{7.21}$$

The velocities (7.20) are given by the formulas

$$W_{x1} = \frac{U_0 \Gamma}{2\pi} w_{x1}, \quad W_{r1} = \frac{U_0 \Gamma}{2\pi} w_{r1}, \tag{7.22}$$

where w_{x1} and w_{r1} are dimensionless functions depending on ξ_0 and ρ_0. From (7.20) and (7.21) we have

$$\left. \begin{aligned} w_{x1} &= -\frac{1}{2} \int_0^{2\pi} \frac{1 - \rho_0 \cos\psi}{\left[\xi_0^2 + 1 + \rho_0^2 - 2\rho_0 \cos\psi\right]^{3/2}} \, d\psi, \\ w_{r1} &= -\frac{\xi_0}{2} \int_0^{2\pi} \frac{\cos\psi \, d\psi}{\left[\xi_0^2 + 1 + \rho_0^2 - 2\rho_0 \cos\psi\right]^{3/2}}. \end{aligned} \right\} \tag{7.23}$$

The dimensionless velocities w_{x1} and w_{r1} can be expressed in terms of complete elliptic integrals of the first and second kinds K and E with modulus

$$k^2 = \frac{4\rho_0}{\xi_0^2 + (1 + \rho_0)^2}, \tag{7.24}$$

where, as is known,

$$K = \int_0^{\frac{\pi}{2}} \frac{d\alpha}{\sqrt{1 - k^2 \sin^2\alpha}}, \quad E = \int_0^{\frac{\pi}{2}} \sqrt{1 - k^2 \sin^2\alpha} \, d\alpha.$$

Detailed tables of the functions $K(k^2)$ and $E(k^2)$ can be found in [46].

After some simple transformations the relations (7.23) yield

$$\left.\begin{aligned}
w_{x1} &= \frac{1}{2\sqrt{\xi_0^2+(1+\rho_0)^2}}\left(\frac{2-k^2}{1-k^2}E-2K-\frac{1}{\rho_0}\frac{k^2}{1-k^2}E\right), \\
w_{r1} &= -\frac{\xi_0}{2\rho_0\sqrt{\xi_0^2+(1+\rho_0)^2}}\left(\frac{2-k^2}{1-k^2}E-2K\right).
\end{aligned}\right\} \tag{7.25}$$

The functions w_{x1} and w_{r1} have the following symmetry properties:

$$w_{x1}(-\xi_0,\ \rho_0)=w_{x1}(\xi_0,\ \rho_0), \qquad w_{r1}(-\xi_0,\ \rho_0)=-w_{r1}(\xi_0,\ \rho_0), \tag{7.26}$$

and

$$w_{r1}(0,\ \rho_0)=0, \qquad w_{x1}(\xi_0,\ 0)=-\frac{\pi}{(\xi_0^2+1)^{3/2}}, \qquad w_{r1}(\xi_0,\ 0)=0, \tag{7.27}$$

The annular vortex considered in this section is very important in the determination of annular-wing motion.

§5. Annular Vortices with Strength Proportional to Cos φ

We now consider the case when the strength of a bound vortex varies according to the law [37]:

$$\Gamma_+ = U_0 r\Gamma\cos\varphi. \tag{7.28}$$

This vortex system will be used in the determination of the translational motion of an annular wing at an angle of attack when it is also rotating.

From (7.16), the velocity field of the bound vortex can be obtained from the relations

$$\left.\begin{aligned}
U_{x2} &= -\frac{U_0 r^2\Gamma}{4\pi}\int_0^{2\pi}\frac{\cos\varphi\,[r-r_0\cos(\varphi-\varphi_0)]\,d\varphi}{[x_0^2+r^2+r_0^2-2rr_0\cos(\varphi-\varphi_0)]^{3/2}}, \\
U_{r2} &= -\frac{U_0 x_0 r^2\Gamma}{4\pi}\int_0^{2\pi}\frac{\cos\varphi\cos(\varphi-\varphi_0)\,d\varphi}{[x_0^2+r^2+r_0^2-2rr_0\cos(\varphi-\varphi_0)]^{3/2}}, \\
U_{\varphi2} &= -\frac{U_0 x_0 r^2\Gamma}{4\pi}\int_0^{2\pi}\frac{\cos\varphi\sin(\varphi-\varphi_0)\,d\varphi}{[x_0^2+r^2+r_0^2-2rr_0\cos(\varphi-\varphi_0)]^{3/2}}.
\end{aligned}\right\} \tag{7.29}$$

The change of variable

$$\varphi - \varphi_0 = \psi,$$

easily shows that these velocities can be given by the expressions

$$U_{x2}=\frac{U_0\Gamma}{2\pi}u_{x2}\cos\varphi_0, \qquad U_{r2}=\frac{U_0\Gamma}{2\pi}u_{r2}\cos\varphi_0, \qquad U_{\varphi2}=\frac{U_0\Gamma}{2\pi}u_{\varphi2}\sin\varphi_0. \tag{7.30}$$

where u_{x2}, u_{r2}, and $u_{\varphi2}$ are functions of the dimensionless parameters ξ_0 and ρ_0. From (7.29) and (7.30) we easily obtain

$$\left.\begin{aligned}
u_{x2} &= -\frac{1}{2}\int_0^{2\pi}\frac{\cos\psi\,(1-\rho_0\cos\psi)\,d\psi}{\left(\xi_0^2+1+\rho_0^2-2\rho_0\cos\psi\right)^{3/2}}, \\
u_{r2} &= -\frac{\xi_0}{2}\int_0^{2\pi}\frac{\cos^2\psi\,d\psi}{\left(\xi_0^2+1+\rho_0^2-2\rho_0\cos\psi\right)^{3/2}}, \\
u_{\varphi2} &= -\frac{\xi_0}{2}\int_0^{2\pi}\frac{\sin^2\psi\,d\psi}{\left(\xi_0^2+1+\rho_0^2-2\rho_0\cos\psi\right)^{3/2}}.
\end{aligned}\right\} \tag{7.31}$$

These dimensionless velocities can also be given in terms of complete elliptic integrals of the first and second kind with the modulus (7.24):

$$\left.\begin{aligned}
u_{x2} &= \frac{1}{2\sqrt{\xi_0^2+(1+\rho_0)^2}}\left[4K+\frac{k^2}{1-k^2}E-8\frac{K-E}{k^2}-\frac{K^2}{\rho_0^2}\left(\frac{E}{1-k^2}-2\frac{K-E}{k^2}\right)\right], \\
u_{r2} &= -\frac{\xi_0}{2\rho_0\sqrt{\xi_0^2+(1+\rho_0)^2}}\left(4K+\frac{k^2}{1-k^2}E-8\frac{K-E}{k^2}\right), \\
u_{\varphi2} &= \frac{2\xi_0}{\rho_0\sqrt{\xi_0^2+(1+\rho_0)^2}}\left(2\frac{K-E}{k^2}-K\right).
\end{aligned}\right\} \tag{7.32}$$

These functions have the following symmetry properties:

$$\left.\begin{aligned}
u_{x2}(-\xi_0,\ \rho_0) &= u_{x2}(\xi_0,\ \rho_0), \quad u_{r2}(-\xi_0,\ \rho_0) = -u_{r2}(\xi_0,\ \rho_0), \\
u_{\varphi2}(-\xi_0,\ \rho_0) &= -u_{\varphi2}(\xi_0,\ \rho_0).
\end{aligned}\right\} \tag{7.33}$$

We also have the relations

$$\left.\begin{aligned}
u_{x2}(\xi_0,\ 0) &= 0, \quad u_{r2}(0,\ \rho_0)=0, \quad u_{\varphi2}(0,\ \rho_0)=0, \\
u_{r2}(\xi_0,\ 0) &= -\frac{\pi}{2}\frac{\xi_0}{\left(\xi_0^2+1\right)^{3/2}}, \quad u_{\varphi2}(\xi_0,\ 0)=-\frac{\pi}{2}\frac{\xi_0}{\left(\xi_0^2+1\right)^{3/2}},
\end{aligned}\right\} \tag{7.34}$$

We now investigate the velocity field of the free vortices corresponding to the bound vortex (7.28). Substituting

$$\frac{d\Gamma_+}{d\varphi}=-U_0 r\Gamma\sin\varphi,$$

in (7.17), as in the previous case we have

$$V_{x2}=0, \quad V_{r2}=\frac{U_0\Gamma}{2\pi}v_{r2}\cos\varphi_0, \quad V_{\varphi2}=\frac{U_0\Gamma}{2\pi}v_{\varphi2}\sin\varphi_0, \tag{7.35}$$

where

$$\begin{aligned}
v_{r2} &= -\frac{1}{2}\int_0^{2\pi}\frac{\sin^2\psi\,d\psi}{1+\rho_0^2-2\rho_0\cos\psi}-\frac{\xi_0}{2}\int_0^{2\pi}\frac{\sin^2\psi\,d\psi}{(1+\rho_0^2-2\rho_0\cos\psi)\sqrt{\xi_0^2+1+\rho_0^2-2\rho_0\cos\psi}}, \\
v_{\varphi2} &= -\frac{1}{2}\int_0^{2\pi}\frac{\rho_0\cos\psi-\cos^2\psi}{1+\rho_0^2-2\rho_0\cos\psi}\,d\psi-\frac{\xi_0}{2}\int_0^{2\pi}\frac{(\rho_0\cos\psi-\cos^2\psi)\,d\psi}{(1+\rho_0^2-2\rho_0\cos\psi)\sqrt{\xi_0^2+1+\rho_0^2-2\rho_0\cos\psi}}.
\end{aligned} \tag{7.36}$$

These functions can be expressed in terms of complete elliptic integrals of the first, second, and third kind. The modulus k^2 of the integrals K and E, as in the previous case, is given by (7.24). In our case, the complete elliptic integral of the third kind will always have a negative value for the first argument, which we will denote by $-n^2$:

$$\Pi\left(-n^2,\ k^2\right) = \int_0^{\pi/2} \frac{d\alpha}{(1-n^2\sin^2\alpha)\sqrt{1-k^2\sin^2\alpha}} ,$$

where

$$n^2 = \frac{4\rho_0}{(1+\rho_0)^2}, \qquad k^2 \leqslant n^2 \leqslant 1. \tag{7.37}$$

Detailed tables of $\Pi\left(-n^2,\ k^2\right)$ can be found in [46].

Using the identities

$$\frac{\sin^2\psi}{1+\rho_0^2-2\rho_0\cos\psi} = \frac{\cos\psi}{2\rho_0} + \frac{1+\rho_0^2}{4\rho_0^2} - \frac{\dfrac{(1-\rho_0^2)^2}{4\rho_0^2}}{1+\rho_0^2-2\rho_0\cos\psi},$$

$$\frac{\rho_0\cos\psi-\cos^2\psi}{1+\rho_0^2-2\rho_0\cos\psi} = \frac{\cos\psi}{2\rho_0} + \frac{1-\rho_0^2}{4\rho_0^2} - \frac{\dfrac{1-\rho_0^4}{4\rho_0^2}}{1+\rho_0^2-2\rho_0\cos\psi},$$

we easily establish the following expressions for the integrals:

$$\int_0^{2\pi} \frac{\sin^2\psi\,d\psi}{1+\rho_0^2-2\rho_0\cos\psi} = \begin{cases} \dfrac{\pi}{\rho_0^2} & \text{for}\quad \rho_0 \geqslant 1, \\[2mm] \pi & \text{for}\quad \rho_0 \leqslant 1, \end{cases} \tag{7.38}$$

and also

$$\int_0^{2\pi} \frac{\rho_0\cos\psi-\cos^2\psi}{1+\rho_0^2-2\rho_0\cos\psi}\,d\psi = \begin{cases} -\pi & \text{for}\quad \rho_0 < 1, \\[2mm] 0 & \text{for}\quad \rho_0 = 1, \\[2mm] \dfrac{\pi}{\rho_0^2} & \text{for}\quad \rho_0 > 1, \end{cases} \tag{7.39}$$

The two remaining integrals in (7.36) can be expressed in terms of elementary functions and elliptic integrals by using the change of variable $\psi = \pi - 2\alpha$.

After some transformations we obtain

$$\left.\begin{aligned} & v_{r2} = -(A+\Delta v_r), \qquad v_{\varphi 2} = -(A_1+\Delta v_\varphi), \\[2mm] & A = \begin{cases} \dfrac{\pi}{2} & \text{for}\quad \rho_0 \leqslant 1, \\[2mm] \dfrac{\pi}{2\rho_0^2} & \text{for}\quad \rho_0 \geqslant 1, \end{cases} \qquad A_1 = \begin{cases} -\dfrac{\pi}{2} & \text{for}\quad \rho_0 < 1, \\[2mm] 0 & \text{for}\quad \rho_0 = 1, \\[2mm] \dfrac{\pi}{2\rho_0^2} & \text{for}\quad \rho_0 > 1, \end{cases} \\[2mm] & \Delta v_\varphi = B_1+\Delta_1 v_\varphi, \quad \Delta_1 v_\varphi = \frac{\xi_0}{\rho_0\sqrt{\xi_0^2+(1+\rho_0)^2}}\left(2\frac{K-E}{k^2}+2\frac{1-n^2}{n^2}K-\rho_0 K\right), \end{aligned}\right\} \tag{7.40}$$

$$B_1 = \text{sign}\left[\xi_0(\rho_0 - 1)\right] \frac{1}{\rho_0} \frac{2-n^2}{n^2} \sqrt{\frac{(1-n^2)(n^2-k^2)}{n^2}} \, \Pi(-n^2,\, k^2),$$

$$\Delta v_r = \frac{\xi_0}{\rho_0 \sqrt{\xi_0^2 + (1+\rho_0)^2}}\left[\left(\frac{2-k^2}{k^2} + \frac{2-n^2}{n^2}\right)K - \frac{2}{n^2}(1-n^2)\,\Pi(-n^2,\, k^2) - \frac{2}{k^2}E\right]. \tag{7.40}$$

The function "sign" is defined as follows:

$$\text{sign } z = \begin{cases} +1 & \text{for } z > 0, \\ -1 & \text{for } z < 0. \end{cases}$$

It is clear from (7.40) that the dimensionless velocities for free vortices consist of two parts, the first of which (A and A_1) is independent of ξ_0 and the second (Δv_r and Δv_φ) changes sign when ξ_0 changes sign:

$$\begin{aligned} \Delta v_r\,(-\xi_0,\ \rho_0) &= -\Delta v_r\,(\xi_0,\ \rho_0), \\ \Delta v_\varphi\,(-\xi_0,\ \rho_0) &= -\Delta v_\varphi\,(\xi_0,\ \rho_0). \end{aligned} \tag{7.41}$$

In the case $\rho_0 = 1$, which is of basic importance in the annular wing problem, Eqs. (7.40) for v_{r2} are greatly simplified:

$$v_{r2} = -\left(\frac{\pi}{2} + \Delta v_r\right), \qquad \Delta v_r = \frac{2\xi_0}{\sqrt{\xi_0^2 + 4}} \frac{K-E}{k^2}. \tag{7.42}$$

We also have the relations

$$\Delta v_r\,(0,\ \rho_0) = 0, \qquad \Delta v_r\,(\xi_0,\ 0) = \frac{\pi}{2} \frac{\xi_0}{\sqrt{1+\xi_0^2}}. \tag{7.43}$$

From the basic theorems of Chapter II, we know that the velocity $v_{\varphi2}$ will have a discontinuity of the first kind at the cylindrical surface $\rho_0 = 1$ occupied by free vortices. We write $v_{\varphi2+}$ and $v_{\varphi2-}$ for the limiting values of the dimensionless velocities when $\rho_0 \to 1$ inside and outside the cylinder. Then

$$\begin{aligned} v_{\varphi2+} &= -\pi - \frac{\xi_0}{\sqrt{\xi_0^2 + 4}}\left(2\,\frac{K-E}{k^2} - K\right), \\ v_{\varphi2-} &= \pi - \frac{\xi_0}{\sqrt{\xi_0^2 + 4}}\left(2\,\frac{K-E}{k^2} - K\right). \end{aligned} \tag{7.44}$$

The velocity at any point of the vortex sheet, i.e., for $\rho_0 = 1$, will be denoted by $v_{\varphi20}$: from (7.36) we can therefore write

$$v_{\varphi20} = -\frac{\xi_0}{\sqrt{\xi_0^2 + 4}}\left(2\,\frac{K-E}{k^2} - K\right). \tag{7.45}$$

From Theorem 3 of Chapter II we have

$$\frac{1}{2}\left(v_{\varphi2+} + v_{\varphi2-}\right) = v_{\varphi20}.$$

this relation can also be obtained from (7.44) and (7.45). The theorem referred to also yields

$$V_{\varphi 2+} - V_{\varphi 2-} = \frac{1}{r}\frac{d\Gamma_+}{d\varphi_0},$$

which also holds in the present case.

The total velocities generated by all the vortices under consideration can now be given by the formulas

$$W_{x2} = \frac{U_0\Gamma}{2\pi}\,w_{x2}\cos\varphi_0, \quad W_{r2} = \frac{U_0\Gamma}{2\pi}\,w_{r2}\cos\varphi_0, \quad W_{\varphi 2} = \frac{U_0\Gamma}{2\pi}\,w_{\varphi 2}\sin\varphi_0, \left.\right\} \quad (7.46)$$
$$w_{x2} = u_{x2}, \qquad\qquad w_{r2} = u_{r2}+v_{r2}, \qquad\qquad w_{\varphi 2} = u_{\varphi 2}+v_{\varphi 2}.$$

§ 6. Vortices of Infinite Span. Chains of Vortices of Infinite Span

In solving our problem for a half plane of infinite span, we consider isolated vortices of an infinitely long filament. We obtain formulas for the calculation of the velocities due to such a filament located at the origin and directed along the Oz axis (Fig. 7.6). The velocity W will be perpendicular to the radius vector r joining the origin to the points (x_0, y_0) at which we are calculating the velocity. Hence for the projections of the disturbed velocity we have

$$W_x = \frac{\Gamma_+}{2\pi}\,\frac{y_0}{x_0^2+y_0^2}, \quad W_y = -\frac{\Gamma_+}{2\pi}\,\frac{x_0}{x_0^2+y_0^2},$$

where Γ_+ is the strength of the vortex filament. We write

$$\Gamma_+ = U_0 b\Gamma, \quad \xi_0 = \frac{x_0}{b}, \qquad \eta_0 = \frac{y_0}{b}, \qquad\qquad (7.47)$$

and then we have

$$W_x = \frac{U_0\Gamma}{2\pi}\,w_x, \quad W_y = \frac{U_0\Gamma}{2\pi}\,w_y,$$
$$w_x = \frac{\eta_0}{\xi_0^2+\eta_0^2}, \qquad w_y = -\frac{\xi_0}{\xi_0^2+\eta_0^2}. \qquad\qquad (7.48)$$

In the consideration of the flow about a profile cascade, the basic singularities are infinite chains of vortices of identical strengths Γ_+ (Figs. 3.5 and 7.7). Let the distance between vortices be t, let the $O\eta$ axis be parallel to the vortex chain, and let the $O\xi$ axis be perpendicular to this chain. The origin O is located at the position of one of the vortices. We take the first axis as the imaginary axis and the second as the real axis; the complex coordinate of the point where we are calculating the velocity is

$$z_0 = \xi_0 + i\eta_0.$$

The conjugate complex velocity of the infinite vortex chain parallel to the imaginary axis is [11]

$$W_\xi - iW_\eta = \frac{-\Gamma_+}{2t}\,\mathrm{ctg}\,\frac{i\pi z_0}{t}. \qquad (7.49)$$

But

$$\mathrm{ctg}\,\frac{i\pi z_0}{t} = \frac{\cos\left(-\frac{\pi\eta_0}{t}+i\frac{\pi\xi_0}{t}\right)}{\sin\left(-\frac{\pi\eta_0}{t}+i\frac{\pi\xi_0}{t}\right)};$$

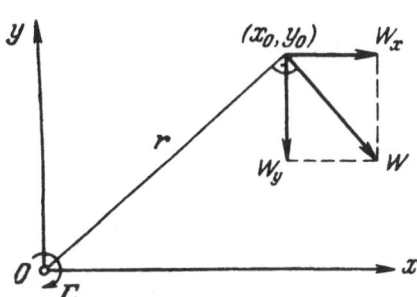

Fig. 7.6. A vortex filament of infinite span.

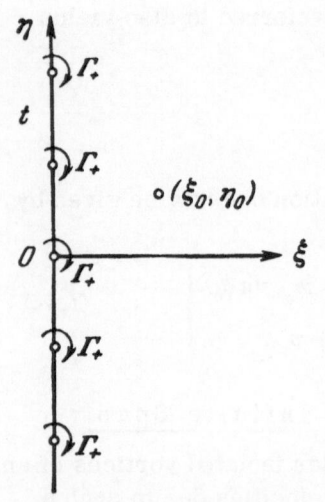

Fig. 7.7. A vortex chain.

now since

$$\cos i\varphi = \operatorname{ch} \varphi, \qquad \sin i\varphi = i \operatorname{sh} \varphi,$$

we have

$$\cot \frac{i\pi z_0}{t} = -\frac{1}{2} \frac{\sin \frac{2\pi \eta_0}{t} + i \operatorname{sh} \frac{2\pi \xi_0}{t}}{\sin^2 \frac{\pi \eta_0}{t} + \operatorname{sh}^2 \frac{\pi \xi_0}{t}}.$$

Separating real and imaginary parts in (7.49) we obtain

$$W_\xi = \frac{\Gamma_+}{4t} \frac{\sin \frac{2\pi \eta_0}{t}}{\sin^2 \frac{\pi \eta_0}{t} + \operatorname{sh}^2 \frac{\pi \xi_0}{t}},$$

$$W_\eta = -\frac{\Gamma_+}{4t} \frac{\operatorname{sh} \frac{2\pi \xi_0}{t}}{\sin^2 \frac{\pi \eta_0}{t} + \operatorname{sh}^2 \frac{\pi \xi_0}{t}}.$$

This relation can also be written

$$\left. \begin{aligned} w_\xi(\xi_0, \eta_0) &= \frac{1}{2} \frac{\sin 2\pi \eta_0}{\sin^2 \pi \bar\eta_0 + \operatorname{sh}^2 \pi \bar\xi_0}, \\ w_\eta(\xi_0, \eta_0) &= -\frac{1}{2} \frac{\operatorname{sh} 2\pi \bar\xi_0}{\sin^2 \pi \bar\eta_0 + \operatorname{sh}^2 \pi \bar\xi_0}, \end{aligned} \right\} \tag{7.50}$$

where

$$\Gamma_+ = U_0 \Gamma t, \quad \bar\xi_0 = \frac{\xi_0}{t}, \quad \bar\eta_0 = \frac{\eta_0}{t}, \quad W_\xi = \frac{U_0 \Gamma}{2} w_\xi, \quad W_\eta = \frac{U_0 \Gamma}{2} w_\eta. \tag{7.51}$$

§7. Broken Bound Vortices of Variable Strength along the Span

In the derivation of the integral equation for a monoplane wing, in the solution of a wing of very small aspect ratio, etc., it is expedient to consider as an auxiliary vortex system an interrupted horseshoe vortex with a variable strength $\Gamma_+(z)$ along its span. From the bound vortex there will originate free vortices formed behind the vortices in the sheet. The span of the vortex will be denoted by l, the sweepback from the leading edge by χ, and the origin O will be located at the midpoint of the segment of length $(l/2) \tan \chi$ (Fig. 7.8). In the same diagram we show the positive directions of the circulation of the bound vortex Γ_+ and an element $d\Gamma_+$ of the sheet of free vortices.

From the Biot-Savart formula for the velocity due to an element from the right half of an attached vortex dl we obtain (Fig. 7.9)

$$\left. \begin{aligned} dU_y &= -\frac{\Gamma_+}{4\pi R^2} \sin \varphi\, dl, \qquad R = (x - x_0)^2 + (z - z_0)^2, \\ \sin(\varphi - \chi) &= \frac{x - x_0}{R}, \qquad \cos(\varphi - \chi) = \frac{z - z_0}{R}. \end{aligned} \right\} \tag{7.52}$$

The velocity is calculated at the point with coordinates $(x_0, 0, z_0)$ in the plane of the vortex, and the coordinates of the element dl are $(x, 0, z)$.

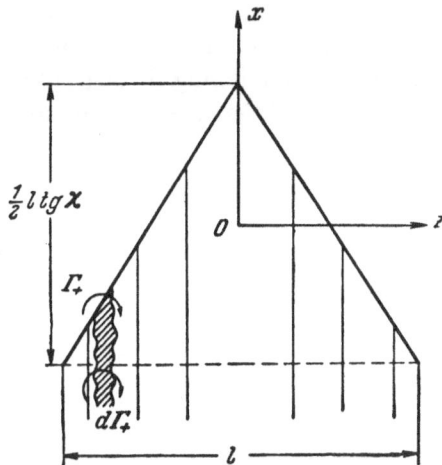

Fig. 7.8. An interrupted bound vortex with variable circulation.

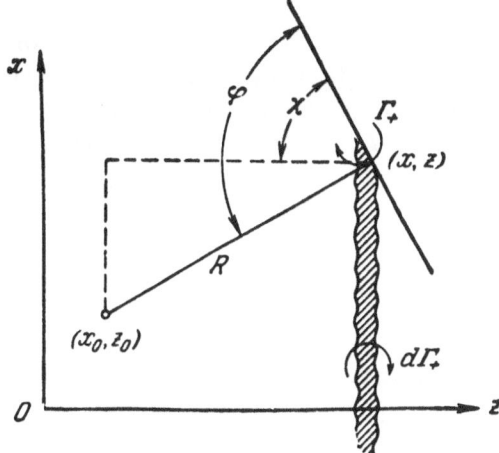

Fig. 7.9. The calculation of the velocity field due to the vortex system shown in Fig. 7.8.

Using the relations

$$\sin \varphi = \frac{x-x_0}{R}\cos\chi + \frac{z-z_0}{R}\sin\chi, \quad dz = dl\cos\chi,$$

we obtain

$$dU_y = -\frac{\Gamma_+}{4\pi}\frac{(x-x_0)+(z-z_0)\,\mathrm{tg}\,\chi}{[(x-x_0)^2+(z-z_0)^2]^{3/2}}\,dz. \tag{7.53}$$

To find the velocity due to an elementary semi-infinite vortex band of width dz (Fig. 7.9), we use Eq. (7.2) which, together with (7.52), yields

$$dV_y = \frac{\dfrac{d\Gamma_+}{dz}\,dz}{4\pi\,(z-z_0)}\left(1 + \frac{x-x_0}{\sqrt{(x-x_0)^2+(z-z_0)^2}}\right). \tag{7.54}$$

The left half of the bound vortex is inclined to the Oz axis at an angle with sign opposite to that of the angle of inclination of the sheet (Fig. 7.8). Hence for the left half the external form of the formula (7.54) remains unchanged, but in (7.53) we must change the sign of χ. Integrating the relations (7.53) and (7.54) for the left and right halves of the vortex with respect to z and adding the results, we obtain the total velocity

$$W_y = -\frac{1}{4\pi}\int_{-\frac{l}{2}}^{0}\frac{\Gamma_+(z)\,[(x-x_0)-\mathrm{tg}\,\chi\,(z-z_0)]\,dz}{[(x-x_0)^2+(z-z_0)^2]^{3/2}} -$$

$$-\frac{1}{4\pi}\int_{0}^{\frac{l}{2}}\frac{\Gamma_+(z)\,[(x-x_0)+\mathrm{tg}\,\chi\,(z-z_0)]\,dz}{[(x-x_0)^2+(z-z_0)^2]^{3/2}} + \tag{7.55}$$

$$+\frac{1}{4\pi}\int_{-\frac{l}{2}}^{\frac{l}{2}}\frac{\dfrac{d\Gamma_+}{dz}}{z-z_0}\left(1+\frac{x-x_0}{\sqrt{(x-x_0)^2+(z-z_0)^2}}\right)dz.$$

We should note that the variables x and z in (7.55) are not independent; x can be obtained as a function of z by writing the equation of the straight lines corresponding to the bound vortex. Since the left part of the vortex is a segment of the straight line passing through the point $z = -l/4$ and inclined at an angle χ, and the right part is a segment of the straight line passing through $z = -l/4$ with angle of inclination $-\chi$, we have (Fig. 7.8)

$$\left.\begin{array}{ll} x = \operatorname{tg}\chi\left(z + \dfrac{l}{4}\right) & \text{for } z \leqslant 0, \\[2mm] x = -\operatorname{tg}\chi\left(z - \dfrac{l}{4}\right) & \text{for } z \geqslant 0. \end{array}\right\} \tag{7.56}$$

Hence instead of (7.55) we can write

$$
\begin{aligned}
W_y = & -\frac{1}{4\pi}\int_{-\frac{l}{2}}^{0} \frac{\Gamma_+(z)\left[\operatorname{tg}\chi\left(z+\frac{l}{4}\right)-x_0-\operatorname{tg}\chi(z-z_0)\right]dz}{\left\{\left[\operatorname{tg}\chi\left(z+\frac{l}{z}\right)-x_0\right]^2+(z-z_0)^2\right\}^{3/2}} - \\[2mm]
& -\frac{1}{4\pi}\int_{0}^{\frac{l}{2}} \frac{\Gamma_+(z)\left[\operatorname{tg}\chi\left(\frac{l}{4}-z\right)-x_0+\operatorname{tg}\chi(z-z_0)\right]dz}{\left\{\left[\operatorname{tg}\chi\left(\frac{l}{4}-z\right)-x_0\right]^2+(z-z_0)^2\right\}^{3/2}} + \\[2mm]
& +\frac{1}{4\pi}\int_{-\frac{l}{2}}^{0} \frac{\frac{d\Gamma_+}{dz}}{z-z_0}\left(1+\frac{\operatorname{tg}\chi\left(z+\frac{l}{4}\right)-x_0}{\sqrt{\left[\operatorname{tg}\chi\left(z+\frac{l}{4}\right)-x_0\right]^2+(z-z_0)^2}}\right)dz + \\[2mm]
& +\frac{1}{4\pi}\int_{0}^{\frac{l}{2}} \frac{\frac{d\Gamma_+}{dz}}{z-z_0}\left(1+\frac{\operatorname{tg}\chi\left(\frac{l}{4}-z\right)-x_0}{\sqrt{\left[\operatorname{tg}x\left(\frac{l}{4}-z\right)-x_0\right]^2+(z-z_0)^2}}\right)dz.
\end{aligned}
\tag{7.57}
$$

We now reduce Eq. (7.57) to dimensionless form; to do this we set

$$\bar{z} = \frac{2z}{l}, \quad \bar{z}_0 = \frac{2z_0}{l}, \quad \bar{x}_0 = \frac{x_0}{b}, \quad \Gamma_+ = U_0 l\Gamma; \tag{7.58}$$

then instead of (7.57) we have

$$
\begin{aligned}
\frac{W_y(x_0, z_0)}{U_0} = & -\frac{1}{8\pi}\left(\frac{l}{b}\right)^2\int_{-1}^{0} \frac{\Gamma\left[\operatorname{tg}\chi\frac{l}{2b}\left(\bar{z}+\frac{1}{2}\right)-\bar{x}_0-\operatorname{tg}\chi\frac{l}{2b}(\bar{z}-\bar{z}_0)\right]d\bar{z}}{\left\{\left[\operatorname{tg}\chi\frac{l}{2b}\left(\bar{z}+\frac{1}{2}\right)-\bar{x}_0\right]^2+\left(\frac{l}{2b}\right)^2(\bar{z}-\bar{z}_0)^2\right\}^{3/2}} - \\[2mm]
& -\frac{1}{8\pi}\left(\frac{l}{b}\right)^2\int_{0}^{1} \frac{\Gamma\left[-\operatorname{tg}\chi\frac{l}{2b}\left(\bar{z}-\frac{1}{2}\right)-\bar{x}_0+\operatorname{tg}\chi\frac{l}{2b}(\bar{z}-\bar{z}_0)\right]d\bar{z}}{\left\{\left[\operatorname{tg}\chi\frac{l}{2b}\left(\bar{z}-\frac{1}{2}\right)+x_0\right]^2+\left(\frac{l}{2b}\right)^2(\bar{z}-\bar{z}_0)^2\right\}^{3/2}} + \\[2mm]
& +\frac{1}{2\pi}\int_{-1}^{0} \frac{\frac{d\Gamma}{d\bar{z}}}{\bar{z}-\bar{z}_0}\left(1+\frac{\operatorname{tg}\chi\frac{l}{2b}\left(\bar{z}+\frac{1}{2}\right)-\bar{x}_0}{\sqrt{\left[\operatorname{tg}\chi\frac{l}{2b}\left(\bar{z}+\frac{1}{2}\right)-\bar{x}_0\right]^2+\left(\frac{l}{2b}\right)^2(\bar{z}-\bar{z}_0)^2}}\right)d\bar{z} + \\[2mm]
& +\frac{1}{2\pi}\int_{0}^{1} \frac{\frac{d\Gamma}{d\bar{z}}}{\bar{z}-\bar{z}_0}\left(1+\frac{-\operatorname{tg}\chi\frac{l}{2b}\left(\bar{z}-\frac{1}{2}\right)-\bar{x}_0}{\sqrt{\left[\operatorname{tg}\chi\frac{l}{2b}\left(\bar{z}-\frac{1}{2}\right)+\bar{x}_0\right]^2+\left(\frac{l}{2b}\right)^2(\bar{z}-\bar{z}_0)^2}}\right)d\bar{z}.
\end{aligned}
\tag{7.59}
$$

CHAPTER VIII

EQUATIONS FOR DETERMINATION OF THE STRENGTH OF A BOUND VORTEX SHEET

§ 1. Profile Cascades and Multiplanes [19, 33]

By using the formulas given in the preceding chapter for an infinite vortex chain, we can obtain expressions for the disturbed velocities in the fluid about a cascade of thin profiles. We replace each profile by a continuous distribution of vortex sheets of strength $\gamma(s)$. We use the coordinate system ξ, η, where the $O\eta$ axis is the front of the cascade (Fig. 1.8 and 1.11). Let $\bar{\xi}_0$ and $\bar{\eta}_0$ be the dimensionless coordinates defined in (7.51) for any point of the plane; then for the disturbed velocity at this point we have

$$\left.\begin{aligned}
W_\xi(\bar{\xi}_0, \bar{\eta}_0) &= \frac{1}{2t} \int \gamma(s)\, w_\xi(\bar{\xi}_0 - \bar{\xi},\, \bar{\eta}_0 - \bar{\eta})\, ds, \\
W_\eta(\bar{\xi}_0, \bar{\eta}_0) &= \frac{1}{2t} \int \gamma(s)\, w_\eta(\bar{\xi}_0 - \bar{\xi},\, \bar{\eta}_0 - \bar{\eta})\, ds,
\end{aligned}\right\} \tag{8.1}$$

since the differential $\gamma(s)ds$ will correspond in dimension to the circulation Γ_+ of Eqs. (7.51). In (8.1), the integration is taken along the length of each profile and the expressions for w_ξ and w_η are given by Eqs. (7.50).

The normal component of the disturbed velocity at the profile surface, from (1.33), can be given by the formula

$$W_n(\bar{\xi}_0, \bar{\eta}_0) = \frac{1}{2t} \int \gamma(s)\, w_n(\bar{\xi}_0 - \bar{\xi},\, \bar{\eta}_0 - \bar{\eta})\, ds, \tag{8.2}$$

where

$$w_n(\bar{\xi} - \bar{\xi}_0,\, \bar{\eta} - \bar{\eta}_0) = w_\xi(\bar{\xi} - \bar{\xi}_0,\, \bar{\eta} - \bar{\eta}_0) \sin(\varepsilon - \beta_\Gamma) + w_\eta(\bar{\xi} - \bar{\xi}_0,\, \bar{\eta} - \bar{\eta}_0) \cos(\varepsilon - \beta_\Gamma). \tag{8.3}$$

We substitute the expression in (8.2) in the boundary condition (1.32) and obtain an equation for $\gamma(s)$:

$$-\frac{1}{2U_0 t} \int \gamma(s)\, w_n(\bar{\xi}_0 - \bar{\xi},\, \bar{\eta}_0 - \bar{\eta})\, ds = \sin(\vartheta - \beta_\Gamma) \cos\varepsilon + \cos(\vartheta - \beta_\Gamma) \sin\varepsilon. \tag{8.4}$$

Using the relation

$$\vartheta - \beta_\Gamma = \alpha + \alpha_0,$$

105

we conclude that $\gamma(s)$ can be expressed in the form (4.45):

$$\gamma(s) = U_0 \left[\gamma_1(s) \sin(\alpha + \alpha_0) + \gamma_2(s) \cos(\alpha + \alpha_0) \right]. \tag{8.5}$$

The integral equations for $\gamma_1(s)$ and $\gamma_2(s)$ are

$$\left. \begin{aligned} \frac{1}{2t} \int \gamma_1(s) \, w_n(\overline{\xi}_0 - \overline{\xi}, \ \overline{\eta}_0 - \overline{\eta}) \, ds &= -\cos\varepsilon, \\ \frac{1}{2t} \int \gamma_2(s) \, w_n(\overline{\xi}_0 - \overline{\xi}, \ \overline{\eta}_0 - \overline{\eta}) \, ds &= -\sin\varepsilon. \end{aligned} \right\} \tag{8.6}$$

For the numerical solution of these equations, we replace the continuous vortex sheet of each profile by m discrete vortices. We denote the number of vortices on each profile by i, and the number of points of this same profile at which the boundary conditions are satisfied by j. Let the coordinates of the first set of points be $\overline{\xi}_i$ and $\overline{\eta}_i$ and the coordinates of the second be $\overline{\xi}_j$ and $\overline{\eta}_j$. The components of the dimensionless velocity generated by the vortex i at the point j will be written $w_{\xi ij}$, $w_{\eta ij}$, w_{nij}, and w_{tij}, where

$$w_{\xi ij} = w_\xi(\overline{\xi}_j - \overline{\xi}_i, \ \overline{\eta}_j - \overline{\eta}_i), \quad w_{\eta ij} = w_\eta(\overline{\xi}_j - \overline{\xi}_i, \ \overline{\eta}_j - \overline{\eta}_i). \tag{8.7}$$

Let Γ_{+i} be the circulation of the i-th vortex; then from (8.5) we have

$$\Gamma_{+i} = 2U_0 t \left[\Gamma_i^{(1)} \sin(\alpha + \alpha_0) + \Gamma_i^{(2)} \cos(\alpha + \alpha_0) \right]. \tag{8.8}$$

where $\Gamma_i^{(1)}$ and $\Gamma_i^{(2)}$ are dimensionless quantities. To find these quantities we have two sets of equations obtained from (8.6) by replacing $\gamma_1(s)ds$ by $2t\,\Gamma_i^{(1)}$ and $\gamma_2(s)ds$ by $2t\Gamma_i^{(2)}$ and converting the integrals into sums:

$$\left. \begin{aligned} \sum_{i=1}^{m} \Gamma_i^{(1)} w_{nij} &= -\cos\varepsilon_j, \\ \sum_{i=1}^{m} \Gamma_i^{(2)} w_{nij} &= -\sin\varepsilon_j. \end{aligned} \right\} \quad j = 1, \, 2, \, \ldots, \, m, \tag{8.9}$$

In the special case of a cascade of flat plates we have $\varepsilon = 0$, $\Gamma_i^{(2)} = 0$, and the first set of equations becomes

$$\sum_{i=1}^{m} \Gamma_i^{(1)} w_{nij} = -1, \quad j = 1, \, 2, \, \ldots, \, m. \tag{8.10}$$

The position of the vortices and the points for which the calculations are performed and at which the boundary conditions are satisfied are chosen according to the following considerations.

Firstly the Chaplygin-Zhukovskii hypothesis must be satisfied. This means that for $m \to \infty$ the velocities generated by the vortices at the trailing edges of a profile must be bounded. Secondly, the sums which replace the integrals in the transition from a continuous distribution of the bound vortex sheet to discrete vortices must, when $m \to \infty$, tend to corresponding Cauchy principal values of the integrals, since these are improper integrals. Without giving a rigorous proof, we only note that both these requirements will be satisfied if the bound vortices and the grid points of our calculation are located as shown in Fig. 8.1.

In this diagram, the chord of the profile is divided into m parts. The vortices are located at the points of the profile contour with coordinates (x_i, y_i) corresponding to one quarter the length of the divisions (measuring from the nose). The points for which calculations are performed and at which the boundary conditions are satisfied are located at the points with coordinates (x_j, y_j), three quarters of the distance along a division.

This arrangement of the singularities ensures that the Chaplygin-Zhukovskii hypothesis concerning the finiteness of the velocities at the trailing edge of the profile is satisfied. Since the trailing edge where the boundary condition concerning smooth flow is satisfied is closer to the tail than any other vortex, it follows that for $m \to \infty$ the disturbed velocities at the tail will be finite. The sums which replace the integrals in (8.6) will in the limit yield the principal values of the integrals, since all the points at which calculations are made are midpoints between neighboring vortices.

From Fig. 1.11, the coordinates x, y, and ξ, η are related by the equations

$$\left. \begin{array}{l} \xi = x \cos \beta_r - y \sin \beta_r, \\ \eta = y \cos \beta_r + x \sin \beta_r. \end{array} \right\} \tag{8.11}$$

It is not difficult to see that the dimensionless coordinates $\bar{x}_i = x_i / b$ and $\bar{x}_j = x_j / b$ of the points will be given by the equations

$$\left. \begin{array}{l} \bar{x}_i = \dfrac{i - \frac{3}{4}}{m}, \qquad \bar{x}_j = \dfrac{j - \frac{1}{4}}{m}, \\ i = 1, 2, \ldots, m; \ j = 1, 2, \ldots, m. \end{array} \right\} \tag{8.12}$$

The coordinates $\bar{y}_i$ and $\bar{y}_j$ of the calculation points are obtained from the equation of the profile contour at the corresponding points $\bar{x}_i$ and $\bar{x}_j$.

A similar conversion from integrals to sums must be used for the calculation of the aerodynamic coefficients from (4.47) and (4.52).

A similar method can be used to solve the problem for a multiplane of infinite span. In this case the flow is not periodic, and this makes the problem more difficult. We consider, for example, a multiplane without stagger made up of plates (Fig. 8.2). Each plane is replaced by m bound discrete vortices located as in Fig. 8.1. The coordinates of the points at which vortices are located and the coordinates of the calculation points will be determined by Eqs. (8.12). For calculating the velocities due to the vortices we use (7.48). The number of unknowns will be larger than in the previous case. We will solve the problem in its linear formulation, using small angles of attack α. Considerations of symmetry reduce the number of unknown functions. We consider the plane through the Ox axis, relative to which the planes are symmetrically located. It is clear that each pair of symmetrically located planes will have the same aerodynamic characteristics and, in particular, the same bound-vortex strength. If there is an even

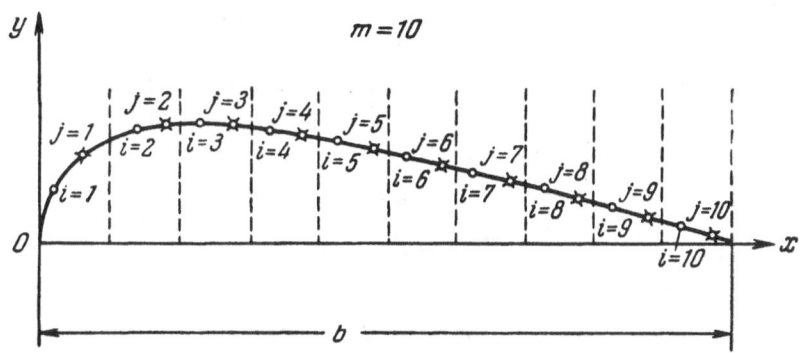

Fig. 8.1. Attached vortices i and points j at which calculations
are performed on a profile in a grid.

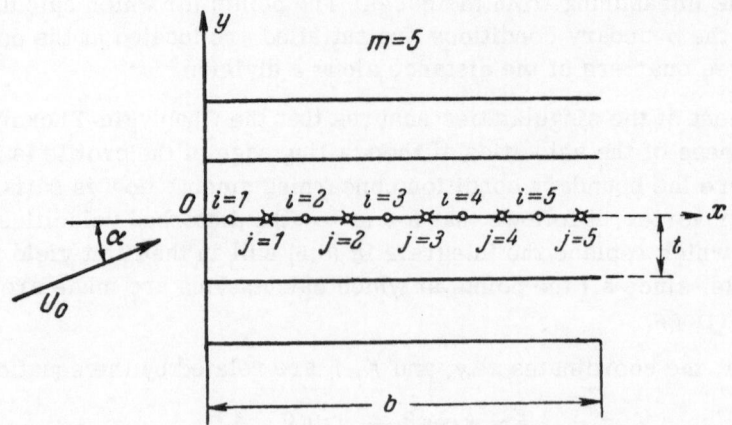

Fig. 8.2. Bound vortices i and calculation points j on one
of the plane profiles of a polyplane.

number of planes, then the number of unknowns will be nm/2, where n is the number of planes
and m the number of vortices on each plane. When n is odd, the number of unknowns is
$\{[(n-1)/2] + 1\}$ m.

The equations for the circulation $\Gamma_{+i\nu}$, of the bound vortices, where ν is the plane
number and i the number of a vortex in a plane, are obtained from the condition from smooth
flow

$$W_y = -U_0 \sin \alpha \cong -U_0 \alpha. \tag{8.13}$$

We set

$$\Gamma_{+i\nu} = U_0 b \Gamma_{i\nu}^a \alpha, \tag{8.14}$$

and then, from (7.48) without any considerations of symmetry, we have

$$\sum_{i=1}^{m} \frac{\Gamma_{i\nu}^a}{2\pi} w_{yij\nu} = -1, \quad j = 1, 2, \ldots, m, \quad \nu = 1, 2, \ldots, n, \tag{8.15}$$

where

$$w_{yij\nu} = -\frac{\bar{x}_j - \bar{x}_i}{(\bar{x}_j - \bar{x}_i)^2 + (\bar{y}_{\nu j} - \bar{y}_{\nu i})^2}. \tag{8.16}$$

By applying the N. E. Zhukovskii theorem "in the small," it is not difficult to find the
aerodynamic characteristics for each plane and then for the multiplane as a whole. For example, for the latter case we have with linear accuracy the relations

$$c_y^a = 2 \sum_{i=1}^{m} \sum_{\nu=1}^{n} \Gamma_{i\nu}^a, \quad m_z^a = 2 \sum_{i=1}^{m} \sum_{\nu=1}^{n} \Gamma_{i\nu}^a \bar{x}_i, \quad \bar{x}_{c.p.} = \frac{m_z^a}{c_y^a}, \tag{8.17}$$

where we have used the total area of all the planes and the chord b as a characteristic area and
characteristic linear dimension respectively.

§ 2. Annular Wings

For the bound vortices on an annular wing, we will use the annular vortices considered in §4 and §5 of Chapter VII, and locate them on a cylindrical surface close to the wing surface. Strictly speaking these vortices should be distributed continuously, and they will form an attached cylindrical vortex sheet. The determination of the strength of the attached vortices will reduce to the successive solution of the integral equations for the three functions of a single variable occurring in (4.20). For the approximations of the linear problem, the free vortices will be parallel to the axis of symmetry Ox of the wing (Fig. 8.3).

The velocities due to the vortex system of an annular wing can be calculated by using the functions w_{r1} and w_{r2} studied above (see §4 and §5 of Chapter VII). To obtain the integral equations referred to above, we must find the radial component W_r of the vortex layer of the form (4.20) and substitute the resulting expression in the boundary condition (1.31). The boundary conditions will be approximately satisfied on the cylindrical surface of the vortex layer.

We consider an elementary bound vortex filament of width dx and circulation $\gamma_+ dx$ (Fig. 8.3) and the corresponding system of free vortices. The dependence of γ_+ on the angle φ of a cylindrical coordinate system is given by (4.20). Since the first term of this equation is independent of φ and the second and third are proportional to $\cos \varphi$, it follows that the relations (7.19)-(7.27) must be used once and (7.28)-(7.46) must be used twice. In the first case, instead of $U_0 r \Gamma$ we must use the quantity $U_0 \alpha_0 \gamma_0 dx$, in the second $U_0 \alpha \gamma_\alpha dx$, and in the third $U_0 \omega_z \gamma_{\omega_z} dx$. We must also take into account the fact that the Ox axes are in opposite directions in Fig. 7.5 and Fig. 8.3, and so:

$$dW_r = \frac{U_0 \alpha_0 \gamma_0\, dx}{2\pi r} w_{r1}(\xi - \xi_0,\ 1) + \frac{U_0 \alpha \gamma_\alpha\, dx}{2\pi r} w_{r2}(\xi - \xi_0,\ 1) \cos \varphi_0 + \frac{U_0 \omega_z \gamma_{\omega_z}\, dx}{2\pi r} w_{r2}(\xi - \xi_0,\ 1) \cos \varphi_0,$$

$$\xi = \frac{x}{r}, \qquad \xi_0 = \frac{x_0}{r}. \tag{8.18'}$$

Let the origin O in Fig. 8.3 be the point in the plane x = const which bisects the chord of the wing section. To obtain the velocities due to the whole vortex system of the wing, we must integrate (8.18') with respect to x from $-b/2$ to $b/2$ or with respect to ξ from $-1/\lambda$ to $1/\lambda$, where λ is the aspect ratio of the annular wing:

$$\frac{W_r}{U_0} = \frac{\alpha_0}{2\pi} \int_{-\frac{1}{\lambda}}^{\frac{1}{\lambda}} \gamma_0(\xi)\, w_{r1}(\xi - \xi_0,\ 1)\, d\xi +$$

$$+ \frac{\alpha}{2\pi} \cos \varphi_0 \int_{-\frac{1}{\lambda}}^{\frac{1}{\lambda}} \gamma_\alpha(\xi)\, w_{r2}(\xi - \xi_0,\ 1)\, d\xi + \frac{\omega_z}{2\pi} \cos \varphi_0 \int_{-\frac{1}{\lambda}}^{\frac{1}{\lambda}} \gamma_{\omega_z}(\xi)\, w_{r2}(\xi - \xi_0,\ 1)\, d\xi. \tag{8.18}$$

We substitute the expression in (8.18) into the boundary condition (1.31). Since it must be an identity in α_0, α, ω_z and φ_0, we obtain three independent integral equations. The equation for γ_0 corresponds to longitudinal motion of the wing; this equation is

$$\frac{1}{2\pi} \int_{-\frac{1}{\lambda}}^{\frac{1}{\lambda}} \gamma_0(\xi)\, w_{r1}(\xi - \xi_0,\ 1)\, d\xi = -1. \tag{8.19}$$

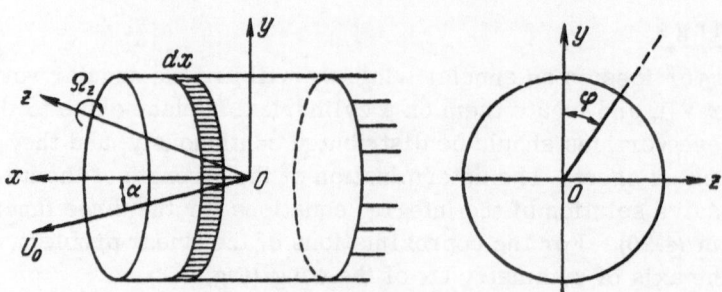

Fig. 8.3. The derivation of the integral equation for an annular wing.

For a cylindrical wing this equation does not exist since the first term in (4.20) is equal to zero.

The circulation γ_α corresponds to lateral flow past a wing, and it is determined by the equation

$$\frac{1}{2\pi} \int_{-\frac{1}{\lambda}}^{\frac{1}{\lambda}} \gamma_\alpha(\xi)\, w_{r2}(\xi - \xi_0,\, 1)\, d\xi = -1. \tag{8.20}$$

Finally, to find the circulation γ_{ω_Z} for a rotating wing we have

$$\frac{1}{2\pi} \int_{-\frac{1}{\lambda}}^{\frac{1}{\lambda}} \gamma_{\omega_z}(\xi)\, w_{r2}(\xi - \xi_0,\, 1)\, d\xi = \frac{x_0}{b}. \tag{8.21}$$

To solve the integral equations (8.19), (8.20), and (8.21) numerically, we convert the integral equations into systems of algebraic equations by using the method described in the previous section for uniform flow. To do this, we replace the functions $\gamma_0 d\xi$, $\gamma_\alpha d\xi$ and $\gamma_{\omega_z} d\xi$ respectively by Γ_{0i}, $\Gamma_{\alpha i}$, and $\Gamma_{\omega_z i}$, and the integrals by sums. If m is the number of bound vortices replacing the wing, then instead of (8.19) we obtain

$$\frac{1}{2\pi} \sum_{i=1}^{m} \Gamma_{0i} w_{r1ij} = -1, \qquad j = 1, 2, \ldots, m. \tag{8.22}$$

The subscript j, as above, denotes the number of the point at which we are carrying out the calculation. Equation (8.21) now becomes

$$\frac{1}{2\pi} \sum_{i=1}^{m} \Gamma_{\alpha i} w_{r2ij} = -1, \qquad j = 1, 2, \ldots, m, \tag{8.23}$$

and instead of (8.22) we have

$$\frac{1}{2\pi} \sum_{i=1}^{m} \Gamma_{\omega_z i} w_{r2ij} = \frac{x_{0j}}{b}, \qquad j = 1, 2, \ldots, m. \tag{8.24}$$

Equations (8.22), (8.23), and (8.24) are independent sets of algebraic equations for the unknowns Γ_{0i}, $\Gamma_{\alpha i}$ and $\Gamma_{\omega_z i}$. Using (4.20), we find that the circulation of each i-th bound vortex will correspond to $\gamma_+ dx$ and will be

$$\Gamma_{+i} = U_0 r \left[a_0 \Gamma_{0i} + \alpha \Gamma_{\alpha i} \cos\varphi + \omega_z \Gamma_{\omega_z i} \cos\varphi \right]. \tag{8.25}$$

The location of the bound vortices and the points at which calculations are performed must be chosen to satisfy the requirements described in the preceding section. For this we divide the annular-wing surface into m equal sections by dividing the chord b of the wing into the same number of equal parts (Fig. 8.4). The length of each part, referred to the radius r of the wing, will be b/mr =2/mλ. One annular vortex will be located in each part at a distance from the leading edge equal to one quarter of the length of the section. The boundary conditions will be satisfied at points of the surface at a distance of three quarters of the length from the leading edge. In Fig. 8.4, the heavy lines indicate the bound vortices, and the crosses and dashed lines indicate the sections where the boundary conditions are satisfied. It is now clear that these conditions need be satisfied at only one point in each section, and then at the remaining points of the section x = const they will be satisfied automatically.

The coefficients in Eqs. (8.22), (8.23), and (8.24) are obtained from (7.19)-(7.27) and (7.28)-(7.46) with the appropriate values of the arguments:

$$w_{r1ij} = w_{r1}(\xi_i - \xi_{0j}, 1), \qquad w_{r2ij} = w_{r2}(\xi_i - \xi_{0j}, 1).$$

The values of the arguments ξ_i and ξ_j as functions of i and j can be given by a table of the following type:

j \ i	1	2	3	—	m
1	$\dfrac{1}{m\lambda}$	$-\dfrac{1}{m\lambda}$	$-\dfrac{3}{m\lambda}$	—	$-\dfrac{2m-3}{m\lambda}$
2	$\dfrac{3}{m\lambda}$	$\dfrac{1}{m\lambda}$	$-\dfrac{1}{m\lambda}$	—	$-\dfrac{2m-5}{m\lambda}$
3	$\dfrac{5}{m\lambda}$	$\dfrac{3}{m\lambda}$	$\dfrac{1}{m\lambda}$	—	$-\dfrac{2m-7}{m\lambda}$
—	—	—	—	—	—
m	$\dfrac{2m-1}{m\lambda}$	$\dfrac{2m-3}{m\lambda}$	$\dfrac{2m-5}{m\lambda}$	—	$\dfrac{1}{m\lambda}$

After the dimensionless circulations Γ_{0i}, $\Gamma_{\alpha i}$ and $\Gamma_{\omega_z i}$ of the bound vortices have been obtained by solving the sets of equations (8.22), (8.23), and (8.24), it is not difficult to calculate the rotational-derivative coefficients for annular wings. As an example we give all the numerical formulas for cylindrical annular wings, for which $\alpha_0 = 0$. In this case the suction force has no effect on the values of these coefficients, and if we set $\alpha_0 = 0$ in the relations (4.27) the bars over m_z^α and $m_z^{\omega_z}$ can be omitted. From (4.30) we also obtain

$$p_\alpha = 2\gamma_\alpha, \qquad p_{\omega_z} = 2\gamma_{\omega_z};$$

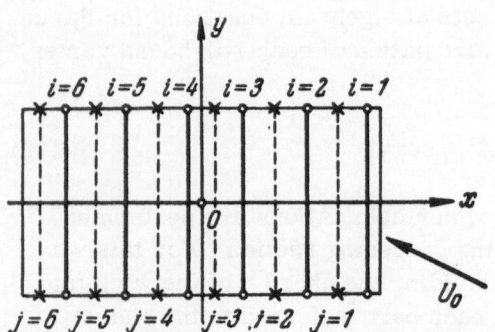

Fig. 8.4. Bound vortices i and calculation sections j on an annular wing for circulating flow.

hence (4.27) can be replaced by

$$c_y^{\alpha} = \pi \int_{-\frac{1}{2}}^{\frac{1}{2}} \gamma_{\alpha} \, d\bar{x}, \qquad m_z^{\alpha} = \pi \int_{-\frac{1}{2}}^{\frac{1}{2}} \gamma_{\alpha} \bar{x} \, d\bar{x},$$
$$c_y^{\omega_z} = \pi \int_{-\frac{1}{2}}^{\frac{1}{2}} \gamma_{\omega_z} \, d\bar{x}, \qquad m_z^{\omega_z} = \pi \int_{-\frac{1}{2}}^{\frac{1}{2}} \gamma_{\omega_z} \bar{x} \, d\bar{x}. \tag{8.26}$$

We must use the discrete vortex method to replace $\gamma_+ dx$ by Γ_+ and integrals by sums in our expressions, and it is easily established that $\gamma_{\alpha} d\bar{x}$ will correspond to $(\lambda/2)\Gamma_{\alpha}i$ and $\gamma \omega_z d\bar{x}$ to the product $(\lambda/2) \Gamma_{\omega_z}i$; hence

$$c_y^{\alpha} = \frac{\pi\lambda}{2} \sum_{i=1}^{m} \Gamma_{\alpha i}, \qquad m_z^{\alpha} = \frac{\pi\lambda}{2} \sum_{i=1}^{m} \Gamma_{\alpha i} \bar{x}_i,$$
$$c_y^{\omega_z} = \frac{\pi\lambda}{2} \sum_{i=1}^{m} \Gamma_{\omega_z i}, \qquad m_z^{\omega_z} = \frac{\pi\lambda}{2} \sum_{i=1}^{m} \Gamma_{\omega_z i} \bar{x}_i. \tag{8.27}$$

The dimensionless coordinates of the bound vortices (Fig. 8.4) are obtained from the formula

$$\bar{x}_i = \frac{x_i}{b} = \frac{2m - 1 - 4\,(i-1)}{4m}$$

§3. Monoplane Wings

We will solve the monoplane-wing problem by replacing the wing by a vortex surface. The lifting vortex layer, in its turn, will be modeled by a series of bound vortex filaments, the number of which can be as large as we please. Then each such filament will be replaced by certain oblique horseshoe vortices with sweepback angles χ chosen to be the same as on the corresponding parts of the filament.

We will use the following notation (Fig. 8.5): k is the number of the strip parallel to the Ox axis with the enumeration from right to left, $1 \le k \le N$; μ is the number of the bound vortex filament running from the leading to the trailing edge, $1 \le \mu \le n$; ν is the number of the line on which the boundary conditions are satisfied, starting also from the leading edge, $1 \le \nu$ n; i is the number of the oblique horseshoe-shaped vortex, going from right to left on each filament in the order of the filament number,

$$1 \leqslant i \leqslant m = nN;$$

j is the number of the point at which the boundary conditions are satisfied, $1 \le j \le m$.

In Fig. 8.5 we show the case when there are 20 bound vortices (m = 20) on the right half of a wing, when four vortex filaments are used (n = 4), and when each of them is replaced by five oblique horseshoe vortices (N = 5). There will be the same number of bound vortices on the left half of a wing that is symmetric in plan.

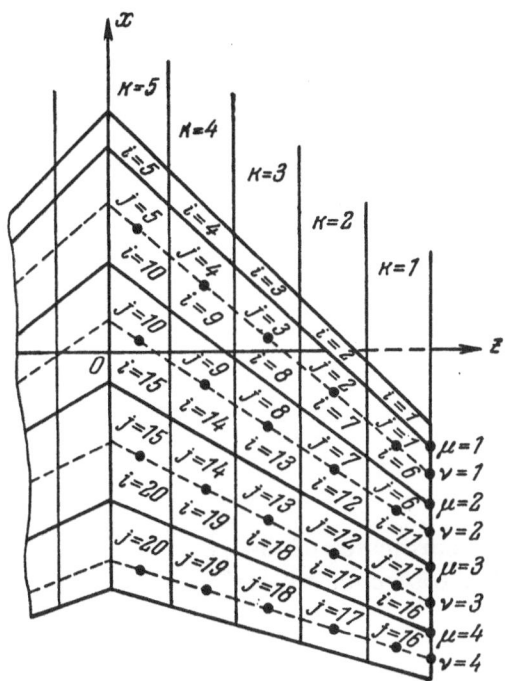

Fig. 8.5. The notation for a monoplane wing replaced by discrete vortices (circulatory flow).

To carry out the calculation, we will consider separately symmetric and antisymmetric circulation distributions along the span. In both cases we will satisfy the boundary conditions on the right half of the wing, taking into account the effect of the left half, and then the boundary conditions will be satisfied automatically on the left half. Hence the numbering of the vortices and the calculation points need only be carried out for the right half of the wing.

Let k_i and k_j be the values of k corresponding to the values of i and j under consideration; then

$$l = k_i + (\mu - 1)N, \qquad j = k_j + (\nu - 1)N. \quad (8.28)$$

These relations are independent of the shape of the wing in plan, and depend only on the method of replacing the vortex layer by discrete vortices (i.e., on the choice of the number of vortices n along the chord and N on the multispan).

The positions of the bound vortices and the calculation points at which the boundary conditions are satisfied must be chosen in the same way as these positions were chosen in the uniform plane flow case. We must first satisfy the Chaplygin-Zhukovskii hypothesis in each strip k = const. This means first, that when $n \to \infty$ the velocities at the trailing edges of the sections of the wing must be bounded. Second, the sums replacing the integrals in the conversion from a continuously distributed vortex layer to oblique horseshoe-shaped vortices must, for $n \to \infty$ and $N \to \infty$, tend to the Cauchy principal values of the integrals, since these are improper integrals.

These conditions are satisfied when the calculation points and the bound vortices are located as shown in Fig. 8.5. The following considerations are essential. First, in each strip k = const the last calculation point must be closer to the tail than the last bound vortex. Second, all the calculation points lie successively between adjacent bound and free vortices.

These points are selected as follows (Fig. 8.6): In the middle of the strip k = const we draw the wing chord of length b_k with its midpoint at the distance Δx_k from the Oz axis. We divide this chord into n equal parts, and then we divide each of these parts into four parts. The points at a distance one quarter the length of each part from its upper boundary will give the positions of the bound vortices (points μ). The points at the same distance from the lower boundaries of the parts are taken as the calculation points (points ν).

The coordinates of the calculation points will be denoted by x_{0j} and z_{0j}, and the coordinates of the midpoint of the i-th oblique horseshoe-shaped bound vortex by x_i, and z_i. From Fig. 8.6 we have

$$\left. \begin{aligned} x_i &= \Delta x_{k_i} + \frac{b_{k_i}}{2} - \frac{1}{4}\frac{b_{k_i}}{n} - (\mu - 1)\frac{b_{k_i}}{n}, \\ x_{0j} &= \Delta x_{k_j} + \frac{b_{k_j}}{2} - \frac{3}{4}\frac{b_{k_j}}{n} - (\nu - 1)\frac{b_{k_j}}{n}, \end{aligned} \right\} \quad (8.29)$$

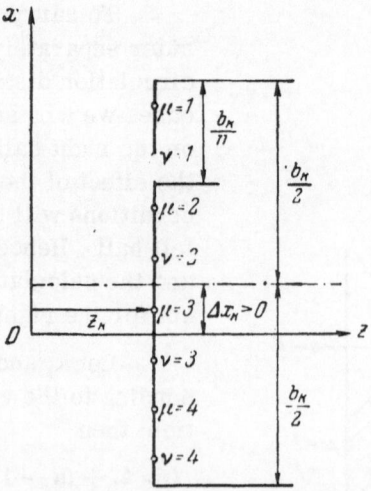

Fig. 8.6. The determination of the coordinates of the vortices (the points μ) and the calculation points ν for circulatory flow.

and so

$$\frac{x_l}{b_{k_l}} = \frac{\Delta x_{k_l}}{b_{k_l}} + \frac{1}{2} - \frac{\mu - \frac{3}{4}}{n}, \qquad \frac{x_{0j}}{b_{k_j}} = \frac{\Delta x_{k_j}}{b_{k_j}} + \frac{1}{2} - \frac{\nu - \frac{1}{4}}{n}. \qquad (8.30)$$

From Fig. 8.5 we see that

$$z_l = \frac{l}{2} - \frac{l_l}{2}(2k_l - 1), \qquad z_{0j} = \frac{l}{2} - \frac{l_j}{2}(2k_j - 1);$$

thus using the fact that, if the multispan is divided into N equal parts,

$$\frac{l_l}{l} = \frac{1}{2N}, \qquad (8.31)$$

we obtain

$$\bar{z}_k = \frac{2z_k}{l} = 1 - \frac{2k - 1}{2N}. \qquad (8.32)$$

The coordinates $\bar{z}_k$ do not depend on the shape of the wing in plan, and for a given value of N they are immediately fixed for all wings.

Using (1.7), we obtain expressions for the calculation of b_k/b and $\Delta x_k/b$ at the midsection of the plane k = const, which characterize the shape of the wing in plan:

$$\left.\begin{aligned} \frac{b_k}{b} &= 1 - \bar{z}_k\left(1 - \frac{1}{\eta}\right), \\ \frac{\Delta x_k}{b} &= -\bar{z}_k\left[\frac{\lambda}{4}\left(1 + \frac{1}{\eta}\right)\operatorname{tg}\chi_0 - \frac{1}{2}\left(1 - \frac{1}{\eta}\right)\right]. \end{aligned}\right\} \qquad (8.33)$$

In (8.32) and (8.33) we must substitute the values of k_i and k_j corresponding to the points i and j under consideration.

For wings with straight edges and constant sweepback (see Fig. 8.5, for example), the angle of sweepback χ will be constant for each bound vortex filament. If we are considering wings with corners at the edges, these corners will give rise to vortex filaments. For wings with curved corners, the vortex filaments will be curvilinear, and the angle χ for the oblique horseshoe vortices must be the mean value of the angles of inclination of the tangent to the filament on each strip k = const. For rectangular wings of different aspect ratios, it is clear that the oblique vortices are represented by the usual horseshoe vortices.

We recall that the circulation about the span for oblique and ordinary horseshoe vortices is constant, while free vortices parallel to the Ox axis arise at the ends of the bound vortices. Hence the vortex scheme of a multiwing, replaced by a system of oblique horseshoe vortices, will have the form shown in Fig. 8.7. Here the free vortices stretching from an attached vor- downstream to infinity are conventionally broken off.

Hence the vortex layer, continuously distributed over the wing in the present method of solving the problem, is replaced by singularities discretely distributed along the chord. More- over the continuous variation of the circulation along the span is approximated by a step-type variation. This method appears to be very convenient for calculations made by an electronic computer. On the one hand this is because it is based on a description of the required flow by the simplest singularities. On the other hand it is essential that the set of algebraic equations to which the solution is reduced have several important properties. The diagonal elements in the coefficient matrix of these equations plays a dominant role, and so the determinant of the system differs considerably from zero, the solution is very stable relative to the original data, etc.

The method is very flexible, and without essential complications it can be generalized to apply to the case of motion of a wing past an interface, the flow about a system of wings, a combination of a wing and a body, etc. It can be conveniently used to determine not only the total, but also the distributed aerodynamic characteristics, both steady and unsteady [31, 36, 43].

The conversion to oblique horseshoe-shaped vortices instead of the usual vortices, as it is carried out in [30], leads to considerable simplification and leads to more accurate results.

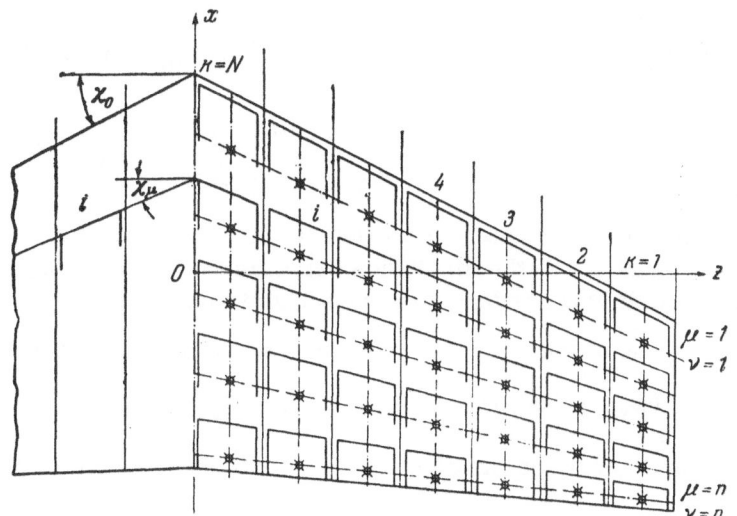

Fig. 8.7. The replacement of a wing by oblique, horseshoe vortices in flow with circulation.

For the replacement of a bound vortex on a swept-back wing by ordinary horseshoe vortices to be sufficiently accurate, the number of horseshoe vortices must be increased when the sweep-back angle χ increases and when the downwash is calculated closer to the vortex. Particularly unsuitable conditions are encountered at the ends of a wing of large taper and sweepback. These difficulties disappear automatically for oblique horseshoe-shaped vortices.

It is also not expedient to use series representations of the bound-vortex-layer strength $\gamma_+(x, z)$. The function $\gamma_+(x, z)$ has singularities close to the leading edge of the wing and its form must be specified, while singularities are usually of the same type as in the plane case [30] which, strictly speaking, is not singular [44]. The same is true of the choice of the law of variation of $\gamma_+(x, z)$ along the span. This law of variation is usually expressed in a form such that close to the ends of a wing the derivative $\partial\gamma_+/\partial z$ is infinite, and this representation is generally not accurate.

We denote by l_i the span of the i-th oblique horseshoe vortex, measured along the Oz axis. The strength of this vortex is

$$\Gamma_{+i} = U_0 l_i \Gamma_i. \tag{8.34}$$

The dimensionless quantity Γ_i can be given by the formula

$$I_i = \Gamma_{\alpha i}\alpha + \Gamma_{\omega_x i}\omega_x + \Gamma_{\omega_z i}\omega_z. \tag{8.35}$$

Let w_{yij} denote the dimensionless velocity due to the oblique horseshoe vortex i at the point j, and let Δw_{yij} be the additional dimensionless velocity at the point j due to the vortex i' on the left multiwing, symmetrically placed relative to the vortex i (Fig. 8.7). When a wing, symmetric in plan, moves with a nonzero angle of attack and rotates about the Oz axis, the loads on the wing and the circulations will also be symmetric. Hence at the relatively symmetric points i and i', $\Gamma_{\alpha i}$ and $\Gamma_{\omega_z i}$ will be identical. The loads and circulations arising when the wing rotates about the Ox axis will be antisymmetric, and so the value of $\Gamma_{\omega_z i}$ at the symmetrically located points i and i' will differ only in sign.

When we take the above into account, we find that the velocities at the point j due to the vortex system are, from (7.1), (7.11), and (8.34), (8.35), equal to

$$W_{yj} = \frac{U_0\alpha}{2\pi}\sum_{i=1}^{m}\Gamma_{\alpha i}(w_{yij}+\Delta w_{yij}) + \frac{U_0\omega_x}{2\pi}\sum_{i=1}^{m}\Gamma_{\omega_x i}(w_{yij}-\Delta w_{yij}) +$$
$$+ \frac{U_0\omega_z}{2\pi}\sum_{i=1}^{m}\Gamma_{\omega_z i}(w_{yij}+\Delta w_{yij}). \tag{8.36}$$

We use this formula in the boundary condition (1.30), and obtain a relation which is an identity in the angle of attack α and the dimensionless angular velocities ω_x and ω_z. Hence the boundary condition yields three independent systems of equations.

The first is for motion of the wing with a nonzero angle of attack without rotation:

$$\sum_{i=1}^{m}(w_{yij}+\Delta w_{yij})\Gamma_{\alpha i} = -2\pi, \qquad j = 1, 2, \ldots, m. \tag{8.37}$$

The second system determines the circulations for a wing rotating about the Oz axis:

$$\sum_{i=1}^{m}(w_{yij}+\Delta w_{yij})\Gamma_{\omega_z i} = 2\pi\frac{x_{0j}}{b}, \qquad j = 1, 2, \ldots, m. \tag{8.38}$$

The third system contains the dimensionless circulations $\Gamma_{\omega x i}$ for a wing rotating about the Ox axis:

$$\sum_{i=1}^{m} (w_{yij} - \Delta w_{yij}) \Gamma_{\omega x i} = -2\pi \frac{z_{0j}}{b}, \qquad j = 1, 2, \ldots, m. \tag{8.39}$$

The coefficients of Eqs. (8.37), (8.38), and (8.39) are calculated from the equations obtained for the velocity field of an oblique horseshoe vortex. A comparison of Fig. 8.7 with Fig. 7.3 shows that in the investigation of the wing under consideration we have used a coordinate system with the opposite directions for the Ox and Oz axes. Moreover the origin of this system was at the midpoint of the bound vortex.

Using the above considerations and (7.4), we obtain the arguments of the functions in (7.10) and (7.11) for the calculation of the velocity at the point j due to the vortex i:

$$\xi_{0ij} = \frac{2(x_i - x_{0j})}{l_i} = 2\left(\frac{b_{k_i}}{b} \frac{x_i}{b_{k_i}} - \frac{x_{0j}}{b_{k_j}} \frac{b_{k_j}}{b}\right) \frac{l}{l_i} \frac{b}{l},$$

$$\zeta_{0ij} = \frac{2(z_i - z_{0j})}{l_i} = \frac{2(z_i - z_{0j})}{l} \frac{l}{l_i}.$$

It remains to find an expression for the sweepback angle of each of the bound vortices. For wings with straight edges without corners, the sweepback angle χ_μ for each vortex filament μ is constant (in Fig. 8.5 all angles $\chi_\mu > 0$). From Fig. 8.5 and 8.6 we obtain the ratio in which the vortex μ divides the root chord:

$$\Delta_\mu = \frac{\frac{b}{4n} + \frac{b}{n}(\mu - 1)}{b} = \frac{\mu}{n} - \frac{3}{4n}.$$

From this relation and (1.6) we obtain an expression for the third argument – the sweepback angle:

$$\operatorname{tg} \chi_\mu = \operatorname{tg} \chi_0 - \frac{4\Delta_\mu}{\lambda} \frac{\eta - 1}{\eta + 1}.$$

Hence (8.31) and (8.32) yield the following equations for ξ_{0ij}, ζ_{0ij}, and χ_μ:

$$\left.\begin{array}{l} \xi_{0ij} = 4 \dfrac{b}{l} N\left(\dfrac{b_{k_i}}{b} \dfrac{x_i}{b_{k_i}} - \dfrac{x_{0j}}{b_{k_j}} \dfrac{b_{k_j}}{b}\right), \\[2mm] \zeta_{0ij} = 2N(\bar{z}_{k_i} - \bar{z}_{0k_j}), \quad \chi_\mu = \operatorname{arctg}\left(\operatorname{tg} \chi_0 - \dfrac{4\Delta_\lambda}{\lambda} \dfrac{\eta - 1}{\eta + 1}\right) \end{array}\right\} \tag{8.40}$$

In the calculation of the velocities Δw_{yij} we must use the fact that in contrast to the vortices i of the right half-span the directions of the Oz axes in Fig. 7.3 and Fig. 8.7 coincide for the vortices i' of the left half-span. Hence we must substitute the same values of ξ_{0ij} and χ_μ in (7.10) and (7.11) as for the vortices i, but different values of the coordinate ζ'_{0ij}. Since the distance from a vortex i' to the symmetrically located vortex i is $2z_i$, where $z_i > 0$, it follows that

$$z_{0j} - z'_i = z_{0j} - z_i + 2z_i = z_{0j} + z_i,$$

and so

$$\zeta'_{0ij} = \zeta_{0ij} + \delta\zeta_{0i} = 2N(\bar{z}_{k_i} + \bar{z}_{0k_j}).$$

All the velocities are calculated from (7.10) and (7.11) and

$$\left.\begin{array}{l} w_{ylj} = w_y(\xi_{0lj}, \zeta_{0lj}, \chi_\mu), \\ \Delta w_{ylj} = w_y(\xi_{0lj}, \zeta_{0lj} + \delta\zeta_{0l}, \chi_\mu). \end{array}\right\} \tag{8.41}$$

Solving the equations (8.37), (8.38), and (8.39) we find Γ_i and Γ_{+i}, after which it is not difficult to calculate all the aerodynamic characteristics both for the wing as a whole and for the sections. We must use (4.6) and (4.11), substituting $\gamma_{z+}dx$ for Γ_{+i}, $\gamma_z^\alpha dx$ for $\Gamma_{\alpha i}l_i$, $\gamma_z^{\omega z}dx$ for $\Gamma_{\omega z i}l_i$ and $\gamma_z^{\omega x}dx$ for $\Gamma_{\omega x i}l_i$. We must also replace dz by l_i and the integrations by sums.

Using (4.6) and (8.31), we obtain the following formulas for the rotational-derivative coefficients:

$$\left.\begin{array}{ll} c_y^\alpha = \dfrac{\lambda}{N^2}\displaystyle\sum_{i=1}^m \Gamma_{\alpha i}, & m_z^\alpha = \dfrac{\lambda}{N^2}\displaystyle\sum_{i=1}^m \Gamma_{\alpha i}\dfrac{x_i}{b_{k_i}}\dfrac{b_{k_i}}{b}, \\[4mm] c_y^{\omega z} = \dfrac{\lambda}{N^2}\displaystyle\sum_{i=1}^m \Gamma_{\omega z i}, & m_z^{\omega z} = \dfrac{\lambda}{N^2}\displaystyle\sum_{i=1}^m \Gamma_{\omega z i}\dfrac{x_i}{b_{k_i}}\dfrac{b_{k_i}}{b}, \\[4mm] m_x^{\omega x} = -\dfrac{\lambda}{2N^2}\dfrac{l}{b}\displaystyle\sum_{i=1}^m \Gamma_{\omega x i}\bar{z}_i. \end{array}\right\} \tag{8.42}$$

Now using (4.11) we obtain an expression for the rotational-derivative coefficients of the sections:

$$\left.\begin{array}{ll} c_{yk_i}'^{\alpha} = \dfrac{1}{N}\dfrac{b}{b_{k_i}}\dfrac{l}{b}\displaystyle\sum_{\mu=1}^n {}_{k_i}\Gamma_{\alpha i}, & m_{zk_i}'^{\alpha} = \dfrac{1}{N}\dfrac{b}{b_{k_i}}\dfrac{l}{b}\displaystyle\sum_{\mu=1}^n {}_{k_i}\Gamma_{\alpha i}\dfrac{x_i}{b_{k_i}}, \\[4mm] c_{yk_i}'^{\omega z} = \dfrac{1}{N}\dfrac{b}{b_{k_i}}\dfrac{l}{b}\displaystyle\sum_{\mu=1}^n {}_{k_i}\Gamma_{\omega z i}, & m_{zk_i}'^{\omega z} = \dfrac{1}{N}\dfrac{b}{b_{k_i}}\dfrac{l}{b}\displaystyle\sum_{\mu=1}^n {}_{k_i}\Gamma_{\omega z i}\dfrac{x_i}{b_{k_i}}, \\[4mm] c_{yk_i}'^{\omega x} = \dfrac{1}{N}\dfrac{b}{b_{k_i}}\dfrac{l}{b}\displaystyle\sum_{\mu=1}^n {}_{k_i}\Gamma_{\omega x i}, & m_{zk_i}'^{\omega x} = \dfrac{1}{N}\dfrac{b}{b_{k_i}}\dfrac{l}{b}\displaystyle\sum_{\mu=1}^n {}_{k_i}\Gamma_{\omega x i}\dfrac{x_i}{b_{k_i}}. \end{array}\right\} \tag{8.43}$$

Here the symbols $\displaystyle\sum_{\mu=1}^n {}_{k_i}$ denote summations with respect to μ with k_i = const., i.e., the

summations are performed for values in the sections $\bar{z}_{k_i}$ = const. The subscript k_i indicates the number of the plane for which the coefficient is being determined.

§4. Apparent Masses of Monoplane Wings

The calculation of apparent masses of monoplane wings is very similar to that described in the previous paragraph. There are two very important differences however. First, the Zhukovskii theorem has a different form "in the small." Second, instead of the Chaplygin-Zhukovskii hypothesis we have here the condition that the vortex strength in each section must vanish. This leads, for example, to the conclusion that there are no free vortices behind the wing. This leads to a change in the position of the bound vortices and a corresponding change of the points at which calculations are performed on the wing.

We again model the wing by a system of oblique bound vortices of span l_i, the strength of each vortex given in the form (8.34) but now with

$$\Gamma_{+i} = U_2 l_i \Gamma_{2i} + U_4 l_i^2 \Gamma_{4i} + U_6 l_i^2 \Gamma_{6i}. \tag{8.44}$$

For a wing that is symmetric in plan, for which Γ_{2i} and Γ_{6i} are even functions of z and Γ_{4i} is an odd function, the speed of the vortex at the point j can be expressed in a form similar to (8.36):

$$W_{yj} = \frac{U_2}{2\pi} \sum_{i=1}^{m} \Gamma_{2i} (w_{yij} + \Delta w_{yij}) + \frac{U_4}{2\pi} \sum_{i=1}^{m} \Gamma_{4i} l_i (w_{ylj} - \Delta w_{ylj}) + \frac{U_6}{2\pi} \sum_{l=1}^{m} \Gamma_{6i} l_i (w_{ylj} + \Delta w_{ylj}). \tag{8.45}$$

The boundary condition is of the form (1.50); substituting (8.45) in this condition, we obtain three independent sets of equations.

The first is for the transverse motion of the wing (in the Oy direction):

$$\sum_{i=1}^{m} (w_{ylj} + \Delta w_{yij}) \Gamma_{2i} = 2\pi, \qquad j = 1, 2, \ldots, m - N. \tag{8.46}$$

The second corresponds to the rotation of the wing about the Ox axis:

$$\sum_{i=1}^{m} (w_{yij} - \Delta w_{yij}) l_i \Gamma_{4i} = -2\pi z_{0j}, \qquad j = 1, 2, \ldots, m - N, \tag{8.47}$$

and the third to rotation about Oz:

$$\sum_{i=1}^{m} (w_{yij} + \Delta w_{yij}) l_i \Gamma_{6i} = 2\pi x_{0j}, \qquad j = 1, 2, \ldots, m - N. \tag{8.48}$$

Since we are considering irrotational flow, in addition to the above conditions at each section k_i = const we also have

$$\sum_{\mu=1}^{n} {}_{k_i} \Gamma_{2i} = 0, \quad \sum_{\mu=1}^{n} {}_{k_i} \Gamma_{6i} = 0, \quad \sum_{\mu=1}^{n} {}_{k_i} \Gamma_{4i} = 0, \qquad k_i = 1, 2, \ldots, N. \tag{8.49}$$

All the formulas of §2, Chapter I remain valid since they refer to geometrical characteristics. The position of the vortices and the grid points are not the same as in the previous section, and so several relations must be rederived.

In the calculation of the apparent mass, the leading edge, the trailing edge, and the lateral edges are qualitatively equivalent. In particular the disturbed velocity will in general be infinite on all these edges. For the same reason, and also for rotational flow, the vortices and the grid points must be located so that the points are midway between the vortices (the crosses in Fig. 8.8). It is easily seen that all the required conditions are met by the model in the diagram. As a basic vortex system we can again use the oblique horseshoe vortex (together with the free vortices) described in §2, Chapter VII. The conditions (8.49) ensure that there are no vortex streets of free vortices behind the wing.

The subscripts i, j, μ, ν, and also n, N, and m have the same meaning as in the previous section. There are two differences however: $1 \le \nu \le n - 1$ and $1 \le j \le (n - 1)N$, while Eqs. (8.28) are unchanged.

The wing is divided along the z axis as follows. The half span is divided into 2N + 1 parts and the corresponding interval lengths are

$$\Delta l = \frac{l}{2(2N+1)}.$$

Let the first strip (k = 1) start at a distance Δl from the terminal chord, and let the width of each strip k = const be $2\Delta l$. The span of each bound vortex is taken to be $2\Delta l$ and so

$$\frac{l_i}{l} = \frac{1}{2N+1}. \tag{8.50}$$

From Fig. 8.8, in which the case N = 4 is illustrated, the coordinate of the midpoint of the vortex i is

$$\bar{z}_i = \frac{l}{2} - \Delta l - \Delta l - 2\,\Delta l\,(k_i - 1) = \frac{l}{2}\left(1 - \frac{2k_i}{2N+1}\right),$$

and so

$$\bar{z}_i = 1 - \frac{2k_i}{2N+1}, \qquad \bar{z}_{0j} = 1 - \frac{2k_j}{2N+1}, \qquad \bar{z} = \frac{2z}{l},$$

where this coordinate depends only on the number k corresponding to i or j; we can thus use the one formula

$$\bar{z}_k = 1 - \frac{2k}{2N+1}. \tag{8.51}$$

We divide each chord b_k corresponding to the middle of the strip k = const as shown in Fig. 8.8 and 8.9. We divide the chord b_k into n equal parts with the midpoint of each segment of length b_k/n at the point through which a bound vortex is generated; at the end of each seg-

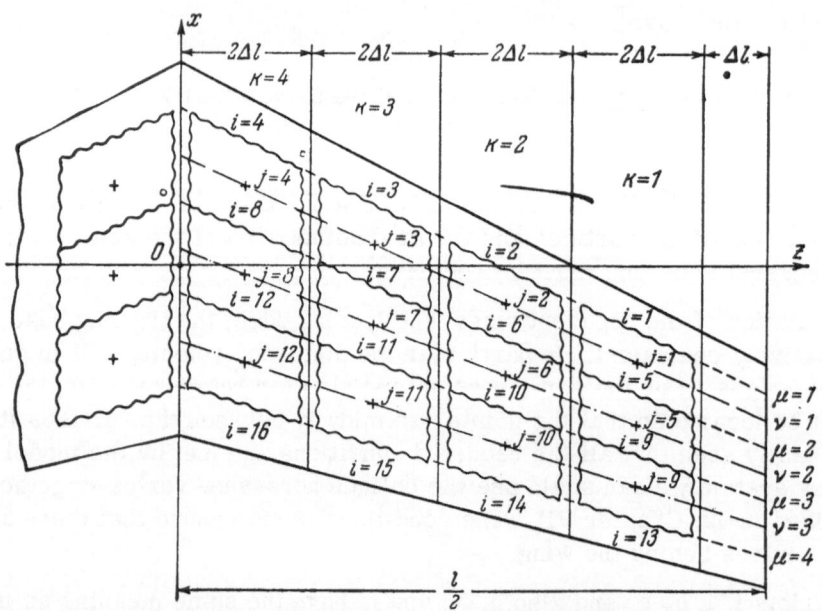

Fig. 8.8. The replacement of a wing by oblique horseshoe vortices
in circulation-free flow.

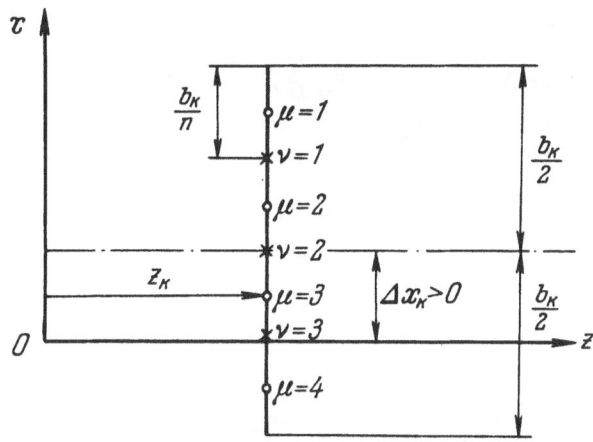

Fig. 8.9. The coordinates of the vortices
(the points μ) and the grid points ν of the
calculation for circulation-free flow.

ment except the last (the calculation is performed from the leading to the trailing edge) the boundary conditions concerning the absence of any circulation are satisfied. Hence in each $k = const$ there will be one less of these conditions than the number of attached vortices.

Let x_i and x_{0j} be the coordinates of the midpoints of the vortices and the points at which the calculations are performed respectively and let Δx_k be the distance from the Oz axis to midpoint of the chord b_k. From Fig. 8.9 we have

$$x_i = \Delta x_{k_i} + \frac{b_{k_i}}{2} - \frac{b_{k_i}}{2n} - \frac{b_{k_i}}{n}(\mu - 1) = \Delta x_{k_i} + \frac{b_{k_i}}{2} - \frac{b_{k_i}}{2n}(2\mu - 1),$$

$$x_{0j} = \Delta x_{k_j} + \frac{b_{k_j}}{2} - \frac{b_{k_j}}{n} - \frac{b_{k_j}}{n}(\nu - 1) = \Delta x_{k_j} + \frac{b_{k_j}}{2} - \frac{b_{k_j}}{n}\nu,$$

and so

$$\frac{x_i}{b_{k_j}} = \frac{\Delta x_{k_i}}{b_{k_i}} + \frac{1}{2} - \frac{2\mu - 1}{2n}, \qquad \frac{x_{0j}}{b_{k_j}} = \frac{\Delta x_{k_j}}{b_{k_j}} + \frac{1}{2} - \frac{\nu}{n}. \tag{8.52}$$

Using the geometrical parameters of the wing and (1.7), we can find x_i and x_{0j} since

$$\left.\begin{array}{l} \dfrac{\Delta x_k}{b} = -\bar{z}_k\left[\dfrac{\lambda}{4}\left(1 + \dfrac{1}{\eta}\right)\operatorname{tg}\chi_0 - \dfrac{1}{2}\left(1 - \dfrac{1}{\eta}\right)\right], \\[2mm] \dfrac{b_k}{b} = 1 - \bar{z}_k\left(1 - \dfrac{1}{\eta}\right), \quad \dfrac{\Delta x_k}{b_k} = \dfrac{\Delta x_k}{b}\dfrac{b}{b_k}. \end{array}\right\} \tag{8.53}$$

We transform (4.93) to the case in which a uniformly distributed layer on a wing is modeled by discrete vortices and the integrals are replaced by sums.

A comparison of the notation (4.81) and (8.44) yields

$$U_2 b \Gamma_2 = U_2 l_i \Gamma_{2i},$$
$$U_4 b^2 \Gamma_4 = U_4 l_i^2 \Gamma_{4i},$$
$$U_6 b^2 \Gamma_6 = U_6 l_i^2 \Gamma_{6i},$$

and hence

$$\Gamma_2 = \Gamma_{2i}\frac{l_i}{b}, \quad \Gamma_4 = \Gamma_{4i}\frac{l_i^2}{b^2}, \quad \Gamma_6 = \Gamma_{6i}\frac{l_i^2}{b^2}.$$

But from (8.50) and (1.7) we have

$$\frac{l_i}{b} = \frac{l_i}{l}\frac{l}{b} = \frac{1}{2N+1}\frac{\lambda}{2}\frac{\eta+1}{\eta} = \frac{\lambda(\eta+1)}{2\eta(2N+1)},$$

$$\frac{l_i^2}{b^2} = \frac{l_i^2}{l^2}\frac{l^2}{b^2} = \frac{\lambda^2(\eta+1)^2}{4\eta^2(2N+1)^2},$$

from which we obtain

$$\Gamma_2 = \Gamma_{2i}\frac{\lambda(\eta+1)}{2\eta(2N+1)}, \quad \Gamma_4 = \Gamma_{4i}\frac{\lambda^2(\eta+1)^2}{4\eta^2(2N+1)^2}, \quad \Gamma_6 = \Gamma_{6i}\frac{\lambda^2(\eta+1)^2}{4\eta^2(2N+1)^2}. \tag{8.54}$$

Replacing dx by b_{k_i}/n, and dz by $l_i = l/(2N+1)$, the summation for a wing symmetric in plan is over the right half of the wing and the results must be doubled. It is more convenient to define the dimensionless circulation not by i, but by the two numbers k_i and μ.

Using the method described above we have, for example,

$$\frac{1}{S}\int_S\int \Gamma_2\, dS = \frac{1}{S}\int_{-\frac{l}{2}}^{\frac{l}{2}}\int_{x_1}^{x_2}\Gamma_2\, dx\, dz \cong \frac{2}{S}\sum_{k_i=1}^{N}\sum_{\mu=1}^{n}{}_{k_i}(\Gamma_{2i})\frac{l_i}{b}\frac{b_{k_i}}{n}\frac{l}{2N+1},$$

where the notation $\displaystyle\sum_{\mu=1}^{n}{}_{k_i}$ indicates that the summation is with respect to μ, i.e, in the sec-

tion k = const. But from (4.81), (4.84), and (8.44) we have

$$\sum_{\mu=1}^{n}{}_{k_i}(\Gamma_{2i}) = \Gamma_2^{k_i 1} + (\Gamma_2^{k_i 1} + \Gamma_2^{k_i 2}) + (\Gamma_2^{k_i 1} + \Gamma_2^{k_i 2} + \Gamma_2^{k_i 3}) + \cdots$$
$$\cdots + (\Gamma_2^{k_i 1} + \Gamma_2^{k_i 2} + \Gamma_2^{k_i 3} + \cdots + \Gamma_2^{k_i n-1})$$

since

$$\Gamma_2^{k_i 1} + \Gamma_2^{k_i 2} + \cdots + \Gamma_2^{k_i n} = 0,$$

where the first superscript is the number k_i of the section and the second is the number of the attached filament in this section.

We introduce the notation

$$R_{2k_i} = -\frac{1}{n}\left[\Gamma_2^{k_i 1} + (\Gamma_2^{k_i 1} + \Gamma_2^{k_i 2}) + (\Gamma_2^{k_i 1} + \Gamma_2^{k_i 2} + \Gamma_2^{k_i 3}) + \cdots + (\Gamma_2^{k_i 1} + \Gamma_2^{k_i 2} + \cdots + \Gamma_2^{k_i n-1})\right],$$

$$R_{4k_i} = -\frac{1}{n}\left[\Gamma_4^{k_i 1} + (\Gamma_4^{k_i 1} + \Gamma_4^{k_i 2}) + (\Gamma_4^{k_i 1} + \Gamma_4^{k_i 2} + \Gamma_4^{k_i 3}) + \cdots + (\Gamma_4^{k_i 1} + \Gamma_4^{k_i 2} + \cdots + \Gamma_4^{k_i n-1})\right],$$

$$R_{6k_i} = -\frac{1}{n}\left[\Gamma_6^{k_i 1} + (\Gamma_6^{k_i 1} + \Gamma_6^{k_i 2}) + (\Gamma_6^{k_i 1} + \Gamma_6^{k_i 2} + \Gamma_6^{k_i 3}) + \cdots + (\Gamma_6^{k_i 1} + \Gamma_6^{k_i 2} + \cdots + \Gamma_6^{k_i n-1})\right],$$

$$L_{6k_i} = -\frac{1}{n}\frac{b_{k_i}}{b}\left[\Gamma_6^{k_i 1}\frac{x_{k_i}^{(1)}}{b_{k_i}} + (\Gamma_6^{k_i 1} + \Gamma_6^{k_i 2})\frac{x_{k_i}^{(2)}}{b_{k_i}} + \right.$$
$$\left. + (\Gamma_6^{k_i 1} + \Gamma_6^{k_i 2} + \Gamma_6^{k_i 3})\frac{x_{k_i}^{(3)}}{b_{k_i}} + \cdots + (\Gamma_6^{k_i 1} + \Gamma_6^{k_i 2} + \cdots + \Gamma_6^{k_i n-1})\frac{x_{k_i}^{(n-1)}}{b_{k_i}}\right], \tag{8.55}$$

in which we use the notation with the subscripts μ and k_i for quantities determined by (8.52) and (8.53), that is,

$$\frac{x_i}{b_{k_i}} = \frac{x_{k_i}^{(\mu)}}{b_{k_i}}.$$

Using (4.93), the notation (8.54) and (8.55), and also the relation

$$S = \frac{bl}{2}\frac{\eta+1}{\eta},$$

we have

$$\left.\begin{array}{ll} k_{22} = \dfrac{\lambda}{(2N+1)^2} \displaystyle\sum_{k_i=1}^{N} R_{2k_i}\dfrac{b_{k_i}}{b}, & k_{44} = \dfrac{\lambda^3(\eta+1)^2}{4\eta^2(2N+1)^3}\displaystyle\sum_{k_i=1}^{N} R_{4k_i}\dfrac{b_{k_i}}{b}\bar{z}_{k_i}, \\[4ex] k_{26} = \dfrac{\lambda^2(\eta+1)}{\eta(2N+1)^3}\displaystyle\sum_{k_i=1}^{N} R_{6k_i}\dfrac{b_{k_i}}{b}, & k_{66} = \dfrac{\lambda^2(\eta+1)}{\eta(2N+1)^3}\displaystyle\sum_{k_i=1}^{N} L_{6k_i}\dfrac{b_{k_i}}{b}. \end{array}\right\} \quad (8.56)$$

§ 5. Apparent Masses of Annular Wings

The apparent masses of cylindrical annular wings are calculated similarly. This problem was investigated by E. P. Kapustina.

We use the customary coordinates shown in Fig. 8.10. We divide the chord of the wing into 2m parts; we locate a bound vortex in the middle of each segment of length b/2m; and we carry out calculations for the end of this segment. The boundary conditions are not satisfied at the end of the last segment (in Fig. 8.11 the circles and heavy lines are vortices and the crosses and dots are grid points for the calculation).

The vortices and grid points are numbered from the leading edge of the wing (x = b/2 is the leading edge), and

$$\left.\begin{array}{ll} x_i = \dfrac{b}{2} - \dfrac{b}{4m} - \dfrac{i-1}{2m}b, & i = 1, 2, \ldots, 2m, \\[3ex] x_{0j} = \dfrac{b}{2} - \dfrac{j}{2m}b, & j = 1, 2, \ldots, 2m-1. \end{array}\right\} \quad (8.57)$$

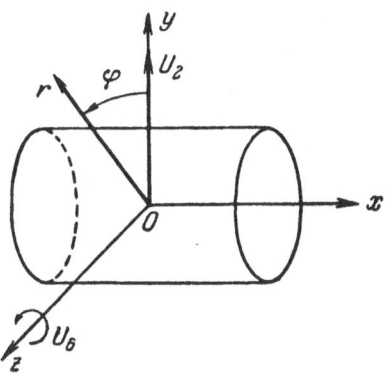

Fig. 8.10. The calculation of apparent masses of annular wings.

From (4.96) the strength of the i-th attached vortex is

$$\Gamma_{+i} = (U_2 r \Gamma_{2i} + U_6 r^2 \Gamma_{6i})\cos\varphi, \quad (8.58)$$

from (7.46) the speed at the point j is

$$\frac{W_{rj}}{\cos\varphi_0} = \frac{U_2}{2\pi}\sum_{i=1}^{2m} w_{r2ij}\Gamma_{2i} + \frac{U_6 r}{2\pi}\sum_{i=1}^{2m} w_{r2ij}\Gamma_{6i}. \quad (8.59)$$

We substitute (8.59) in the boundary condition (1.51), and in addition to the condition for circulation-free flow we obtain two independent sets of equations.

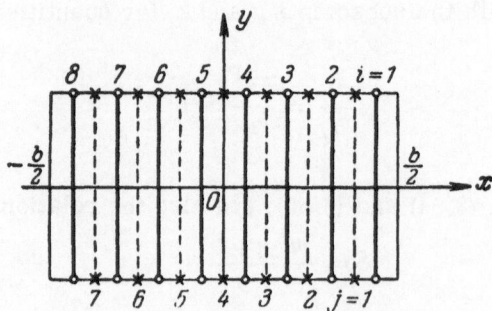

Fig. 8.11. The bound vortices i and the sections j for which calculations are performed on an annular wing with circulation-free flow (2m = 8).

The first set is used for the determination of Γ_{2i}:

$$\sum_{i=1}^{2m} w_{r2ij}\Gamma_{2i} = 2\pi, \quad \sum_{i=1}^{2m}\Gamma_{2i} = 0, \quad j = 1, 2, \ldots, 2m-1, \tag{8.60}$$

and the second is for the determination of Γ_{6i}:

$$\sum_{i=1}^{2m} w_{r2ij}\Gamma_{6i} = 2\pi\frac{x_{0j}}{r}, \quad \sum_{i=1}^{2m}\Gamma_{6i} = 0, \quad j = 1, 2, \ldots, 2m-1. \tag{8.61}$$

From considerations of symmetry the number of unknowns in (8.60) and (8.61) can be halved. We must take into account the fact that for progressive oscillations the vortices symmetrically located relative to the plane x = 0 (Fig. 8.10) have identical strengths but opposite signs. After a rotation about Oz such a pair of vortices will become identical, and

$$\Gamma_{2i} = -\Gamma_{2,2m-i+1}, \quad \Gamma_{6i} = \Gamma_{6,2m-i+1}.$$

When the system (8.60) and (8.61) have been solved, the coefficients of the apparent mass are easily determined. Replacing integrals by sums in (4.99) we have

$$k_{22} = -\frac{\pi\lambda}{8m}\sum_{i=1}^{2m}\Gamma_{2i}, \quad k_{66} = -\frac{\pi\lambda^2}{16m}\sum_{i=1}^{2m}\Gamma_{6i}\frac{x_i}{b}, \quad k_{26} = -\frac{\pi\lambda^2}{16m}\sum_{i=1}^{2m}\Gamma_{6i}. \tag{8.62}$$

Using the definitions of Γ_2 and Γ_6, we can expand the expressions for the sums (8.62) as follows:

$$\sum_{i=1}^{2m}\Gamma_{2i} = \Gamma_{21} + (\Gamma_{21}+\Gamma_{22}) + (\Gamma_{21}+\Gamma_{22}+\Gamma_{23}) + \ldots + (\Gamma_{21}+\Gamma_{22}+\ldots+\Gamma_{22m-1}),$$

$$\sum_{i=1}^{2m}\Gamma_{6i}\frac{x_i}{b} = \Gamma_{61}\frac{x_1}{b} + (\Gamma_{61}+\Gamma_{62})\frac{x_2}{b} + (\Gamma_{61}+\Gamma_{62}+\Gamma_{63})\frac{x_3}{b} + \ldots + (\Gamma_{61}+\Gamma_{62}+\ldots+\Gamma_{62m-1})\frac{x_{2m-1}}{b},$$

$$\sum_{i=1}^{2m}\Gamma_{6i} = \Gamma_{61} + (\Gamma_{61}+\Gamma_{62}) + (\Gamma_{61}+\Gamma_{62}+\Gamma_{63}) + \ldots + (\Gamma_{61}+\Gamma_{62}+\ldots+\Gamma_{62m-1}). \tag{8.63}$$

§ 6. The Effect of Flow Boundaries

The above numerical method of solution for a monoplane wing is easily generalized to be applicable to more complex problems. As an example we consider the motion of a monoplane wing close to an interface parallel to the plane of the wing. This problem was investigated in detail by G. A. Yakovlev.

We assume that the interface is the boundary between two media and that the wing is moving in one of them. The surface can be either free and deformable (a boundary between air and water) or rigid (an air–earth or water–ice boundary, etc.). The mathematical formulation of the problem is similar to that in an unbounded medium. The only difference is in the existence of an extra condition on this surface (see also §2, Chapter VI).

The extra boundary conditions for these two types of surface are different. A fixed (undeformable) interface must be a flow surface, therefore when a wing moves close to this surface the normal components of the disturbed velocity on the surface must be zero. The boundary condition at a free surface is the requirement that the pressures on the two sides of the boundary be equal, since a liquid surface cannot maintain a pressure difference (the weight of the liquid itself is neglected). A free surface will be deformed by the passage of a wing. To the degree of accuracy of the linear theory, the deformations will be assumed to be small and the boundary condition to be that for an undisturbed surface. Thus the boundary condition for a free surface reduces to the condition that the pressure disturbance due to the motion of the wing be zero. From the linearized Bernoulli equation (1.21) we see that for this condition to be satisfied it is sufficient that the velocity disturbance W_X parallel to the surface be zero.

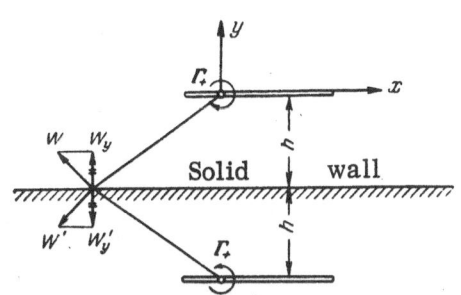

Fig. 8.12. The replacement of a solid wall by a second wing.

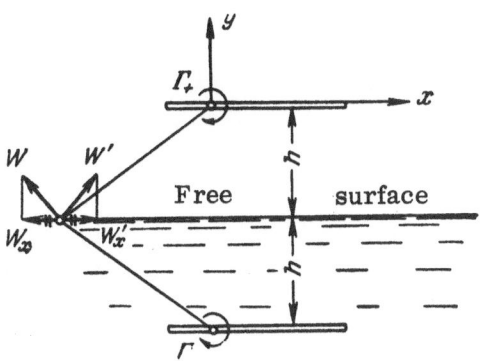

Fig. 8.13. The replacement of a free surface by a second wing.

From (7.3), (7.5), and (7.6) we conclude that the velocity disturbances due to horseshoe vortices possess the following symmetry properties:

$$\left. \begin{aligned} W_x(x_0,\, y_0,\, z_0) &= -W_x(x_0,\, -y_0,\, z_0), \\ W_y(x_0,\, y_0,\, z_0) &= W_y(x_0,\, -y_0,\, z_0), \\ W_z(x_0,\, y_0,\, z_0) &= -W_z(x_0,\, -y_0,\, z_0). \end{aligned} \right\} \quad (8.64)$$

Similar equations hold for oblique horseshoe vortices, and also for the vortices of systems that are combinations of both types of vortices.

Fig. 8.12 shows how we can model the motion of a monoplane wing of arbitrary shape in plan close to a solid wall (close to the earth, for example). By replacing the effect of a wall by the mirror image of the wing in the wall, we ensure that the extra boundary conditions will be satisfied. In this case the strengths of the vortices of the auxiliary wing are equal in magnitude but opposite in sign to the strengths of the vortices of the original wing.

The modeling of a free surface is shown in Fig. 8.13. If the vortices on the mirror image have the same magnitude and sign as the vortices on the original wing, then the tangential components of the velocity disturbances will vanish.

Hence the problem is reduced to the calculation of the flow past a wing in an infinite perfect medium in which there is a second wing at a distance 2h from the first, where h is the distance from the wing to the interface. The strengths of the vortices for the auxiliary wing in terms of those for the actual wing are as shown. The method is applicable to both circulational and circulation-free flow.

All the equations derived in the present chapter for the calculation of flow about a wing remain in force. It is only necessary to take account of the velocities generated by the auxiliary wing in the determination of the coefficients of Eqs. (8.37), (8.38), (8.39) or in (8.46), (8.47), and (8.48).

Strictly speaking the effect of the auxiliary wing causes a change not only of the normal velocities but also of the tangential velocities on the original wing.

The apparent mass is calculated from the exact (nonlinear) scheme. It follows that we must take account of the change in the tangential velocity expressed in Zhukovskii's theorem (3.9) by W_{00l} and W_{00T}.

The reader can easily carry out all the necessary transformations.

CHAPTER IX

SOME EXACT SOLUTIONS OF THE EQUATIONS AND LIMITING VALUES OF THE AERODYNAMIC CHARACTERISTICS

§ 1. Profile Cascades and Multiplanes

Exact solutions of the equations we have obtained can only be found in a few special cases, and in general these equations must be solved numerically. The exact solutions are of interest not only in principle, but also in practice. This is because when systematic calculations are made, it is very important to have some exact check points or regions, to know certain trends in the behavior of the functions to be calculated, to know limiting values, etc. Thus the availability of exact solutions is very useful in all general investigations.

One of the most thoroughly studied forms of exact solution is that for a cascade made up of plates, i.e., the simplest shapes possible. The first solution of this problem was given by S. A. Chaplygin [7], and a detailed investigation can be found in the monographs [12, 16]. The moment characteristics of cascades of plates are investigated in [45]. We limit our discussion in presenting final results to the lift coefficients c_y and the longitudinal moment m_z of a cascade of plates.

Equations (4.47) and (4.53) show that these coefficients are given in terms of the angle of attack α by the formulas

$$c_y = c_y^\alpha \sin \alpha, \quad m_z = -\frac{m_z^\alpha}{2} \sin 2\alpha - \frac{m_z^{\alpha\alpha}}{2} \sin^2 \alpha, \qquad (9.1)$$

since $m_{z0} = 0$ for a plate cascade. The moment M_z is calculated relative to the leading edge; the positive direction of this moment is shown in Fig. 1.8.

Since the camber $\overline{f} = 0$ in the case under consideration, we have $c_y = 0$ when the velocity U_0 is directed along the chord of the plates, and so

$$\alpha_0 = 0, \quad \beta = \beta_\Gamma. \qquad (9.2)$$

(see Fig. 1.7).

The dependence of the coefficient c_y^α on the density τ of the cascade and the gap β_Γ is shown in Fig. 9.1. It should be noted that a change of sign of β_Γ has no effect on c_y^α.

The coefficients $m_z^\alpha/2$ and $m_z^{\alpha\alpha}/2$ as functions of the gap τ and the angle β_Γ can be obtained from the graphs in Fig. 9.2 and 9.3. It is clear from these graphs that $m_z^\alpha/2$ is independent of β_Γ, and $m_z^{\alpha\alpha}/2$ has the same sign as β_Γ.

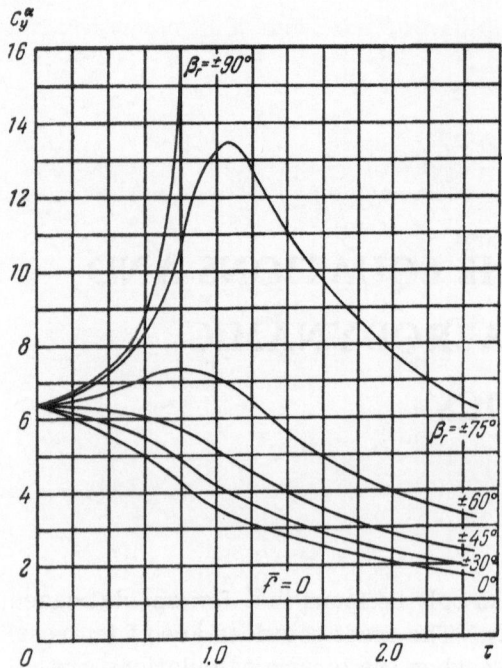

Fig. 9.1. The dependence of c_y^α on the density τ of a grid and the gap β_Γ (exact solution for a plate cascade).

It is useful to keep in mind that for very dense cascades consisting of infinitely thin plates of arbitrary shape an analytic expression is easily found for the lift. In fact for very dense cascades the flow behind maintains the direction it had on leaving the cascade. Since the flow does not separate, its direction on leaving the cascade is the same as that of the tangent to the profile of each plate at the trailing edge. The velocity Γ_+ of circulation about the profile and hence the lift Y are easily calculated from a knowledge of this direction [3].

We use the symbol α_1 for the angle between the tangent to the profile at the trailing edge and the $O\xi$ axis which is perpendicular to the front of the grid (Fig. 9.4). For a dense cascade the velocity behind the grid, both at infinity and at any finite distance, will be parallel to this tangent in the limiting case $\bar{t} \to 0$. Behind a dense cascade we thus have the velocity triangle shown in Fig. 9.4, and so the sine law yields

$$\frac{-U_{2\eta}}{\sin(\vartheta - \alpha_1)} = \frac{U_0}{\sin\left(\frac{\pi}{2} + \alpha_1\right)}.$$

The disturbed velocity $U_{2\eta}$ generated by the bound vortices of the cascade is easily expressed in terms of the circulation Γ_+. From (3.10) and considerations of symmetry (see Fig. 3.5) we have

$$U_{2\eta} = -\frac{\Gamma_+}{2t},$$

and so

$$\Gamma_+ = 2U_0 t \frac{\sin(\vartheta - \alpha_1)}{\cos\alpha_1}. \tag{9.3}$$

For a cascade of profiles the Zhukovskii theorem (3.18) yields

$$Y = \rho U_0 \Gamma_+ = 2\rho U_0^2 t \frac{\sin(\vartheta - \alpha_1)}{\cos\alpha_1},$$

and it follows that for $\vartheta - \alpha_1 = 0$, we have the lift force Y = 0. This means that $\vartheta - \alpha_1 = \alpha$, where α is the angle of attack reckoned from the direction of zero lift. Converting to the lift coefficient of the profile, we obtain the limiting formulas for profile cascades for $\bar{t} \to 0$:

$$c_y = c_y^\alpha \sin\alpha, \qquad c_y^\alpha = \frac{4\bar{t}}{\cos\alpha_1}. \tag{9.4}$$

Here α_1 is the angle between the tangent at the profile trailing edge and the $O\xi$ axis (Fig. 9.4), where this tangent gives the direction of motion for zero lift; i.e., it is in the direction of the aerodynamic chord of a very dense cascade of thin profiles.

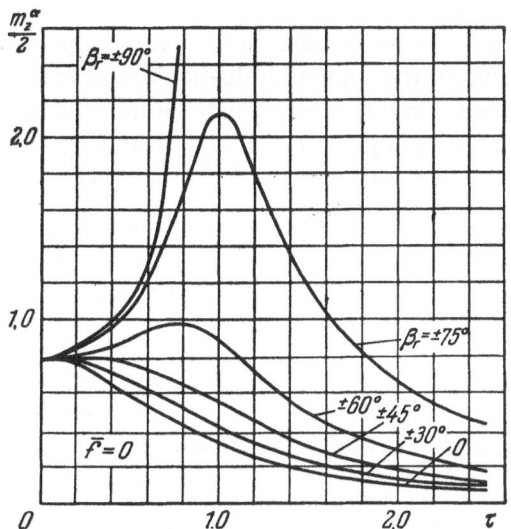

Fig. 9.2. The dependence of $m_z^\alpha/2$ on the density τ of a cascade and the gap β_Γ (exact solution for a plate cascade).

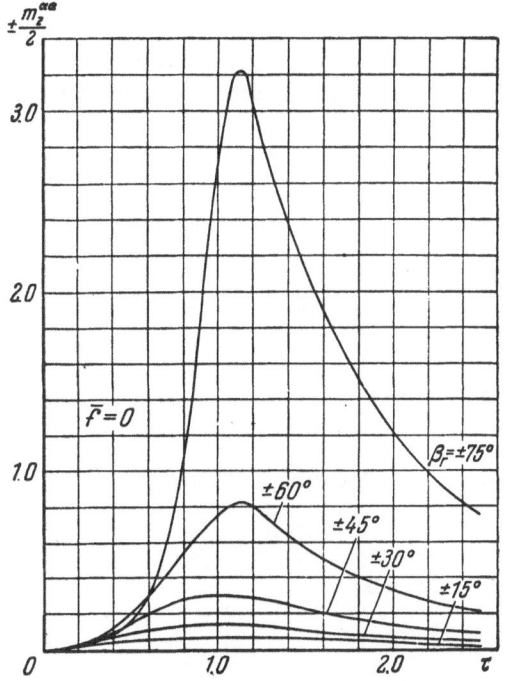

Fig. 9.3. The coefficient $m_z^{\alpha\alpha}/2$ as a function of the cascade density τ and the angle β_Γ (exact solution for a plate cascade).

Flow past a very dense plate cascade can be constructed by using a chain of discrete vortices corresponding to the leading edges of the profiles. The velocities generated by such a chain (Fig. 7.7) can be calculated from (7.50) with $\bar{\xi}_0 \to \infty$ and $\bar{\eta} \to \infty$:

$$w_\xi = 0,$$

$$w_\eta = \begin{cases} -1 & \text{for} \quad \bar{\xi} > 0, \\ 1 & \text{for} \quad \bar{\xi} < 0. \end{cases}$$

For the dimensional velocities, (7.51) yields

$$W_\xi = 0,$$

$$W_\eta = \begin{cases} -\dfrac{\Gamma_+}{2t} & \text{for} \quad \bar{\xi} > 0, \\ \dfrac{\Gamma_+}{2t} & \text{for} \quad \bar{\xi} < 0. \end{cases}$$

Hence the velocity is constant in the half-space to the left of the chain and in the half-space to the right of the chain, and it is discontinuous on the chain line. This result is naturally in complete agreement with Fig. 3.5.

We locate a similar chain along the leading edges of the plates; it is clear that we can satisfy the smooth-flow condition by a suitable choice of Γ_+. Since

$$W_\xi = 0, \quad \varepsilon = 0, \quad \beta_\Gamma = \beta,$$

the boundary condition (1.32) can be written

$$W_\eta \cos\beta = -U_0 \sin\alpha,$$

or

$$-\frac{\Gamma_+}{2t} \cos\beta = -U_0 \sin\alpha,$$

and so

$$\Gamma_+ = 2U_0 t \frac{\sin\alpha}{\cos\beta}.$$

This again leads to Eqs. (9.4), since $\alpha_1 = \beta$.

Among other results we thus find that the center of pressure of a very dense plate cascade is at the leading edge for any angle of attack.

We now find the aerodynamic characteristics of a very dense multiplane. For simplicity we confine ourselves to the case $\beta_\Gamma = 0$ (a multiplane without a stagger).

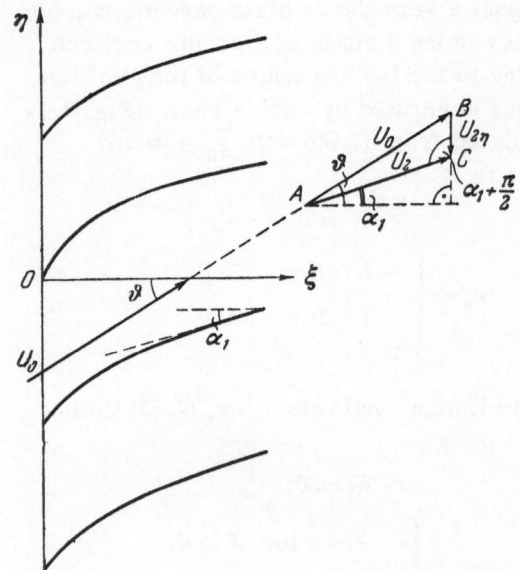

Fig. 9.4. The determination of the lift of a very dense cascade of profiles.

For a given number of plates n = const and $t \to 0$, the aerodynamic load on the interior plates (Fig. 8.2) due to the mutual influence between the plates is zero. In the limit, a multiplane is converted into a single profile and it will have the aerodynamic properties of an isolated profile.

We refer the lift coefficient of the whole multiplane to the area of all n plates [see (8.17)], and so for any M_∞ and $\beta_\Gamma = 0$ we can write

$$\lim_{\bar{t} \to 0} c_y^a(n, \bar{t}, M_\infty) = \frac{c_y^a(1, \infty, M_\infty)}{n},$$

$$\lim_{\bar{t} \to 0} \bar{x}_F(n, \bar{t}, M_\infty) = \bar{x}_F(1, \infty, M_\infty).$$

In particular for $M_\infty = 0$ (an incompressible medium) we have

$$\lim_{\bar{t} \to 0} c_y^a(n, \bar{t}) = \frac{2\pi}{n}, \qquad \lim_{\bar{t} \to 0} \bar{x}_F(n, \bar{t}) = 0,25. \qquad (9.5)$$

§2. Wings of Infinite Aspect Ratio

We first give all the data required for a plate of infinite span. In [15, 16] solutions are given for the case of harmonic vibrations of such a plate in an incompressible medium in which the dimensionless frequency is arbitrary. The relevant rotational derivative coefficients c_y^α, m_z^α, $c_y^{\omega_z}$ and $m_z^{\omega_z}$ corresponding to an angle of attack α that is constant in time and to a dimensionless angular velocity ω_z, can be derived from these solutions. It is sufficient to obtain these coefficients as functions of the dimensionless frequency p* and then to find the limits of these functions when p* $\to$ 0.

In the standard coordinate system with centering $\bar{x}_T = 0.25$ we have

$$c_y^a = 2\pi, \quad m_y^a = 0, \quad c_y^{\omega z} = \pi, \quad m_z^{\omega z} = -\frac{\pi}{8}.$$

We also give these coefficients for the centering $\bar{x}_T = 0.50$, for which the origin coincides with the mean chord of the wing:

$$c_y^a = 2\pi, \quad m_z^a = \frac{\pi}{2}, \quad c_y^{\omega z} = \frac{\pi}{2}, \quad m_z^{\omega z} = 0. \qquad (9.6)$$

Conversion from one center to another is performed by using the conversion formula (4.75).

For high subsonic velocities for which $M_\infty \leq 1$, the relations (5.9) yields ($\bar{x}_T = 0.50$):

$$c_{y\,com}^a = \frac{2\pi}{\sqrt{1 - M_\infty^2}}, \quad m_{z\,com}^a = \frac{\frac{\pi}{2}}{\sqrt{1 - M_\infty^2}}, \quad c_{y\,com}^{\omega z} = \frac{\frac{\pi}{2}}{\sqrt{1 - M_\infty^2}}, \qquad (9.7)$$

$$m_{z\,com}^{\omega z} = 0.$$

Analogous results can be obtained for $\overline{x}_T = 0.25$. It is clear from (9.7) in particular that

$$c_{y+}^{\omega z} = m_{z-}^{\alpha}$$

which is one of the consequences of the reversibility theorem, holds in this case, and that $c_{y+}^{\omega z} = c_{y-}^{\omega z} = c_y^{\omega z}$, and $m_{z+}^{\alpha} = m_{z-}^{\alpha} = m_z^{\alpha}$.

The results we have obtained have the following interpretation. For motion with the parameters α = const and ω_z = const for $\overline{x}_T = 0.5$, the action of the flow on the plate reduces to two forces. One of these forces Y_1, proportional to α, acts at a distance $x_F = 0.25b$ from the leading edge of the wing, while the other Y_2, proportional to ω_z, acts at a distance $x_{\omega_z} = 0.50b$ from the leading edge.

It should be stressed that the force Y_2 and the value of x_{ω_z} depend on the centering $\overline{x}_T$. At the same time the force Y_1 and the coordinate x_F, characterizing the position of the focus of the profile relative to the nose, are independent of $\overline{x}_T$ in steady translational flow.

We also give the apparent masses of rectangular wings of infinite span (plates) [38]. Uniform motion of a plate is characterized by three apparent masses λ_{22}, λ_{66} and λ_{26}. Let the wing span be unity and let the apparent masses [see (4.92)] be

$$k_{22} = \frac{\lambda_{22}}{\rho b^2}, \qquad k_{66} = \frac{\lambda_{66}}{\rho b^4}, \qquad k_{26} = \frac{\lambda_{26}}{\rho b^3}.$$

Then for the center $\overline{x}_T = 0.5$ (the origin at the midpoint of the chord) we have

$$k_{22} = \frac{\pi}{4}, \qquad k_{66} = \frac{\pi}{128}, \qquad k_{26} = 0.$$

In some cases the relation between the coefficients k_{22} of rectangular wings of different aspect ratios determined by elementary considerations can be useful.

To find this relation we note that the apparent mass λ_{22} determines the lift Y for translational vibrations in the direction of the Oy axis. Since the flow and the lift are identical for rectangular wings with aspect ratios λ and $1/\lambda$, we have

$$\lambda_{22}(\lambda) = \lambda_{22}\left(\frac{1}{\lambda}\right) = \lambda_{22}.$$

From the relations

$$k_{22}(\lambda) = \frac{\lambda_{22}}{\rho S b}, \qquad k_{22}\left(\frac{1}{\lambda}\right) = \frac{\lambda_{22}}{\rho S l}$$

we obtain an equation that holds for any λ:

$$k_{22}(\lambda) = \lambda k_{22}\left(\frac{1}{\lambda}\right). \tag{9.8}$$

For very small λ we thus obtain the simple formula

$$k_{22} = \frac{\pi\lambda}{4}. \tag{9.9}$$

Elementary considerations yield the rotational-derivative coefficients and the apparent masses of annular wings of large elongation for which the diameter D is much larger than the chord b. Here we can assume that each meridional section of the wing φ = const acts under conditions analagous to the conditions for a profile of a monoplane wing of infinite aspect ratio. The difference in this and other cases will be caused only by the difference in positions of the sections relative to the xOyz coordinates.

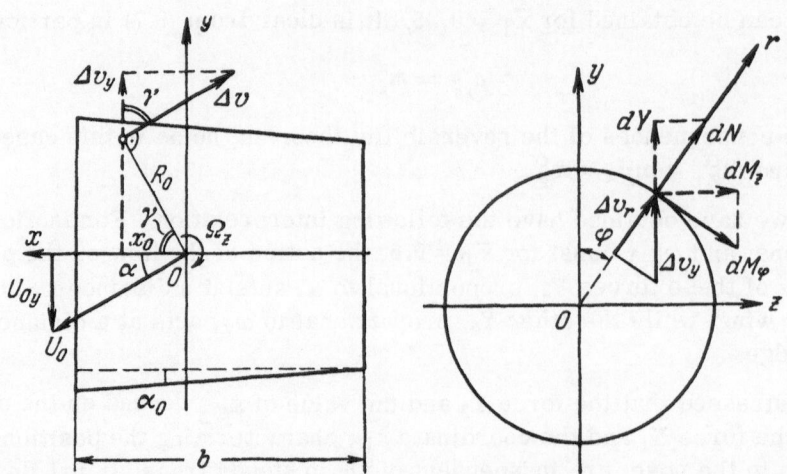

Fig. 9.5. The calculation of characteristics of an annular wing
of very large aspect ratio.

It is clear from Fig. 9.5 that, if Δv_y is the component of any supplementary velocity at any point of an annular wing, then the variation of the local angle of attack is caused by the projection of this component

$$\Delta v_r = \Delta v_y \cos \varphi. \tag{9.10}$$

We thus conclude that for translational motion, the angles of attack of the sections $\varphi = const$ are

$$\alpha'(\varphi) = \alpha_0 + \alpha \cos \varphi. \tag{9.11}$$

Let the wing rotate with an angular velocity Ω_z; then (Fig. 9.5)

$$\Delta v_y = \Delta v \cos \gamma = \Omega_z R_0 \cos \gamma = \Omega_z x_0.$$

when

$$\Delta v_r = \Omega_z \cos \varphi x_0, \tag{9.12}$$

where x_0 is the coordinate of the point where the velocity is calculated. Neglecting the effect of tangential components in each section, we conclude that the rotation of an annular wing about the Oz axis generated the same velocity in the sections as would be generated by a rotation of the section about an axis tangential to the surface in the plane $x = 0$ with an angular velocity

$$\Omega'(\varphi) = \Omega_z \cos \varphi. \tag{9.13}$$

To find the lift dY and the moment dM_z of a section about the Oz axis, we must also take into account the relations

$$dY = dN \cos \varphi, \quad dM_z = dM_\varphi \cos \varphi, \tag{9.14}$$

where dN is the normal force and dM_φ the moment relative to the transverse axis of a section with area $brd\varphi$. To find the lift Y and the longitudinal moment M_z, we must express dN and dM_φ in terms of the apparent mass or in terms of the rotational derivatives of a plate of infinite span, as was done above. Using (9.11), (9.13) and (9.14) and integrating the resulting expressions with respect to φ from 0 to 2π, we obtain expressions for Y and M_z for an annular wing of very large aspect ratio. It is easily seen that these expressions have values that are $\pi/2$ times the corresponding lift and longitudinal moment acting on a similarly loaded rectangular wing (i.e., a wing with span D and chord b). This wing has a very large aspect ratio λ.

If we use the area Db of the corresponding rectangular wing for the characteristic area S of the annular wing, we arrive at the following conclusion. All apparent masses and rotational derivatives of an annular wing of very large aspect ratio are $\pi/2$ times the corresponding parameters of an equivalently loaded rectangular wing (with the same center).

§ 3. The Side-Slipping Wing

We now investigate steady translational motion of a so-called side-slipping wing, i.e., a rectangular wing of infinite span whose leading edge is not perpendicular to the direction of motion. The angle χ between the normal velocity component U_{0n} and the velocity of motion of the wing U_0 is usually called the angle of side slip (Fig. 9.6).

The aerodynamic characteristics of a side-slipping wing can easily be expressed in terms of the profile characteristics, if it is noted that the disturbances will have no tangential component U_{0t} for such a wing in a perfect medium. Hence all the aerodynamic forces and moments acting on a side-slipping wing, and the pressure as well, are functions of the effective velocity U_{0n} and the effective Mach number $\mathbf{M}_{n\infty}$.

From Fig. 9.6 we see that

$$U_{0n} = U_0 \cos \chi, \qquad \mathbf{M}_{n\omega} = \mathbf{M}_\omega \cos \chi, \qquad \alpha = \alpha_n \cos \chi, \tag{9.15}$$

where α_n is the angle of attack corresponding to the normal section of the wing.

We will use the characteristics of a side-slipping wing in the analysis of the aerodynamics of sweptback wings. As parameters for such wings we will use the velocity U_0 and the velocity head $q_\infty = \rho_\infty U_0^2/2$, where

$$q_\infty = \frac{q_{n\infty}}{\cos^2 \chi}, \qquad q_{n\infty} = \frac{\rho_\infty U_{0n}^2}{2}. \tag{9.16}$$

Hence the aerodynamic characteristics of a side-slipping wing will differ from those of a normal wing (a profile) by the coefficients in the formulas

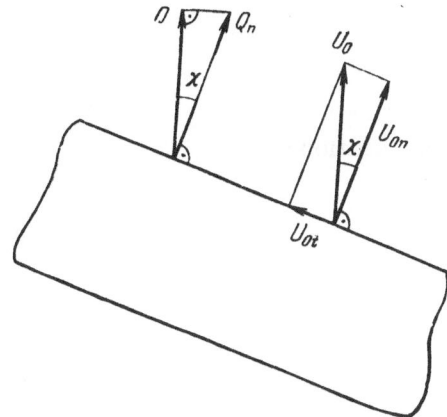

$$c_y = c_{yn} \cos^2 \chi, \qquad m_z = m_{zn} \cos^2 \chi. \tag{9.17}$$

It follows from the above that the coordinates of the center of pressure and the focus of a side-slipping wing are independent of the angle of side slip and are equal to their values for a profile.

The suction force Q is perpendicular to the leading edge and its component in the direction of U_0 is $Q \cos \chi$, and so we have

Fig. 9.6. A side-slipping wing.

$$c_q = c_{qn} \cos^3 \chi. \tag{9.18}$$

According to Zhukovskii's theorem, the lift on a normal wing of infinite span (a profile), both for a low and for high subsonic velocities, is perpendicular to the velocity of motion of the wing. Using the fact that the suction force is in the plane of a plate and the pressure is perpendicular to a plate, for low angles of attack we have

$$c_n = c_y^\alpha \alpha, \qquad c_q = c_y^\alpha \alpha^2. \tag{9.19}$$

We can thus easily find the suction-force coefficient for a side-slipping wing for $\mathbf{M}_\infty < 1$. Since

$$c_q = D \frac{c_y^2}{\pi \lambda},$$

it follows from (9.19) that

$$D = \frac{\pi \lambda}{c_y^\alpha}. \tag{9.20}$$

In an incompressible medium $(\mathbf{M}_\infty = 0)$ we have

$$c_y^2 = 2\pi \alpha_n,$$

and thus from (9.15) and (9.17)

$$c_y = 2\pi \alpha_n \cos^2 \chi = 2\pi \alpha \cos \chi,$$

i.e., for a side-slipping wing

$$c_y = c_y^\alpha \alpha, \qquad c_y^\alpha = 2\pi \cos \chi, \tag{9.21}$$

and these relations can be expressed in the form

$$\frac{c_y^\alpha}{c_{yn}^\alpha} = \cos \chi. \tag{9.22}$$

For high subsonic velocities we use the conversion described in Chapter V and obtain for a profile the equation

$$c_{yn} = \frac{2\pi \alpha_n}{\sqrt{1 - \mathbf{M}_{n\infty}^2}},$$

thus using (9.15) and (9.17) we have for a side-slipping wing the equations

$$c_y = c_y^\alpha \alpha, \qquad c_y^\alpha = \frac{2\pi}{\sqrt{\dfrac{1}{\cos^2 \chi} - \mathbf{M}_\infty^2}}, \tag{9.23}$$

and also

$$\frac{c_y^\alpha}{c_{yn}^\alpha} = \frac{\sqrt{1 - \mathbf{M}_\infty^2}}{\sqrt{\dfrac{1}{\cos^2 \chi} - \mathbf{M}_\infty^2}}. \tag{9.24}$$

The suction-force factor (9.20) in a compressible gas is

$$D_n = \frac{\Lambda}{2}\sqrt{1 - M_\infty^2}, \qquad D = \frac{\Lambda}{2}\sqrt{\frac{1}{\cos^2\chi} - M_\infty^2},$$ (9.25)

and thus

$$\frac{D}{D_n} = \frac{\sqrt{\dfrac{1}{\cos^2\chi} - M_\infty^2}}{\sqrt{1 - M_\infty^2}},$$ (9.26)

and in an incompressible medium ($M_\infty = 0$) we have

$$\frac{D}{D_n} = \frac{1}{\cos\chi}.$$ (9.27)

§4. Annular Wings of Very Small Aspect Ratio

The integral equations for an annular wing are greatly simplified when λ tends to zero. These integral equations can be solved in closed form if terms of the order of the first power of λ are retained.

We first investigate the behavior, when $\lambda \to 0$, of velocities generated by discrete annular vortices of strength proportional to $\cos\varphi$ and the corresponding system of free vortices. We consider only those questions that are of interest for the problem under consideration, that is, we study the velocities u_{r2}, v_{r2}, $w_{r2} = u_{r2} + v_{r2}$ for $\rho_0 = r_0/r = 1$ and a fixed coordinate x_0 and $\lambda \to 0$.

From (7.31) and (7.36), for $\rho_0 = 1$ we have

$$\left.\begin{aligned} u_{r2} &= -\frac{\xi_0}{2}\int_0^{2\pi}\frac{\cos^2\psi\,d\psi}{\left(\xi_0^2 + 2 - 2\cos\psi\right)^{3/2}}, \qquad \xi_0 = \frac{x_0}{r}, \\ v_{r2} &= -\frac{1}{2}\int_0^{2\pi}\cos^2\frac{\psi}{2}\left(1 + \frac{\xi_0}{\sqrt{\xi_0^2 + 2 - 2\cos\psi}}\right)d\psi. \end{aligned}\right\}$$ (9.28)

It is convenient to change from the variable ξ_0 to $\overline{x}_0 = x_0/b$ and use the relation $\xi_0 = \overline{x}_0(2/\lambda)$; then (9.28) becomes

$$\left.\begin{aligned} u_{r2} &= -\frac{\overline{x}_0}{8}\lambda^2\int_0^{2\pi}\frac{\cos^2\psi\,d\psi}{\left(\overline{x}_0^2 + \lambda^2\sin^2\frac{\psi}{2}\right)^{3/2}}, \\ v_{r2} &= -\frac{1}{2}\int_0^{2\pi}\cos^2\frac{\psi}{2}\left(1 + \frac{\overline{x}_0}{\sqrt{\overline{x}_0^2 + \lambda^2\sin^2\frac{\psi}{2}}}\right)d\psi. \end{aligned}\right\}$$ (9.29)

For any fixed value of $\overline{x}_0$ and $\lambda \to 0$, the velocity generated by a bound vortex $u_{r2} \to 0$. From the second equation in (9.29) we easily conclude that the velocities generated by the free vortices have two different limiting values depending on the sign of $\overline{x}_0$. If $\overline{x}_0 > 0$, then when

$\lambda \to 0$ we have

$$\frac{\bar{x}_0}{\sqrt{\bar{x}_0^2 + \lambda^2 \sin^2 \frac{\psi}{2}}} \to +1$$

while if $\bar{x} < 0$, we have

$$\frac{\bar{x}_0}{\sqrt{\bar{x}_0^2 + \lambda^2 \sin^2 \frac{\psi}{2}}} \to -1.$$

Hence if $\bar{x}_0 < 0$, $v_{r2} = 0$, and if $x_0 > 0$,

$$v_{r2} = -\int_0^{2\pi} \cos^2 \frac{\psi}{2} \, d\psi = -\pi,$$

and for $\rho = 1$ and $\lambda \to 0$ the total dimensionless velocity is

$$w_{r2}(\bar{x}_0) = \begin{cases} 0 & \text{for} \quad \bar{x}_0 < 0, \\ -\pi & \text{for} \quad \bar{x}_0 > 0. \end{cases} \tag{9.30}$$

If we consider only cylindrical annular wings, the problem reduces to that of the solution of Eqs. (8.20) and (8.21), in which it is also convenient in the subsequent analysis to change from the variable ξ to the variable $\bar{x}$. In the general case, the equations can be written

$$\int_{-\frac{1}{2}}^{\frac{1}{2}} \gamma_a w_{r2}(\bar{x} - \bar{x}_0, 1) \, d\bar{x} = -\pi\lambda, \qquad \int_{-\frac{1}{2}}^{\frac{1}{2}} \gamma_{\omega_z} w_{r2}(\bar{x} - \bar{x}_0, 1) \, d\bar{x} = \pi\lambda\bar{x}_0, \tag{9.31}$$

where from (9.30) we have

$$w_{r2}(\bar{x} - \bar{x}_0, 1) = \begin{cases} 0 & \text{for} \quad \bar{x} - \bar{x}_0 < 0, \\ -\pi & \text{for} \quad \bar{x} - \bar{x}_0 > 0. \end{cases} \tag{9.32}$$

when $\lambda \to 0$.

Using (9.32), we see that when $\lambda \to 0$ Eqs. (9.31) become

$$\int_{\bar{x}_0}^{\frac{1}{2}} \gamma_a \, d\bar{x} = \lambda, \qquad \int_{\bar{x}_0}^{\frac{1}{2}} \gamma_{\omega_z} \, d\bar{x} = -\lambda\bar{x}_0. \tag{9.33}$$

It is easily seen, however, that in the limiting case $\lambda \to 0$ it is impossible to satisfy the boundary conditions by replacing the wing by a continuous vortex sheet. In addition to a continuous vortex distribution we must here use discrete vortices as well.

We will show that the first equation in (9.33) has an exact solution if we locate a discrete annular vortex at the leading edge ($\bar{x} = 0.5$) and set $\gamma_\alpha = 0$. If this is done, then by using (8.23), we obtain an equation for Γ_α:

$$\frac{1}{2\pi} \Gamma_a w_{r2} = -1,$$

or, since from (9.30) $w_{r_2} = -\pi$ at all points of the surface of an annular wing, we have

$$\Gamma_a = 2, \quad \bar{x} = 0.5. \tag{9.34}$$

We can also obtain an exact solution of the second equation in (9.33) but here, in addition to a discrete vortex at the leading edge, we must also introduce a vortex sheet. From (8.24) and (9.33), a combination of these singularities yields, for $\lambda \to 0$, the equation

$$\frac{1}{\lambda} \int_{\bar{x}_0}^{\frac{1}{2}} \gamma_{\omega_z} d\bar{x} + \frac{1}{2} \Gamma_{\omega_z} = -\bar{x}_0. \tag{9.35}$$

This equation has the solution $\gamma_{\omega_z} = $ const, since in this case

$$\frac{1}{\lambda} \gamma_{\omega_z} \left(\frac{1}{2} - \bar{x}_0\right) + \frac{1}{2} \Gamma_{\omega_z} = -\bar{x}_0,$$

and so from the fact that this is an identity in $\bar{x}_0$, we have

$$\gamma_{\omega_z} = \lambda, \quad \Gamma_{\omega_z} = -1, \quad \bar{x} = 0.5, \tag{9.36}$$

where $\bar{x} = 0.5$ is the coordinate of the additional discrete vortex.

We substitute the resulting values of γ_α and γ_{ω_z} in (8.26) and determine the rotational derivatives corresponding to the distribution of singularities:

$$(c_y^\alpha)_1 = 0, \quad (m_z^\alpha)_1 = 0, \quad (c_y^{\omega z})_1 = \pi\lambda, \quad (m_z^{\omega z})_1 = 0.$$

For the known values of Γ_α and Γ_{ω_z}, using (8.27) we obtain

$$(c_y^\alpha)_2 = \pi\lambda, \quad (m_z^\alpha)_2 = \frac{\pi\lambda}{2}, \quad (c_y^{\omega z})_2 = -\frac{\pi\lambda}{2}, \quad (m_z^{\omega z}) = -\frac{\pi\lambda}{4}.$$

from the discrete vortices.

Adding the results, we obtain the following rotational derivatives for an annular wing of very small aspect ratio with the center $\bar{x}_T = 0.5$:

$$c_y^\alpha = \pi\lambda, \quad m_z^\alpha = \frac{\pi\lambda}{2}, \quad c_y^{\omega z} = \frac{\pi\lambda}{2}, \quad m_z^{\omega z} = -\frac{\pi\lambda}{4}. \tag{9.37}$$

We should note that, in the limiting case under consideration, the Chaplygin-Zhukovskii hypothesis reduces to the requirement that there be no discrete vortex at the trailing edge of the wing.

The consequences of the reversibility theorem hold in the case we have considered; in particular we have

$$c_y^{\omega z} = m_z^\alpha.$$

when $\bar{x}_T = 0.5$.

§ 5. Rectangular Wings of Very Small Aspect Ratio

Results similar to those obtained in the previous section also hold for monoplane wings of very small aspect ratio. It is clear from (7.59) that the velocity due to the fundamental vortex system used to obtain the flow about a monoplane wing depends not only on l/b, but also on the product $(l/b) \tan \chi$. Hence the results depend on the value of this parameter when $l/b \to 0$. For example $(l/b) \tan \chi$ = const for all triangular wings with straight leading edges, and this expression is independent of the aspect ratio λ. In the analysis of results of numerical calculations, the case χ = const, $l/b \to 0$ is of basic interest; for simplicity we consider only rectangular wings of very small aspect ratio.

We first study a rectangular wing with arbitrary aspect ratio $\lambda = l/b$, using a standard coordinate system with origin O at the midpoint of the root chord. We replace the wing by a continuously distributed vortex layer γ_{z+} and the corresponding free vortices. An elementary vortex band of the bound layer of width dx and length l, together with the free vortices, can be considered as a vortex system of the type (7.57) with $\chi = 0$. Replacing Γ_+ by γ_{z+} dx and x_0 by $(x_0 - x)$ in this formula (see Fig. 7.8) and integrating resulting equation with respect to x from $-b/2$ to $b/2$, we obtain the following expression for the disturbed velocities:

$$W_{y1}(x_0, z_0) = \frac{1}{4\pi} \int_{-\frac{b}{2}}^{\frac{b}{2}} \int_{-\frac{l}{2}}^{\frac{l}{2}} \frac{\gamma_{z+}(x_0 - x)\, dz\, dx}{[(x_0 - x)^2 + (z_0 - z)^2]^{3/2}} +$$

$$+ \frac{1}{4\pi} \int_{-\frac{b}{2}}^{\frac{b}{2}} \int_{-\frac{l}{2}}^{\frac{l}{2}} \frac{\frac{d\gamma_{z+}}{dz}}{z - z_0} \left[1 - \frac{x_0 - x}{\sqrt{(x_0 - x)^2 + (z_0 - z)^2}} \right] dz\, dx. \qquad (9.38)$$

It will be shown below that, for a wing of very small aspect ratio, we must consider discrete vortices in addition to a distributed vortex layer. For a bound vortex of span l on the Oz axis, we have

$$W_{v2}(x_0, z_0) = \frac{1}{4\pi} \int_{-\frac{l}{2}}^{\frac{l}{2}} \frac{\Gamma_+ x_0\, dz}{[x_0^2 + (z - z_0)^2]^{3/2}} + \frac{1}{4\pi} \int_{-\frac{l}{2}}^{\frac{l}{2}} \frac{\frac{d\Gamma_+}{dz}}{z - z_0} \left(1 - \frac{x_0}{\sqrt{x_0^2 + (z - z_0)^2}} \right) dz. \qquad (9.39)$$

The relations (9.38) and (9.39) are reduced to dimensionless form by writing

$$\left. \begin{aligned} \gamma_{z+} &= U_0 \lambda \left(\gamma_\alpha \alpha + \gamma_{\omega_z}\omega_z + \gamma_{\omega_x}\omega_x \right), \\ \Gamma_+ &= U_0 l \left(\Gamma_\alpha \alpha + \Gamma_{\omega_z}\omega_z + \Gamma_{\omega_x}\omega_x \right), \\ \bar{z} &= \frac{2z}{l}, \quad \bar{z}_0 = \frac{2z_0}{l}, \quad \bar{x}_0 = \frac{x_0}{b}. \end{aligned} \right\} \qquad (9.40)$$

Then the disturbed velocities W_{y1} and W_{y2} can be given by the formulas

$$\left. \begin{aligned} W_{y1} &= W_{y\alpha}^{(1)}\alpha + W_{y\omega_z}^{(1)}\omega_z + W_{y\omega_x}^{(1)}\omega_x, \\ W_{y2} &= W_{y\alpha}^{(2)}\alpha + W_{y\omega_z}^{(2)}\omega_z + W_{y\omega_x}^{(2)}\omega_x. \end{aligned} \right\} \qquad (9.41)$$

From (9.38) and (9.40) we have

$$\frac{1}{U_0} W_y^{(1)} = \frac{\lambda^2}{8\pi} \int\limits_{-\frac{1}{2}}^{\frac{1}{2}} \int\limits_{-1}^{1} \frac{\gamma(\bar{x}_0 - \bar{x}) \, d\bar{z} \, d\bar{x}}{\left[(\bar{x} - \bar{x}_0)^2 + \frac{\lambda^2}{4} (\bar{z} - \bar{z}_0)^2 \right]^{3/2}} +$$

$$+ \frac{1}{2\pi} \int\limits_{-\frac{1}{2}}^{\frac{1}{2}} \int\limits_{-1}^{1} \frac{\frac{d\gamma}{dz}}{\bar{z} - \bar{z}_0} \left(1 + \frac{\bar{x} - \bar{x}_0}{\sqrt{(\bar{x} - \bar{x}_0)^2 + \frac{\lambda^2}{4} (\bar{z} - \bar{z}_0)^2}} \right) d\bar{z} \, d\bar{x}. \tag{9.42}$$

where $W_y^{(1)}$ and γ must have the same subscripts $(\alpha, \omega_z, \text{ or } \omega_x)$.

With the same condition on the subscripts of $W_y^{(2)}$ and Γ, we obtain from (9.39) and (9.40) the relation

$$\frac{W_y^{(2)}}{U_0} = \frac{\lambda^2}{8\pi} \int\limits_{-1}^{1} \frac{\Gamma \bar{x}_0 \, d\bar{z}}{\left[\bar{x}_0^2 + \frac{\lambda^2}{4} (\bar{z} - \bar{z}_0)^2 \right]^{3/2}} + \frac{1}{2\pi} \int\limits_{-1}^{1} \frac{\frac{d\Gamma}{dz}}{\bar{z} - \bar{z}_0} \left(1 - \frac{\bar{x}_0}{\sqrt{\bar{x}_0^2 + \frac{\lambda^2}{4} (\bar{z} - \bar{z}_0)^2}} \right) d\bar{z}. \tag{9.43}$$

Thus when $\lambda \to 0$ we have

$$\frac{W_y^{(2)}}{U_0} = \begin{cases} 0 & \text{for } \bar{x}_0 > 0, \\ \frac{1}{\pi} \int\limits_{-1}^{1} \frac{\frac{d\Gamma}{dz}}{\bar{z} - \bar{z}_0} \, d\bar{z} & \text{for } \bar{x}_0 < 0. \end{cases} \tag{9.44}$$

If we use this property of the velocities in (9.42), for $\lambda \to 0$ we obtain

$$\frac{W_y^{(1)}}{U_0} = \frac{1}{\pi} \int\limits_{\bar{x}_0}^{\frac{1}{2}} \int\limits_{-1}^{1} \frac{\frac{d\gamma}{dz} \, d\bar{z} \, d\bar{x}}{\bar{z} - \bar{z}_0}. \tag{9.45}$$

We now easily find the integral equations for the strength of the distributed and discrete vortices replacing a wing of small elongation. To do this we substitute

$$\frac{W_y}{U_0} = \frac{W_{y1} + W_{y2}}{U_0}.$$

in the boundary condition (1.30). Now using (9.41), (9.44), and (9.45), we obtain three integral equations that can be solved independently.

The aerodynamic characteristics of a wing in translational motion with a constant angle of attack α are obtained by the solution of the equation

$$\frac{1}{\pi} \int\limits_{-1}^{1} \frac{\frac{d\Gamma_\alpha}{dz}}{\bar{z} - \bar{z}_0} \, d\bar{z} + \frac{1}{\pi} \int\limits_{\bar{x}_0}^{\frac{1}{2}} \int\limits_{-1}^{1} \frac{\frac{d\gamma_\alpha}{dz} \, d\bar{z} \, d\bar{x}}{\bar{z} - \bar{z}_0} = -1. \tag{9.46}$$

The first term of this equation does not contain the coordinate $\bar{x}_0$ explicitly, but it follows from (9.44) that the induced velocities vanish everywhere in front of a bound vortex. Equation (9.46) corresponds to the case when there is a discrete bound vortex at the leading edge of a wing. It is only in this case that the velocities generated by this vortex system will be given by the expression in (9.46) over the whole surface of the wing. It is easily seen that a discrete bound vortex can only exist at the leading edge. If such a vortex were at any point between the leading and trailing edges, the induced velocities would be discontinuous at this point, and this is not possible since the boundary values (1.30) are continuous. The Chaplygin-Zhukovskii condition also shows that there can be no discrete vortex at the trailing edge, since the vortex-layer strength must be zero at the trailing edge for any finite aspect ratio λ.

It is easily seen that

$$\Gamma_\alpha = \Gamma_{\alpha 0} \sqrt{1 - \bar{z}^2}, \qquad \gamma_\alpha = 0. \tag{9.47}$$

is the exact solution of (9.46).

Hence the strength of a discrete bound vortex varies along the span according to an elliptic law.

In fact, writing

$$\bar{z} = -\cos\theta, \qquad \bar{z}_0 = -\cos\theta_0. \tag{9.48}$$

we have the relation

$$\Gamma_{\alpha 0} \int_0^\pi \frac{\cos\theta \, d\theta}{\cos\theta - \cos\theta_0} = \pi. \tag{9.49}$$

But it is known [13, 14] that

$$\int_0^\pi \frac{\cos n\theta}{\cos\theta - \cos\theta_0} \, d\theta = \pi \, \frac{\sin n\theta_0}{\sin\theta_0}. \tag{9.50}$$

It thus follows that the expressions (9.47) give an exact solution of Eq. (9.46), and from (9.49) we find that $\Gamma_{\alpha 0} = 1$ or

$$\Gamma_\alpha = \sqrt{1 - \bar{z}^2}, \qquad \gamma_\alpha = 0. \tag{9.51}$$

We now consider the integral equation for the rotation of a wing about Oz with constant angular velocity. The boundary condition (1.30) leads in this case to the relation

$$\frac{1}{\pi} \int_{-1}^1 \frac{\dfrac{d\Gamma_{\omega_z}}{d\bar{z}}}{\bar{z} - \bar{z}_0} \, d\bar{z} + \frac{1}{\pi} \int_{\bar{x}_0}^{\frac{1}{2}} \int_{-1}^1 \frac{\dfrac{d\gamma_{\omega_z}}{d\bar{z}} \, d\bar{z} \, d\bar{x}}{\bar{z} - \bar{z}_0} = \bar{x}_0. \tag{9.52}$$

It is easily seen that the solution of this equation describes a vortex layer whose strength is independent of $\bar{x}_0$ and which is distributed along the span of the wing elliptically, and also a bound discrete vortex at the leading edge of the wing with a circulation varying along the span elliptically. We express the required functions Γ_{ω_z} and γ_{ω_z} in the form

$$\Gamma_{\omega_z} = \Gamma_{\omega_z 0} \sqrt{1 - \bar{z}^2}, \qquad \gamma_{\omega_z} = \gamma_{\omega_z 0} \sqrt{1 - \bar{z}^2}; \tag{9.53}$$

then using the new variables (9.48) we transform Eq. (9.52) into the equation

$$\frac{\gamma_{\omega_z 0}}{\pi} \left(\frac{1}{2} - \bar{x}_0 \right) \int_0^\pi \frac{\cos \theta \, d\theta}{\cos \theta_0 - \cos \theta} + \frac{\Gamma_{\omega_z 0}}{\pi} \int_0^\pi \frac{\cos \theta \, d\theta}{\cos \theta_0 - \cos \theta} = \bar{x}_0,$$

or according to (9.50)

$$-\gamma_{\omega_z 0} \left(\frac{1}{2} - \bar{x}_0 \right) - \Gamma_{\omega_z 0} = \bar{x}_0.$$

Since this is an identity, valid for all $\bar{x}_0$, we have

$$\gamma_{\omega_z 0} = 1, \qquad \Gamma_{\omega_z 0} = -\frac{1}{2}.$$

The exact solution of (9.52) is thus

$$\Gamma_{\omega_z} = -\frac{1}{2} \sqrt{1 - \bar{z}^2}, \qquad \gamma_{\omega_z} = \sqrt{1 - \bar{z}^2}. \tag{9.54}$$

Finally the third integral equation for a wing of very small aspect ratio rotating with constant angular velocity about the Ox axis is

$$\frac{1}{\pi} \int_{-1}^1 \frac{\dfrac{d\Gamma_{\omega_x}}{d\bar{z}}}{\bar{z} - \bar{z}_0} \, d\bar{z} + \frac{1}{\pi} \int_{\bar{x}_0}^{\frac{1}{2}} \int_{-1}^1 \frac{\dfrac{d\gamma_{\omega_x}}{d\bar{z}} \, d\bar{z} \, d\bar{x}}{\bar{z} - \bar{z}_0} = -\bar{z}_0 \frac{\lambda}{2}. \tag{9.55}$$

We will show that the solution of this equation is given by the functions

$$\Gamma_{\omega_x} = \Gamma_{\omega_x 0} \bar{z} \sqrt{1 - \bar{z}^2}, \qquad \gamma_{\omega_x} = 0. \tag{9.56}$$

In fact if we change to the variables (9.48), we find that (9.55) becomes

$$\frac{\Gamma_{\omega_x 0}}{\pi} \int_0^\pi \frac{\cos 2\theta \, d\theta}{\cos \theta - \cos \theta_0} = \frac{\lambda}{2} \cos \theta_0,$$

or, if we use (9.50),

$$\Gamma_{\omega_x 0} = \frac{\lambda}{4}.$$

Hence the exact solution of (9.55) is given by the functions

$$\Gamma_{\omega_x} = \frac{\lambda}{4} \bar{z} \sqrt{1 - \bar{z}^2}, \qquad \gamma_{\omega_x} = 0. \tag{9.57}$$

It remains to find the rotational derivatives of a rectangular wing of very small aspect ratio. To find those for a continuously distributed vortex layer we must use formulas of the type (4.6), and we must add the extra terms due to the discrete bound vortices.

Comparing the notation (4.2) with the notation (9.40) we see that (4.6) in the present case takes the form

$$(c_y^\alpha)_1 = \frac{2\lambda}{S} \int \int \gamma_\alpha \, dx \, dz, \qquad (m_z^\alpha)_1 = \frac{2\lambda}{Sb} \int \int \gamma_\alpha x \, dx \, dz,$$

$$(c_y^{\omega_z})_1 = \frac{2\lambda}{S} \int \int \gamma_{\omega_z} \, dx \, dz, \qquad (m_z^{\omega_z})_1 = \frac{2\lambda}{Sb} \int \int \gamma_{\omega_z} x \, dx \, dz,$$

$$(m_x^{\omega_x})_1 = -\frac{2\lambda}{Sb} \int \int \gamma_{\omega_x} z \, dx \, dz,$$

the range of integration being the whole wing. Changing to dimensionless variables, we have

$$(c_y^\alpha)_1 = \lambda \int_{-1}^{1} \int_{-\frac{1}{2}}^{\frac{1}{2}} \gamma_\alpha \, d\bar{x} \, d\bar{z}, \qquad (m_z^\alpha)_1 = \lambda \int_{-1}^{1} \int_{-\frac{1}{2}}^{\frac{1}{2}} \gamma_\alpha \bar{x} \, d\bar{x} \, d\bar{z},$$

$$(c_y^{\omega_z})_1 = \lambda \int_{-1}^{1} \int_{-\frac{1}{2}}^{\frac{1}{2}} \gamma_{\omega_z} \, d\bar{x} \, d\bar{z}, \qquad (m_z^{\omega_z})_1 = \lambda \int_{-1}^{1} \int_{-\frac{1}{2}}^{\frac{1}{2}} \gamma_{\omega_z} \bar{x} \, d\bar{x} \, d\bar{z}, \tag{9.58}$$

$$(m_x^{\omega_x})_1 = -\frac{\lambda^2}{2} \int_{-1}^{1} \int_{-\frac{1}{2}}^{\frac{1}{2}} \gamma_{\omega_x} \bar{z} \, d\bar{x} \, d\bar{z}.$$

To find the extra terms due to the discrete bound vortices, we replace $\int_{-b/2}^{b/2} \gamma_+ \, dx$ by Γ_+

and use the fact that $\bar{x} = 0.5$ for a discrete vortex (for the center being used). Thus replacing

$\int_{-1/2}^{1/2} \gamma_\alpha \, d\bar{x}$ by Γ_α, $\int_{-1/2}^{1/2} \gamma_\alpha \bar{x} \, d\bar{x}$ by $\frac{1}{2} \Gamma_\alpha$, etc. in (9.58), we obtain

$$(c_y^\alpha)_2 = \lambda \int_{-1}^{1} \Gamma_\alpha \, d\bar{z}, \qquad (m_z^\alpha)_2 = \frac{\lambda}{2} \int_{-1}^{1} \Gamma_\alpha \, d\bar{z},$$

$$(c_y^{\omega_z})_2 = \lambda \int_{-1}^{1} \Gamma_{\omega_z} \, d\bar{z}, \qquad (m_z^{\omega_z})_2 = \frac{\lambda}{2} \int_{-1}^{1} \Gamma_{\omega_z} \, d\bar{z}, \tag{9.59}$$

$$(m_x^{\omega_x})_2 = -\frac{\lambda^2}{2} \int_{-1}^{1} \Gamma_{\omega_x} \bar{z} \, d\bar{z}.$$

Substituting the solutions obtained for (9.51), (9.54), and (9.57) in (9.58), we obtain the rotational derivatives due to a continuous vortex layer:

$$(c_y^\alpha)_1 = 0, \quad (m_z^\alpha)_1 = 0, \quad (c_y^{\omega_z})_1 = \frac{\pi\lambda}{2}, \quad (m_z^{\omega_z})_1 = 0, \quad (m_x^{\omega_x})_1 = 0. \tag{9.60}$$

From (9.59) we obtain for these same rotational derivatives for the discrete vortices the relations

$$(c_y^\alpha)_2 = \frac{\pi\lambda}{2}, \quad (m_z^\alpha)_2 = \frac{\pi\lambda}{4}, \quad (c_y^{\omega_z})_2 = -\frac{\pi\lambda}{4}, \quad (m_z^{\omega_z})_2 = -\frac{\pi\lambda}{8}, \quad (m_x^{\omega_x})_2 = -\frac{\pi\lambda^3}{64}. \tag{9.61}$$

For $\bar{x}_T = 0.5$ the total rotational derivatives for rectangular wings of very small aspect ratio are

$$c_y^\alpha = \frac{\pi\lambda}{2}, \quad m_z^\alpha = \frac{\pi\lambda}{4}, \quad c_y^{\omega z} = \frac{\pi\lambda}{4}, \quad m_z^{\omega z} = -\frac{\pi\lambda}{8}, \quad m_x^{\omega x} = -\frac{\pi\lambda^3}{64}. \tag{9.62}$$

It should be noted that, in practice, instead of the coefficient $m_x^{\omega x}$ we usually determine the coefficient $m_{x1}^{\omega x1}$ which is introduced in the following way. The characteristic linear dimension for the coefficient is taken to be the span of the wing l, and the characteristic linear dimension for ω_{x1} is taken to be the half-span $l/2$:

$$m_{x1} = \frac{M_x}{qSl}, \quad \omega_{x1} = \frac{\Omega_x l}{2U_0}, \quad m_{x1} = m_{x1}^{\omega x1}\omega_{x1}. \tag{9.63}$$

Hence converting from $m_x^{\omega x}$ to $m_{x1}^{\omega x1}$ as in §6, Chapter IV, we easily obtain

$$m_{x1}^\omega = m_x^\omega \frac{8}{\lambda^2} \frac{1}{\left(1+\frac{1}{\eta}\right)^2}, \tag{9.64}$$

or, for a rectangular wing,

$$m_{x1}^\omega = \frac{2m_x^\omega}{\lambda^2}. \tag{9.65}$$

For very small λ, (9.62) and (9.65) thus yield

$$m_{x1}^\omega = -\frac{\pi\lambda}{32}. \tag{9.66}$$

The consequences of the reversibility of the relation between the coefficients $c_y^{\omega z}$ and m_z^α for direct and reversed wings hold, since from (9.62),

$$c_y^{\omega z} = m_z^\alpha.$$

CHAPTER X

PROFILE CASCADES AND MULTIPLANES

§1. Some Problems in Methods of Calculation

It has already been shown (§1, Chapter VIII), that the problem of a thin-profile cascade reduces to that of the independent solution of two sets of linear algebraic equations (8.9). The number of equations in each of these systems is equal to the number of vortices m replacing the profile cascade. The coefficients of the equations are the dimensionless velocities due to an infinite chain of vortices.

The linearized multiplane problem leads to a single set of algebraic equations (8.15). The coefficients of these equations can be calculated more simply than in the preceding case by using (8.16) for the dimensionless velocity due to a single vortex of infinite span. However the number of equations considerably increases, since the flow is not periodic in this case. If m is the number of bound vortices on each plane and n is the number of planes, then in general there are mn equations. In the case of a multiplane without stagger, elementary considerations of symmetry permit the reduction of the number of equations to approximately one half (to exactly one half if the number of planes is even and to $\{[(n-1)/2] + 1\}$ m equations if n is given).

A very important property of the system of equations under consideration is the stability of their solutions. Small changes in the coefficients and right-hand sides of these equations lead to small changes in the solutions. This is because, in the principal determinants of the systems, the elements that have the largest absolute values are near the main diagonal on which i = j. This fact is not due only to chance — the points with j = i are directly behind the bound vortex with the number i (see Fig. 8.1).

In calculations we must choose a sufficient number m of discrete bound vortices to ensure the required accuracy. It is, of course, very important to demonstrate the convergence of the process when m → ∞ and to have an estimate of the error for various values of m both for the system under consideration and for the other systems considered in the present monograph. Since no such proofs or estimates exist for the general case, we must as usual confine ourselves to an analysis of practical convergence by a comparison of solutions for various, increasing values of m. The values of m can naturally be chosen automatically if a computer is used. This considerably increases the machine time needed for the calculations, however, and overloads the operational memory of the machine (the calculation of the coefficients and the solution of the system are carried out simultaneously). Hence the number of discrete vortices required is usually determined separately for several typical variants.

It is interesting to note that, for the simplest case (an isolated profile, plates of infinite span), the above method of calculation yields exact values of c_y^α and m_z^α for any number m of vortices starting with m = 1. The influence of the number m of bound vortices on the accuracy of the coefficients can be illustrated by graphs of the type in Fig. 10.1. These graphs are somewhat formal, since the number m can actually take only integral values. It is clear from

145

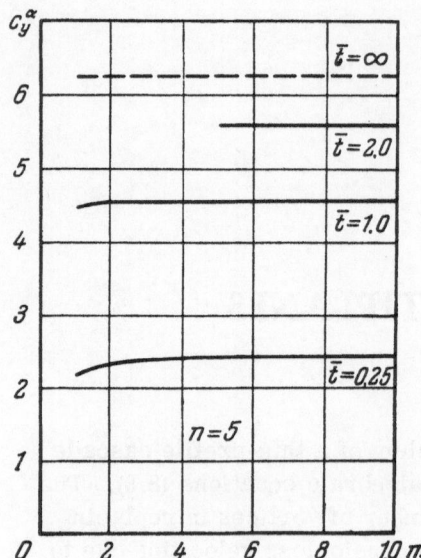

Fig. 10.1. The effect of the number of vortices m on the derivative c_y^α for multiplanes (n = 5).

Fig. 10.1 that the accuracy of the calculations only varies slightly when m is increased beyond 5, especially for large $\bar{t}$.

To obtain high accuracy for dense profiles (for small $\bar{t}$) or for dense multiplanes (for large n and small $\bar{t}$), especially in the moment characteristics, we must use a large number of vortices. As an example we give the results of calculations to determine the effect of the value of m on the increment Δ of the coefficients c_y^α due to a change of unity in the value of m. The first variant corresponds to a cascade with the geometrical parameters

$$\bar{t} = 2, \quad \beta_\Gamma = 45°, \quad \bar{x}_f = 0.5, \quad \bar{f} = 0.01$$

and the second to a cascade with parameters

$$\bar{t} = 0.5, \quad \beta_\Gamma = 30°, \quad \bar{x}_f = 0.5, \quad \bar{f} = 0.15.$$

Results are given with five digits after the decimal point. Somewhat lower accuracy was obtained in the calculations of the moment characteristics (for the same value of m).

				$\Delta \cdot 10^5$		
	m	3	5	6	10	14
1	c_y^α	2.74500	—02	—02	—01	0
2	c_y^α	2.00800	—18	—04	—02	0

We also note that exact solutions for a plate cascade (Figs. 9.1, 9.2, and 9.3) and the asymptotic formula (9.4) are useful in checking numerical results.

§2. Lift Coefficients of Profiles in Cascades

Results of calculations for a long series of thin-profile cascades by the method described above are given in [19]. We give below the equations and graphs needed for the determination of the lift of profile cascades.

We have

$$c_y = c_y^\alpha \sin \alpha, \quad \alpha = \alpha_\Gamma - \alpha_0, \quad \alpha_0 = \beta - \beta_\Gamma, \tag{10.1}$$

and so it is sufficient to determine β and c_y^α as functions of the geometrical parameters $\tau = b/t$ and β_Γ of the cascade and f and $\bar{x}_f$ of the profile. The position of a profile relative to the flow is determined by the geometrical angle of attack α_Γ

We give the required data for a parabolic profile, for which the maximum curvature is at the midpoint of the chord, and which has the equation

$$\frac{y}{b} = \text{const} \frac{x}{b}\left(1 - \frac{x}{b}\right). \tag{10.2}$$

(see Fig. 1.8).

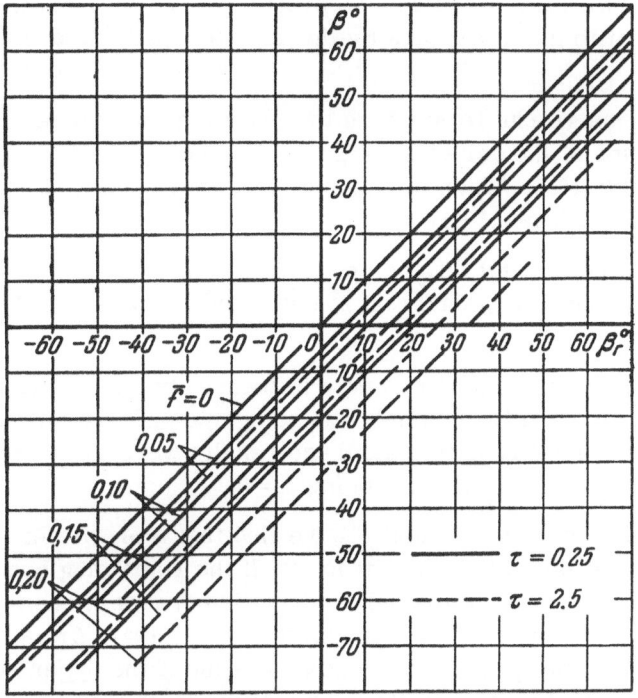

Fig. 10.2. The relation between the aerodynamic and geometrical stagger for a profile cascade.

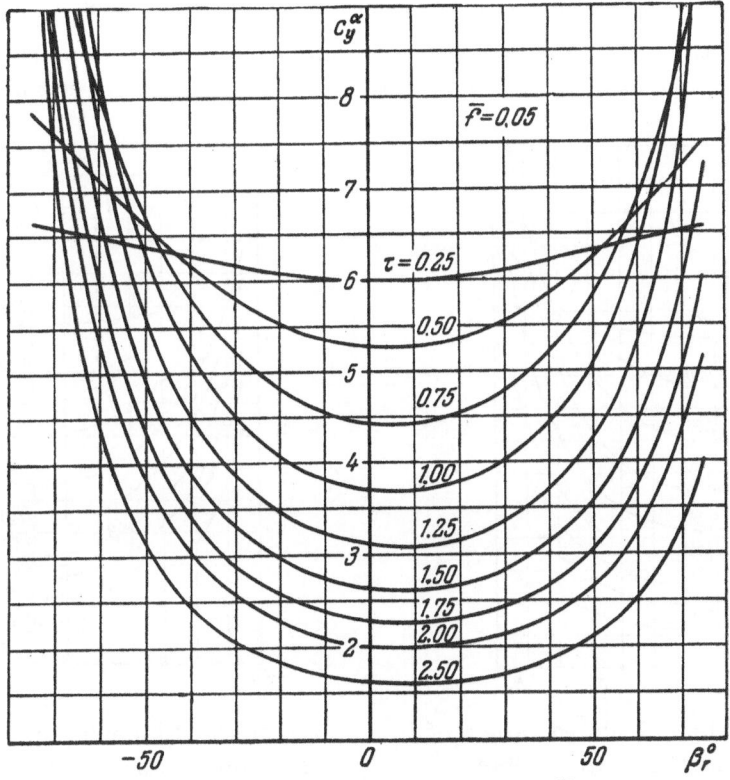

Fig. 10.3 Numerically calculated values of c_y^α for a cascade of profiles with a camber $\overline{f} = 0.05$.

Since $\overline{x}_f = 0.5$ for all these profiles, the profile geometry is determined by the camber $\overline{f}$ and the position of a profile in the cascade by the density $\tau = 1/\overline{t}$ and the geometrical stagger β_Γ.

Figure 10.2 contains graphs from which the aerodynamic stagger β can be determined from β_Γ, $\overline{f}$ and τ. Values of the derivative c_y^α as a function of β_Γ for various values of $\overline{f}$ and τ are shown in Figs. 10.3-10.6. When the geometrical angle of attack is known, the lift coefficient c_y is easily obtained from (10.1).

It is interesting to note that β depends practically linearly on β_Γ, while the angle of inclination of this line depends very weakly on $\overline{f}$ and τ and is close to $\pi/4$ which is obtained for a plate cascade ($\overline{f} = 0$).

The value of the derivative c_y^α is strongly influenced by the position of the profile in the cascade (or by the density τ and the geometrical stagger β_Γ). The curvature $\overline{f}$ has relatively little effect on c_y^α. We should stress that all these results were obtained for infinitely thin profiles (the relative thickness of a profile being $\overline{c} = 0$).

By using appropriate values of τ and β_Γ, we find that a profile in a cascade has a greater lift than an isolated profile (for which the lift is c_y^α) if the remaining conditions, in particular the geometrical angles of attack, are the same. This is because there is positive interference between the profiles in a cascade when the magnitude of the angle β_Γ is sufficiently large (Fig. 1.7). When the lift is positive, the pressure on most of the upper surface of a profile is lower except for a region at the rear, while the pressure on the lower surface is raised. When β_Γ is small, the mutual effect of the profiles leads to a lowering of the pressure difference on a profile and to a drop in c_y^α. For large β_Γ, this interference is positive.

Relative to the effect of the density τ cascades can be divided into three classes: high-density cascades, low-density cascades, and medium-density cascades.

The name "high-density cascade" is usually applied to configurations for which the direction of flow far behind the cascade is practically independent of the direction of flow in front of it.

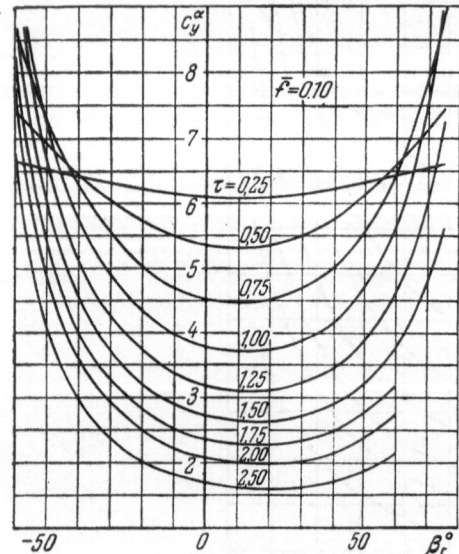

Fig. 10.4. Numerically calculated values of c_y^α for a cascade of profiles with camber $\overline{f} = 0.10$.

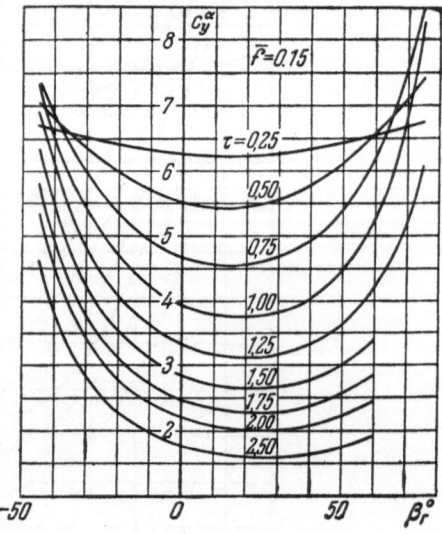

Fig. 10.5. Numerically calculated values of c_y^α for a cascade of profiles with camber $\overline{f} = 0.15$.

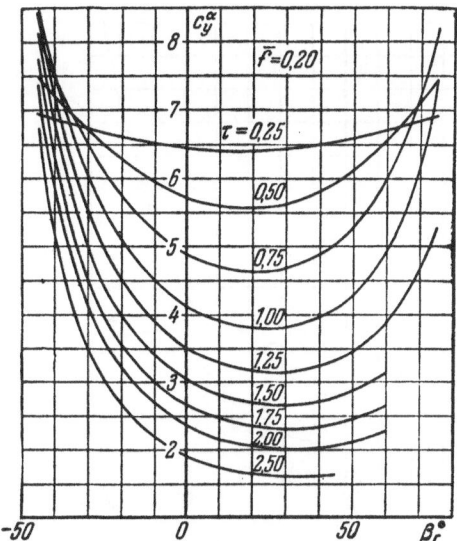

Fig. 10.6. Numerically calculated values of c_y^α for a cascade of profiles with relative camber $\bar{f} = 0.20$.

In this case the asymptotic formula (9.4) holds. The results mentioned show that this is true of practically all cascades for which $\tau \geq 1.2 - 1.4$.

A "low-density cascade" is one for which the direction of flow behind depends strongly on the direction of the oncoming flow. The derivative c_y^α for such a cascade does not differ greatly from 2π, which is its value for an isolated profile. For average values of β_Γ, such grids are those for which $\tau \leq 0.5 - 0.7$.

Medium-density cascades are, of course, intermediate between high- and low-density cascades.

As was shown in the derivation of the asymptotic formula (9.4), the aerodynamic chord of a high-density cascade coincides with the tangent to the profile at the trailing edge. For medium-density cascades the aerodynamic chord will be close to the bisector of the angle between this tangent and the geometrical chord. This follows from Chaplygin's theorem, according to which this is an exact result for smooth flow past an infinitely thin isolated profile [6, 8, 13, 14]. The aerodynamic chord of a medium-density cascade is intermediate relative to the two extreme cases.

In the analysis of the relations shown in Figs. 10.3-10.6, it is useful to compare them with the exact theory for plate cascades (Fig. 9.1). We note that the calculated results are in good agreement with the results of theory.

§ 3. Moment Coefficients of Profiles in Cascades

From (4.53) the longitudinal moment m_z relative to the nose of a profile is given by the formula

$$m_z = - m_{z0} - \frac{m_z^\alpha}{2} \sin 2\alpha - \frac{m_z^{\alpha\alpha}}{2} \sin^2 \alpha. \tag{10.3}$$

The positive direction is shown in Fig. 4.6.

Figures 10.7-10.16 show the dependence of m_{z0}, $m_z^\alpha / 2$ and $m_z^{\alpha\alpha} / 2$ on the parameters τ and β_Γ, which determine the position of a profile in the cascade for various values of the relative camber $\bar{f}$ of the profile. We consider profiles of the same shape as in the preceding section, and in particular $\bar{x}_f = 0.5$ for all profiles. The method of finding the angle of attack α. is the same as that described above.

Equation (10.3) shows that m_{z0} is equal to minus the pitching moment m_z for the aerodynamic angle of attack $\alpha = 0$; thus for $c_y = 0$ we have

$$m_{z0} = - m_z|_{\alpha = 0}. \tag{10.4}$$

It is also useful to remember that the absolute value of m_{z0} is equal to the moment in a cascade relative to the focus F (for more detail see §4, Chapter IV):

$$m_F = - m_{z0}. \tag{10.5}$$

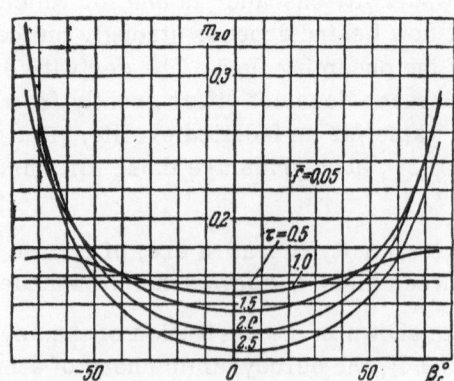

Fig. 10.7. Calculated values of m_{Z0} for cascades of profiles with camber $\overline{f} = 0.05$.

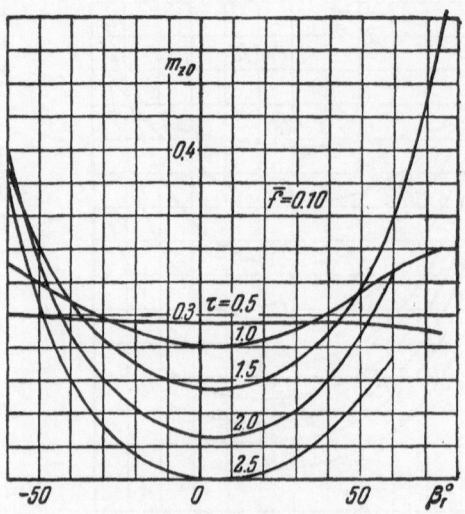

Fig. 10.8. Calculated values of m_{Z0} for grids of profiles with camber $\overline{f} = 0.10$.

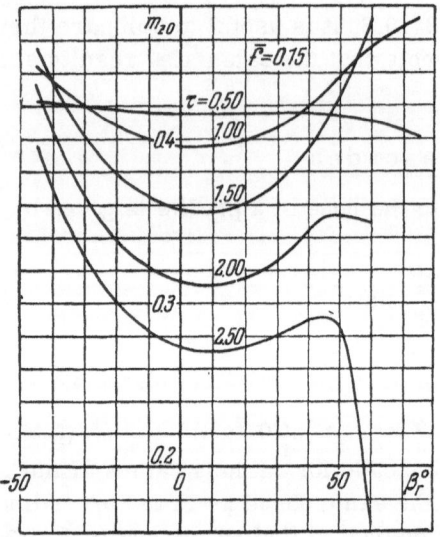

Fig. 10.9. Calculated values of m_{Z0} for grids of profiles with camber $\overline{f} = 0.10$.

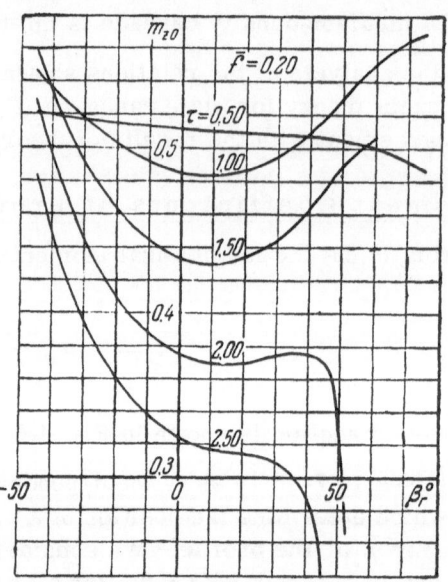

Fig. 10.10. Calculated values of m_{Z0} for grids of profiles with camber $\overline{f} = 0.15$.

For low-density cascades, m_{Z0} is nearly equal to this same quantity for the corresponding isolated profile. In this case m_{Z0} depends only insignificantly on the angle of geometrical stagger β_Γ (Figs. 10.7–10.10), and this is evidence for the unimportance of interference between planes for $\alpha = 0$.

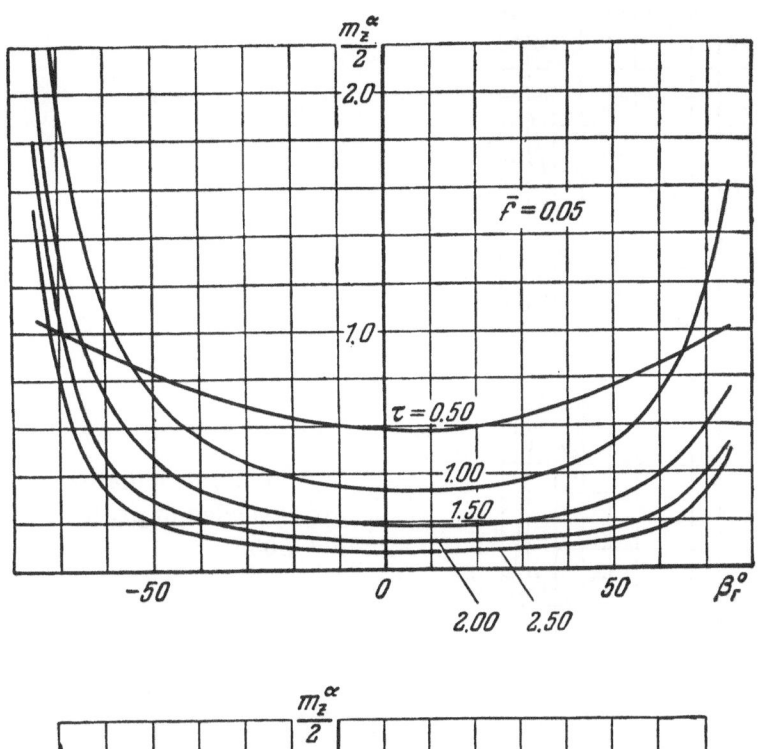

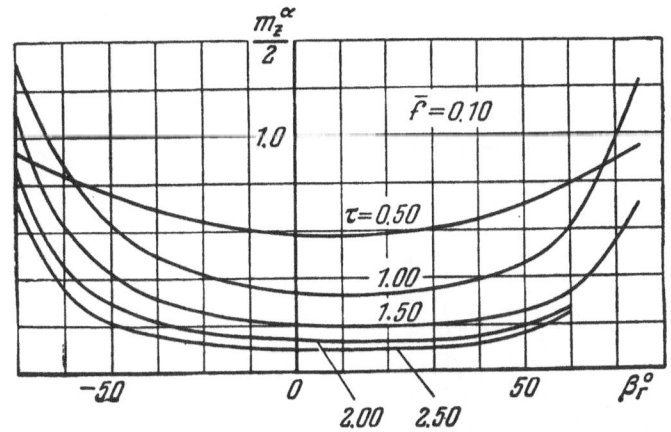

Fig. 10.11 Calculated values of $m_z^\alpha/2$ for grids of profiles with cambers $\overline{f} = 0.05$ and $\overline{f} = 0.10$.

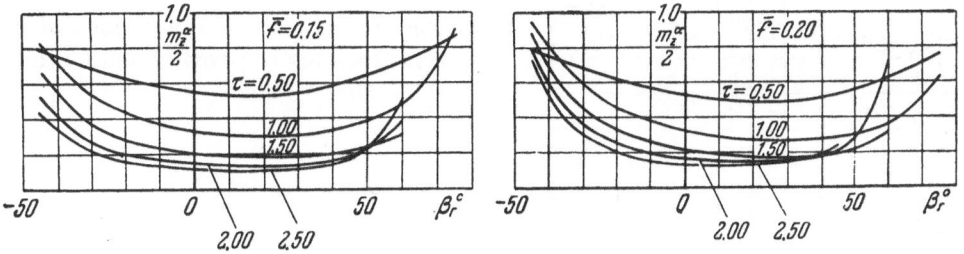

Fig. 10.12. Calculated values of $m_z^\alpha/2$ for cascades of profiles with camber $\overline{f} = 0.15$ and $\overline{f} = 0.20$.

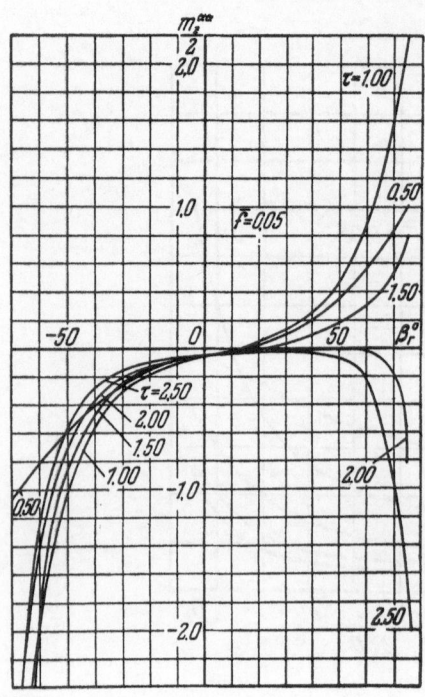

Fig. 10.13. Calculated values of $m_z^{\alpha\alpha}/2$ for cascades of profiles with camber $\bar{f} = 0.05$.

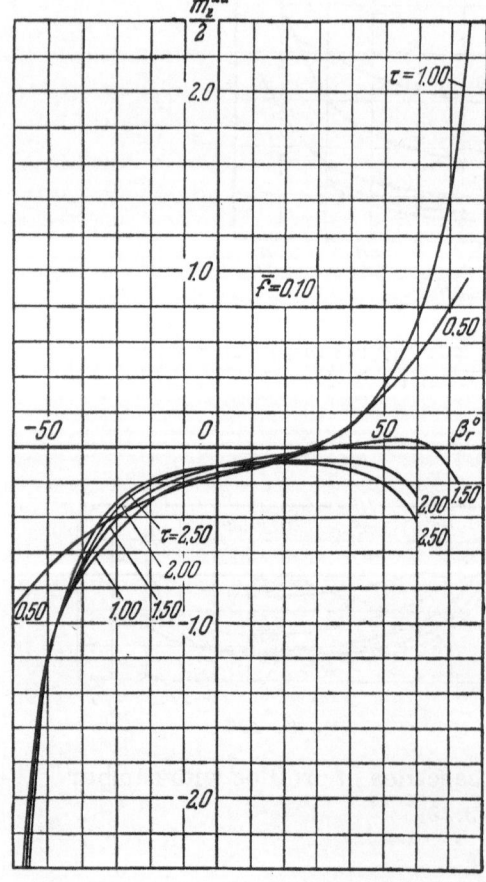

Fig. 10.14. Calculated values of $m_z^{\alpha\alpha}/2$ for cascades of profiles with camber $\bar{f} = 0.10$.

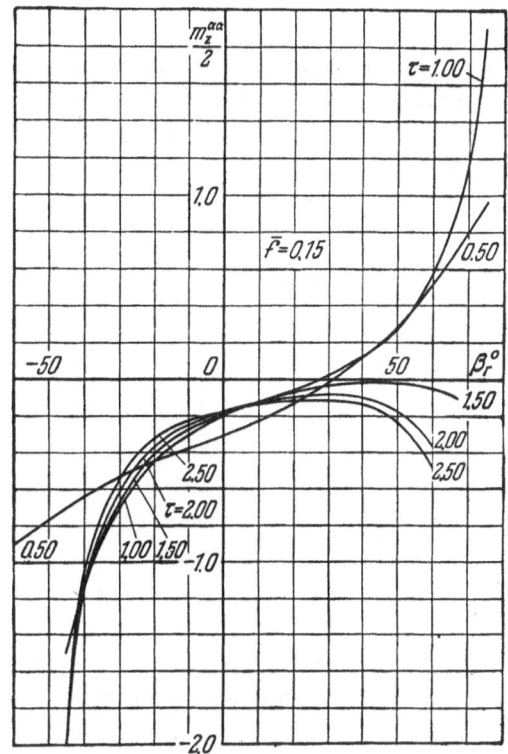

Fig. 10.15. Calculated values of $m_z^{\alpha\alpha}/2$ for cascades of profiles with camber $\bar{f} = 0.15$.

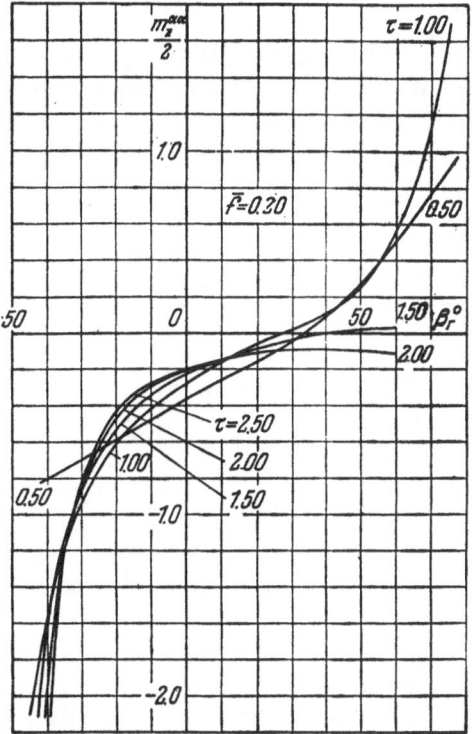

Fig. 10.16. Calculated values of $m_z^{\alpha\alpha}/2$ for cascades of profiles with camber $\bar{f} = 0.20$

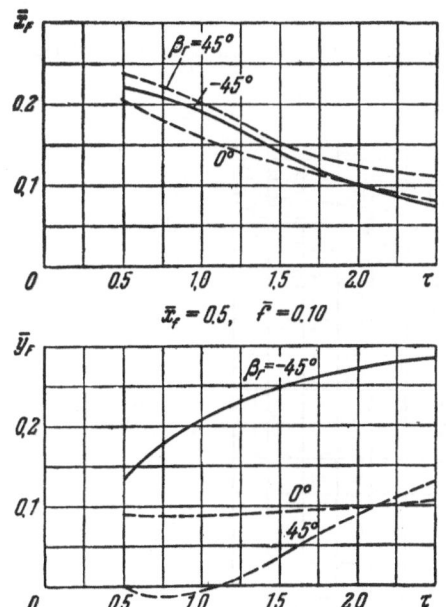

Fig. 10.17. Coordinates of the focus of profile cascades with $\bar{x}_f = 0.5$, $f = 0.10$.

In the case of high-density cascades, for which this interference is very strong, the transition from one value of β_Γ to another with $\tau = $ const can even lead to a change of sign of m_{z0}. This coefficient depends strongly on the camber $\bar{f}$ of the profile, and $m_{z0} = 0$ for $\bar{f} = 0$.

A good representation of the features of the other moments can be obtained by considering the coordinates of the focus F of a profile in a cascade for $\bar{x}_f = 0.5$, and $\bar{f} = 0.10$. Figure 10.17 shows how the density τ and the geometrical stagger influence the coordinates $\bar{x}_F$ and $\bar{y}_F$. It is interesting to note that, for various β_Γ, the focus moves forward relative to the profile when τ increases. If $\bar{x}_F = 0.25$ for an isolated profile, then for a high-density cascade it reaches a value of the order of 0.1 and smaller (see also §1, Chapter IX). Hence for high-density cascades the load, which varies with the angle of attack, is concentrated near the leading edge to a still greater degree than for an isolated profile.

Results obtained from exact theory for a plate cascade are shown in Figs. 9.2 and 9.3. The calculated results shown in Figs. 10.7-10.17 are in good agreement with these theoretical results.

§ 4. Multiplanes of Infinite Span

The multiplane of infinite span has been investigated by many authors from many points of view. We will not give a detailed bibliography of works on this subject, but will only mention a very elegant theorem concerning the stability parabola for a multiplane proved by M. V. Keldysh (for more detail see the monograph [14]). The reader can find other methods of investigating polyplanes in [9, 16, 17, 18] and elsewhere.

The method developed in the present monograph can also be used in the numerical solution of this problem for any thin profiles both in the linearized and exact formulations (in the latter case we refer to an incompressible medium).

Figures 10.18 and 10.19 show results obtained by N. G. Lavrenko for a series of multiplanes made up of plates for $\beta_\Gamma = 0$ (a multiplane of this type is illustrated in Fig. 8.2). We consider the linear case and have

$$c_y = c_y^a \alpha, \qquad m_z = m_z^a \alpha, \qquad \bar{x}_F = \frac{m_z^a}{c_y^a}, \qquad (10.6)$$

for the whole multiplane, where $\bar{x}_F = x_F/b$ and x_F is the distance from the nose of the profile to the focus. The coefficients under consideration thus characterize the lift and moment properties of the multiplane as a whole. For a characteristic dimension we use the area nb of the whole plane (since the flow is uniform, the span is taken to be unity), and so the derivative c_y^α determines the mean lift properties of a profile in a multiplane.

The limiting values of c_y^α and $\bar{x}_F$ in the diagrams when $\bar{t} \to 0$ were obtained from (9.5); corresponding results are plotted for a plate grid ($n = \infty$) for $\beta_\Gamma = 0$.

It is clear from Fig. 10.18 that when a profile is in a multiplane without stagger an increase in the density τ (or a decrease in the relative step $\bar{t}$) leads to a fall in the lifting properties of the multiplane. When $\bar{t}$ is given, this fall becomes greater when the number of planes n increases. Calculations show that the interference between the planes, disadvantageous from the point of view of lift properties, has a greater influence on the inner planes than on the outer

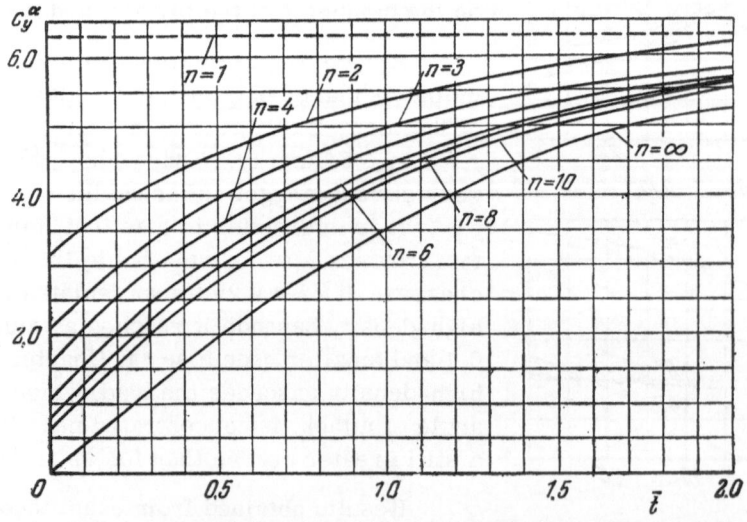

Fig. 10.18. The dependence of the derivative c_y^α on the number n of planes and the relative step $\bar{t}$ for $\beta_\Gamma = 0$.

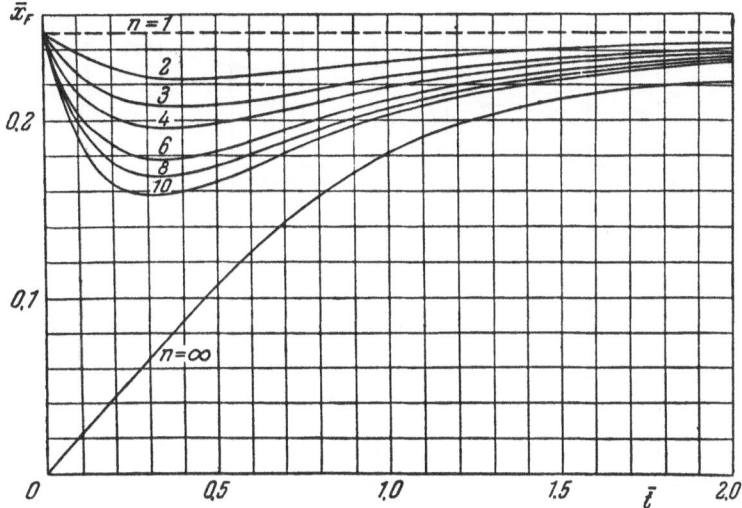

Fig. 10.19. Dependence of the focus coordinate of a multi-plane on n and $\bar{t}$ for $\beta_\Gamma = 0$.

planes. The difference between lift properties of internal planes is much smaller than the difference between lift properties of internal and outer planes.

It was noted above that for profile cascades an increase in τ (a decrease in $\bar{x}_t$) leads to a decrease in the focus coordinate $\bar{x}_F$ and to a displacement of the focus F towards the nose of the profile. This same tendency is also observed initially (for large $\bar{\tau}$) with multiplanes (Fig. 10.19), while for a fixed relative step $\bar{t}$ this displacement is greater when the number n of planes is larger.

For very small $\bar{t}$, however, it follows from (9.5) that the focus of a multiplane approaches the point $\bar{x}_F = 0.25$ when $\bar{t}$ decreases.

Multiplanes can naturally be considered as airfoils intermediate between isolated profiles (n = 1) and profile cascades (n = ∞). It is plain from Figs. 10.18 and 10.19 that c_y^α and $\bar{x}_F$ approach their limiting values in different ways.

§ 5. The Effect of the Mach Number

The approximate numerical method described in §4 of Chapter V can be used to calculate the aerodynamic characteristics of a profile cascade or a multiplane for high subsonic velocities. This method, as already noted, is based on linearized theory and is valid up to larger Mach numbers M_∞ when the disturbances caused by multiplane or profile cascade are smaller.

It should be noted that the aerodynamic characteristics for an incompressible medium in the case under consideration can be calculated from nonlinear theory (as described above for a profile cascade). Hence the accuracy of the original data is higher than the accuracy of the calculated results.

The calculations are particularly convenient when the results of systematic calculations for a series of profile cascades or multiplanes with various geometrical parameters are available. In this case the aerodynamic characteristics of the transformed cascade or multiplane are obtained directly, and the calculation of the influence of the Mach number becomes elementary.

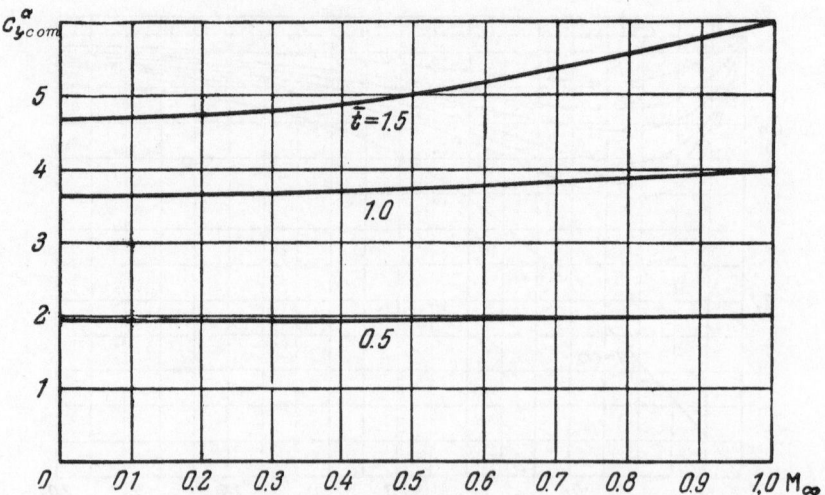

Fig. 10.20. The effect of M_∞ on the derivative c_y^α for a cascade of plates for various values of $\bar{t}$ and $\beta_\Gamma = 0$.

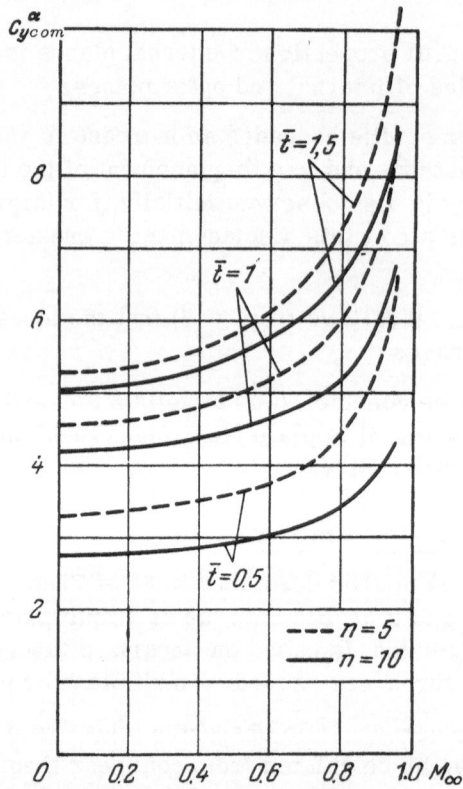

Fig. 10.21. The effect of M_∞ on the derivative c_y^α for $\beta_\Gamma = 0$, n = 5, n = 10 and various values of $\bar{t}$.

Figure 10.20 contains graphs showing the effect of the compressibility of air on the derivative c_y^α for a cascade of plates with a geometrical stagger $\beta_\Gamma = 0$ and various values of the relative step $\bar{t}$. When $M_\infty \to 1$, the values of c_y^α are calculated by using the asymptotic formula (9.4) and the relation (5.36).

Analagous graphs for multiplanes with n = 5 and n = 10 planes for the same values of $\bar{t}$ are shown in Fig. 10.21.

The influence of M_∞ becomes smaller when the disturbances due to the cascade or multiplane are weaker.

CHAPTER XI

ANNULAR WINGS

§ 1. Some Problems in Methods of Calculation

In the solution of the linearized problem for annular wings, the dependence of the circulation of the bound vortices on the angle φ of the cylindrical coordinate system is known exactly. It is thus necessary to study only one problem: How many of these vortices must be located along the chord of the wing to ensure the required accuracy in the calculated rotational-derivative coefficients?

Calculations were carried out for cylindrical annular wings of various elongations λ with from one to six bound vortices. The results are given in the following tables.

$\lambda = 0.5$

m	c_y^{α}	$m_z^{\alpha} = c_y^{\omega_z}$	$m_z^{\omega_z}$
1	1.52	0.380	0.095
2	1.53	0.530	—0.127
3	1.54	0.565	—0.178
4	1.54	0.578	—0.197
5	1.54	0.583	—0.205
6	1.54	0.586	—0.209

$\lambda = 1.0$

m	c_y^{α}	$m_z^{\alpha} = c_y^{\omega_z}$	$m_z^{\omega_z}$
1	2.87	0.718	0.179
2	2.89	0.886	—0.148
3	2.90	0.917	—0.210
4	2.90	0.926	—0.233
5	2.90	0.931	—0.243
6	2.90	0.933	—0.248

$\lambda = 1.5$

m	c_y^{α}	$m_z^{\alpha} = c_y^{\omega_z}$	$m_z^{\omega_z}$
1	3.96	0.990	0.247
2	3.98	1.13	—0.126
3	3.99	1.15	—0.197
4	3.99	1.16	—0.220
5	3.99	1.17	—0.232
6	3.99	1.17	—0.238

$\lambda = 2.0$

m	c_y^{α}	$m_z^{\alpha} = c_y^{\omega_z}$	$m_z^{\omega_z}$
1	4.80	1.20	0.300
2	4.82	1.31	—0.099
3	4.82	1.32	—0.173
4	4.83	1.33	—0.200
5	4.83	1.34	—0.212
6	4.83	1.34	—0.218

$\lambda = 2.5$

m	c_y^{α}	$m_z^{\alpha} = c_y^{\omega_z}$	$m_z^{\omega_z}$
1	5.46	1.36	0.341
2	5.47	1.45	—0.074
3	5.47	1.46	—0.151
4	5.47	1.47	—0.178
5	5.47	1.47	—0.191
6	5.47	1.47	—0.198

$\lambda = 3.0$

m	c_y^{α}	$m_z^{\alpha} = c_y^{\omega_z}$	$m_z^{\omega_z}$
1	5.98	1.49	0.374
2	5.99	1.56	—0.052
3	5.99	1.57	—0.132
4	5.99	1.58	—0.159
5	5.99	1.58	—0.172
6	5.99	1.58	—0.179

We see that the smallest number of bound vortices is required for the calculation of c_y^α and the greatest number for the calculation of $m_z^{\omega z}$. If we are interested only in the lift characteristics of annular wings, we need use only one or two bound vortices for any aspect ratio λ.

The consequences of the reversibility theorem (§3. Chapter VI) can be used as a second check of the correctness of calculations. In the present case, for a center $\overline{x}_T = 0.5$, this theorem yields

$$m_z^\alpha = c_y^{\omega z} \qquad (11.1)$$

for cylindrical annular wings. However the system of algebraic equations and formulas used in the calculation of m_z^α and $c_y^{\omega z}$ have the interesting property that for any number of vortices m the relation (11.1) is exactly satisfied and there is no need to check its accuracy.

There is still another method of checking calculations, based on the asymptotic formulas for c_y^α, m_z^α, $c_y^{\omega z}$ and $m_z^{\omega z}$ which hold when $\lambda \to 0$ (see §4, Chapter IX). We will use this last method below.

§2. Aerodynamic Characteristics of Annular Wings

Calculated rotational-derivative coefficients for cylindrical annular wings are shown graphically in Figs. 11.1-11.3. The only geometrical parameter of a cylindrical annular wing is the aspect ratio $\lambda = D/b$ (D is the diameter of the wing and b the chord). On all the graphs there is a dotted straight line emanating from the origin which gives the asymptotic solution for $\lambda \to 0$ [Eq. (9.37)].

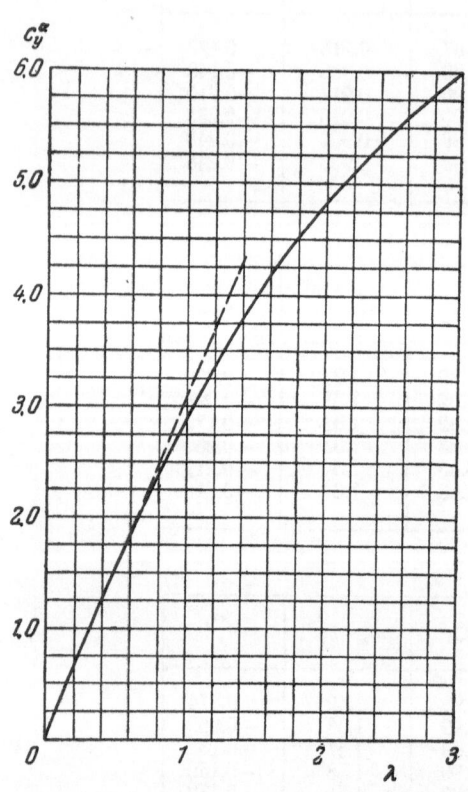

Fig. 11.1. Dependence of c_y^α on the aspect ratio of an annular wing.

The derivative c_y^α is independent of the position of the origin, and the values of the coefficients m_z^α, $c_y^{\omega z}$ and $m_z^{\omega z}$ are for the center $\overline{x}_T = 0.5$ and a standard coordinate system (see Fig. 8.3, for example).

An increase in the aspect ratio λ leads to a monotonic increase in the lift of an annular wing, and the derivative c_y^α increases with increasing λ. This is due, first, to the fact that a decrease in aspect ratio leads to an increase in the adverse effect of the mutual interaction between the lower part of the upper section of the wing and the upper part of the lower section (an effect qualitatively similar to that observed in multiplanes and profile cascades for small β_Γ). Second, a decrease of aspect ratio leads to an increase in the downwash due to the free vortices on the wing surface, which leads on the whole to a decrease of the angle of attack and to a consequent decrease of lift.

The quantity $m_z^{\omega z}$ first increases with increasing λ and then decreases; it has a maximum for a value of λ close to 1 (for a given center). This type of relation is understandable if we recall that for the center $\overline{x}_T = 0.5$ we have $m_z^{\omega z} = 0$ for a plate of infinite span, and the characteristics of an annular wing of large aspect ratio are $\pi/2$ times those of a plate of infinite span (§2, Chapter IX).

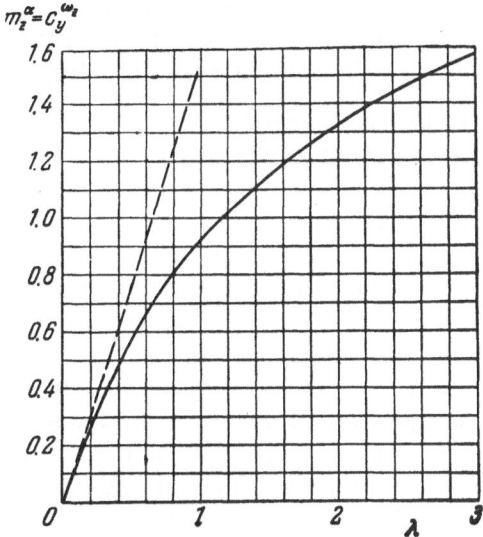

Fig. 11.2. Dependence of $m_z^\alpha = c_y^{\omega_z}$ on the aspect ratio of an annular wing.

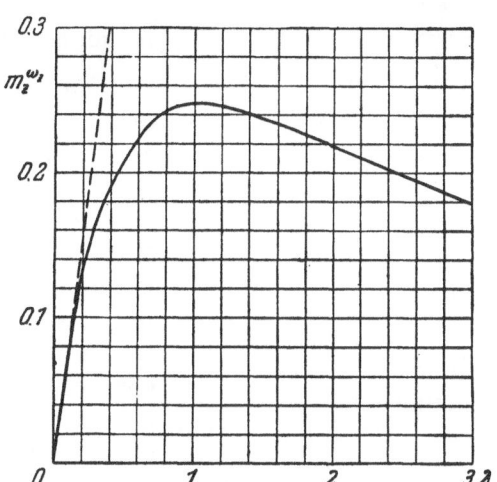

Fig. 11.3. Dependence of $m_z^{\omega z}$ on the aspect ratio of an annular wing.

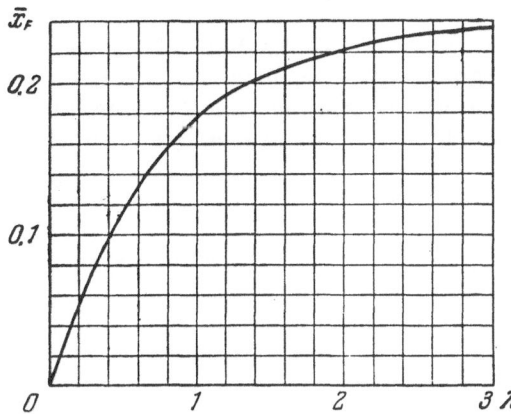

Fig. 11.4. The variation of the coordinate of the focus of an annular wing as a function of the aspect ratio λ.

The coordinate of the focus F of a cylindrical annular wing can be found by using the graphs in Fig. 11.4. Here we use the usual notation: $\bar{x}_F = x_F/b$, where x_F is the distance from the section $x = x_F$ to the section passing through the leading edge of the wing ($x = 0.5b$ in this coordinate system). The focus of an annular wing of very small aspect ratio lies in the plane of the leading edges ($\bar{x}_F = 0$). When λ increases it moves back, and when $\lambda \to \infty$ its coordinate tends to $\bar{x}_F = 0.25$.

The information given is sufficient to determine the over-all characteristics of an annular wing for arbitrary motion, including curvilinear motion.

Since rotation of this type of wing about the Ox axis does not generate any disturbances in the medium, we can solve the problem for:

1) rectilinear motion of an annular wing ($\alpha = $ const, $\omega_z = 0$);
2) circular motion for $\alpha = 0$ and $\omega_z = $ const;
3) combinations of these types of motion.

We should stress that the information we have given is not sufficient for an investigation of a combination of translational motion with $U_0 = $ const and rotational motion, or translational motion with $\alpha \neq $ const [31, 32].

§3. A Comparison of Annular Wings and Rectangular Wings with the Same Over-All Dimensions

In many practical problems it is often important or even essential to compare different airfoils with the same over-all dimensions. In this case the optimum airfoil among all airfoils

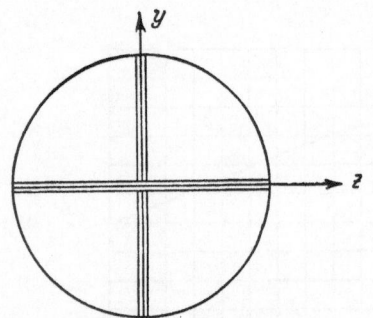

Fig. 11.5. Annular and cruciform wings with the same
over-all dimensions.

with the same over-all dimensions is that with the best aerodynamic characteristics, and it is most important to obtain
the highest lift possible and the best possible lift–drag ratio.

We will compare the main aerodynamic characteristics
of annular wings and rectangular wings with the same over-
all dimensions, i.e., we will assume that the chords of all the
wings under consideration are identical and the span of the
rectangular wing is equal to the diameter D of an annular wing.
It is useful to consider, in addition to rectangular wings,
cruciform wings with the same over-all dimensions consisting of two similar rectangular wings at right angles (Fig. 11.5).
There is very little difference between the aerodynamic
characteristics of cruciform and annular wings, and so a
comparison of these wings is quite natural. In the absence
of slip only one plane of a cruciform wing will generate lift,
while both will generate drag. Hence there is no difference between rectangular and cruciform
wings relative to c_y^α, m_z^α, $c_y^{\omega_z}$ and $m_z^{\omega_z}$.

It was established in §2 of Chapter IX that, for very large aspect ratio λ, the values of all
these parameters for annular wings are $\pi/2$ times their values for rectangular wings. According to §4 and §5 of Chapter IX, this ratio becomes 2 for very small values of λ.

By using numerical results for intermediate values of λ, we can obtain graphs of the ratios of corresponding coefficients for annular and rectangular wings as functions of λ (Fig. 11.6).
One important conclusion that can be drawn from these graphs is that the foci F of annular and
rectangular (or cruciform) wings are close to one another for identical values of λ. Linear
theory shows that they tend to coincidence when $\lambda \to 0$ and $\lambda \to \infty$, and their positions differ
somewhat for intermediate cases.

We now compare the maximum lift-drag ratio k_{max} of annular, rectangular, and cruciform wings. If

$$c_x = c_{x0} + Ac_y^2,$$

it follows from (4.58) and (4.60) that

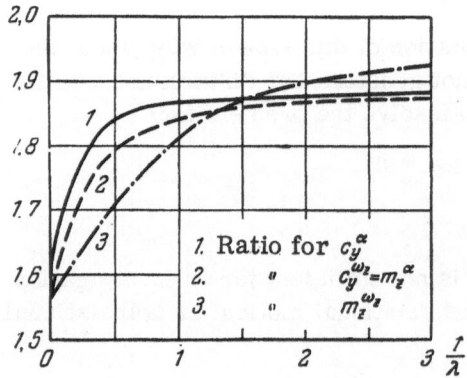

Fig. 11.6. Ratios of annular-wing coefficients to the same coefficients
for a cruciform wing with the same
over-all dimensions

$$k_{max} = \frac{1}{2} \sqrt{\frac{1}{Ac_{x0}}}.$$

For a profile, we denote by $c_{x\infty}$ the value of
that part of the drag that is independent of c_y,
and we assume that the profile is the same for
all wings. Then we can assume that c_{x0} for a
rectangular wing is approximately equal to $c_{x\infty}$
for a cruciform wing (twice as large as for an
annular wing) and π times $c_{x\infty}$ for a rectangular
wing.

From (4.34) and (4.17), the coefficient A is
equal to $1/2\pi\lambda$ for an annular wing and equal to
$B/\pi\lambda$ for a for a monoplane wing if the wings are
rounded and if their leading edges generate a
suction force.

In the case of sharp leading edges, which do not generate any suction force, A is equal to $1/c_y^\alpha$ for all wings.

When there is no suction force we thus have for annular, rectangular, and cruciform wings respectively the formulas

$$k_{\max 1} = \frac{1}{2}\sqrt{\frac{2\lambda}{c_{x\infty}}}, \quad k_{\max 2} = \frac{1}{2}\sqrt{\frac{\pi\lambda}{Bc_{x\infty}}}, \quad k_{\max 3} = \frac{1}{2}\sqrt{\frac{\pi\lambda}{2Bc_{x\infty}}}. \tag{11.2}$$

When suction force exists we have

$$\bar{k}_{\max 1} = \frac{1}{2}\sqrt{\frac{c_{y1}^\alpha}{\pi c_{x\infty}}}, \quad \bar{k}_{\max 2} = \frac{1}{2}\sqrt{\frac{c_{y2}^\alpha}{c_{x\infty}}}, \quad \bar{k}_{\max 3} = \frac{1}{2}\sqrt{\frac{c_{y3}^\alpha}{2c_{x\infty}}}. \tag{11.3}$$

The relations (11.2) and (11.3) are useful in clarifying the role of suction force in raising $k_{\max}$, in carrying out comparisons between the maximum lift-drag ratios of different airfoils, etc.

For example for a comparison of the maximum lift-drag ratios of annular and cruciform wings without suction force we have

$$\frac{\bar{k}_{\max 1}}{\bar{k}_{\max 3}} = \sqrt{\frac{c_{y1}^\alpha}{c_{y3}^\alpha}\frac{2}{\pi}}. \tag{11.4}$$

We thus conclude that when $\lambda \to 0$, which leads to $c_{y1}^\alpha/c_{y3}^\alpha \to 2$, this ratio tends to $\sqrt{4/\pi}$; when $\lambda \to \infty$, which leads to $c_{y1}^\alpha/c_{y3}^\alpha \to \pi/2$, we obtain $\bar{k}_{\max 1}/k_{\max 3} \to 1$.

We should note that all the above analysis refers to isolated wings. No account has been taken in it of either the mutual interaction between a wing and the body or of the effect of the bracket by which an annular wing is fixed to the body.

§4. The Effect of the Mach Number

By using the method described in §3, Chapter V, we can easily convert the rotational-velocity coefficients of an annular wing of any aspect ratio to values for high subsonic velocities. The results of the previous section permit this to be done rather simply. When $\mathbf{M}_\infty \to 1$ and consequently the reduced elongation $\lambda_M = \lambda\sqrt{1-\mathbf{M}_\infty^2}$ tends to zero, the limiting equations of

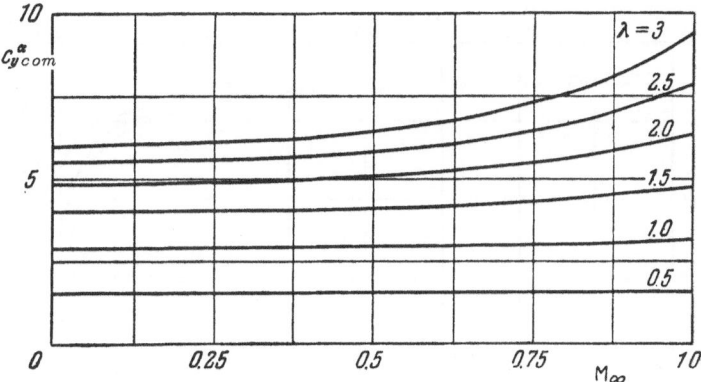

Fig. 11.7. The influence of the value of $\mathbf{M}_\infty$ on c_y^α for annular wings of various aspect ratios.

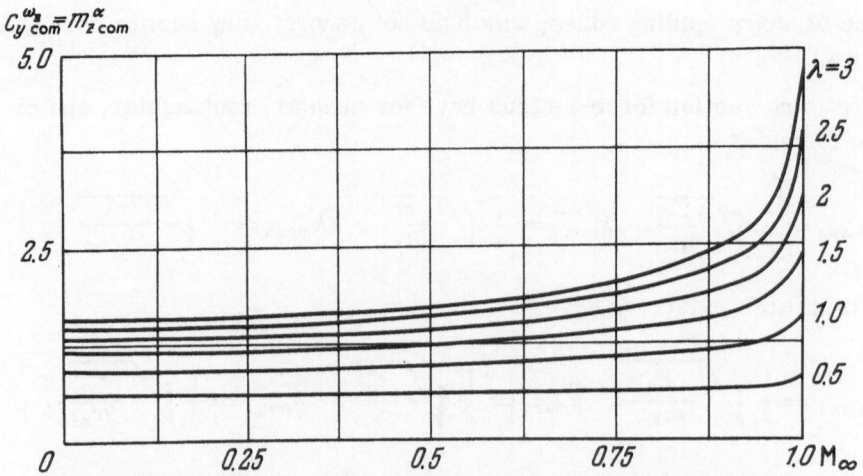

Fig. 11.8. The influence of the value of $\mathbf{M}_\infty$ on the coefficient $m_z^\alpha = c_y^{\omega z}$ for annular wings of various aspect ratios.

§ 4, Chapter IX come into play. Figures 11.7–11.9 show the dependence of c_y^α, $m_z^\alpha = c_y^{\omega z}$, and $m_z^{\omega z}$ on $\mathbf{M}_\infty$ for various aspect ratios λ of an annular wing for fixed center $(\overline{x}_T = 0.5)$. These coefficients were obtained for $\mathbf{M}_\infty = 1$ from the asymptotic formulas.

All the rotational-derivative coefficients increase when the Mach number increases, and the relative rate of increase becomes greater for larger aspect ratios. For wings of very small elongation ($\lambda = 0.5$), the effect of the value of $\mathbf{M}_\infty$ is negligible. This is to be expected, since the effect of air compressibility for a given $\mathbf{M}_\infty$ will become weaker when the disturbances due to the body become weaker. A decrease in aspect ratio when the transverse dimensions of the body decreases relative to the longitudinal dimensions leads to a weakening of these disturbances.

The dimensionless coordinate of the focus of an annular wing in a compressible flow is

$$\overline{x}_{F\,com} = \overline{x}_{FM},$$

where $\overline{x}_{FM}$ is obtained for a transformed wing with aspect ratio $\lambda_M = \lambda \sqrt{1 - \mathbf{M}_\infty^2}$ in an incompressible medium.

As has already been noted, the focus moves forward towards the leading edge of the wing when the aspect ratio decreases (Fig. 11.4). Hence from the linear theory, when $\mathbf{M}_\infty$ increases at subsonic velocities the focus moves forward. The calculations are left to the reader.

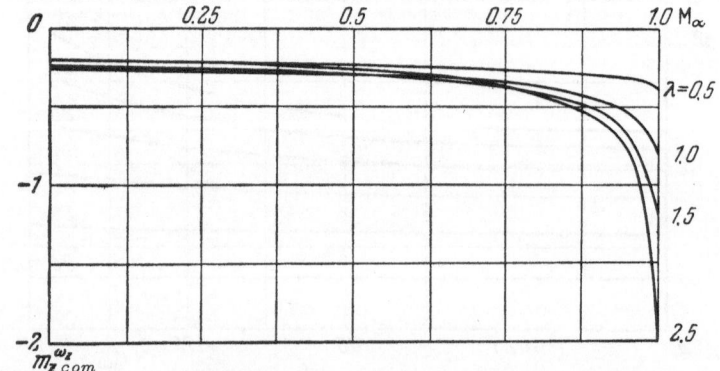

Fig. 11.9. The effect of the value of $\mathbf{M}_\infty$ on the coefficient $m_z^{\omega z}$ for annular wings of various aspect ratios.

§ 5. A Comparison of Some Calculated and Experimental Results

It is very important to determine how well the linear theory considered above, based on an airfoil scheme, is confirmed by experiment. If the calculations are checked mathematically and their accuracy is satisfactorily verified, such a comparison might yield evidence of the validity of the hypotheses on which the theory is based. Such comparisons would also indicate the range of applicability of the theory.

It should be noted that experimental results should also not be considered to be absolutely reliable. On the one hand there are many sources of experimental errors which are very difficult to estimate, especially when it is a question of the effect of the various types of support used for holding a model. On the other hand it is difficult to model any phenomenon completely accurately in a wind tunnel, especially for large angles of attack when separation, depending on the structure of the boundary layer, plays a decisive role. A simultaneous analysis of both theoretical and experimental data is therefore often very fruitful. In such an analysis particular attention should be focused on regions in which the results disagree and there is no known reason for this disagreement.

Below we compare the theoretical results obtained above with experimental results obtained by L. N. Kravchenko and the author for low Mach numbers. The models used in the experiments had various relative thicknesses (from 1% to 3%), since annular wings of high aspect ratio were obtained by truncating part of a wing of lower aspect ratio. The leading edges of all wings were rounded. The moment characteristics of annular wings are given for the case when the Oz axis is in a plane passing through the leading edge.

Figures 11.10–11.12 contain graphs of experimental results obtained for cylindrical annular wings ($\alpha_0 = 0$) of various aspect ratios. Calculated results are plotted in the same

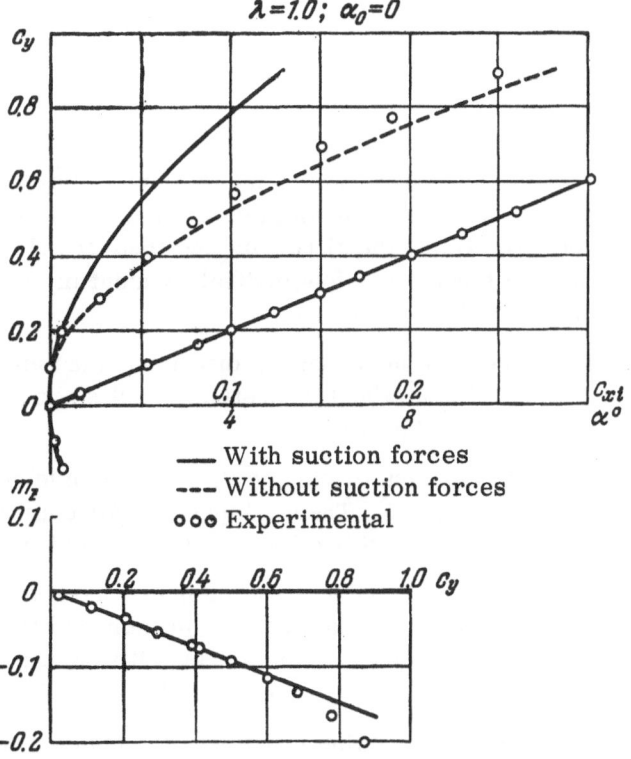

Fig. 11.10. A comparison of theoretical and experimental results for a cylindrical annular wing with $\lambda = 1.0$.

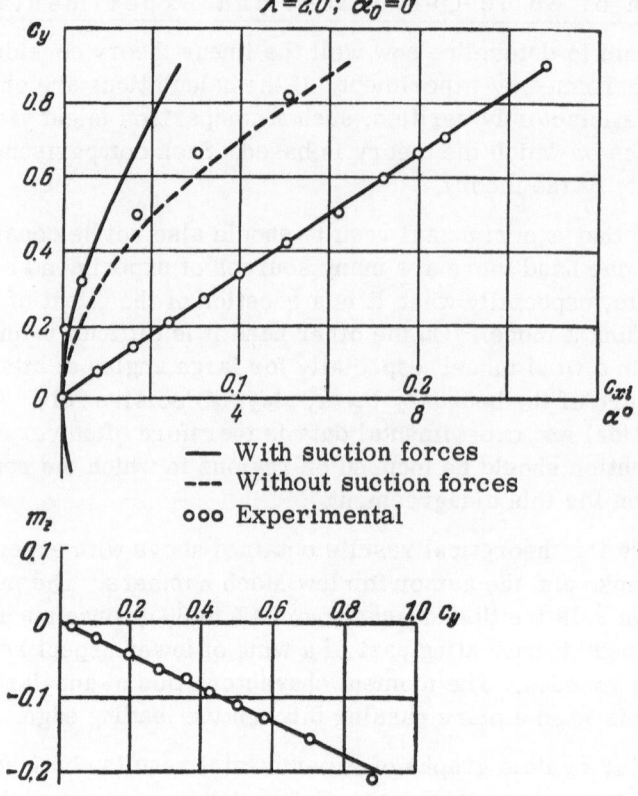

Fig. 11.11. A comparison of theoretical and experimental results for a cylindrical annular wing with $\lambda = 2.0$.

diagrams, the induced drag being calculated from two formulas — one taking account of and the other not taking account of the suction force. The smaller the relative thickness $\bar{c}$, the better should be the agreement between experimental results and theoretical results obtained without taking account of the suction force. The longitudinal moment m_z is theoretically independent of the magnitude of the suction force when $\alpha_0 = 0$.

The graphs confirm our position and show that the functions $c_y(\alpha)$ and $m_z(\alpha)$ are linear for low angles of attack, and the experimental and theoretical results are in good agreement in the range of angles of attack under consideration.

The influence of the taper α_0 of an annular wing on its aerodynamic characteristics can be judged from the graphs in Fig. 11.13-11.15. These graphs are for conical annular wings of the same aspect ratio with $\alpha_0 = -2°$ (the vertex of the cone is in front, see Fig. 11.16).

The conical shape has the greatest influence on the moment characteristics. To explain the results we consider Fig. 11.16. The cone angle α_0 influences both the moment due to the normal forces and the moment due to the suction forces. It is clear from (4.27) that the latter coefficient is equal to $(\alpha_0 \lambda /2) c_y$, i.e., it is negative for $\alpha_0 < 0$.

We consider the forces (including the suction forces) on the upper section (the meridional half-plane $\varphi = 0$) and the lower section ($\varphi = \pi$) of a conical wing. In comparison with a cylindrical wing with the same aspect ratio, there is here an extra negative moment due to the longitudinal suction pressures c_t^l.

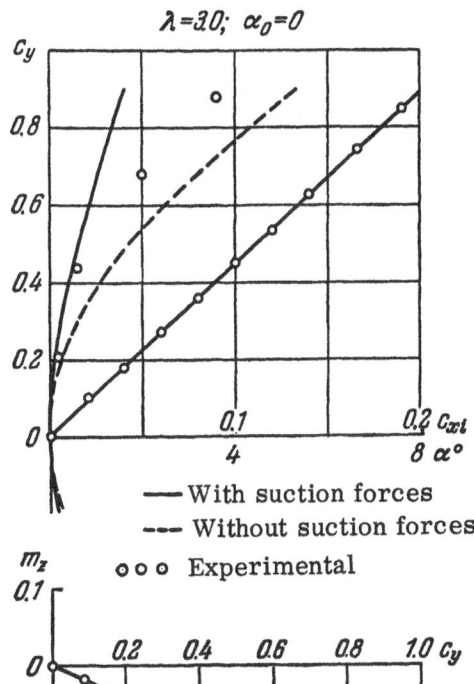

Fig. 11.12. A comparison of theoretical and experimental results for a cylindrical annular wing with $\lambda = 3.0$.

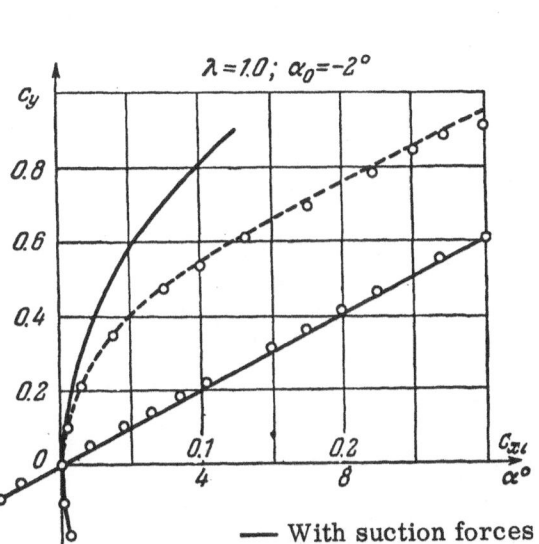

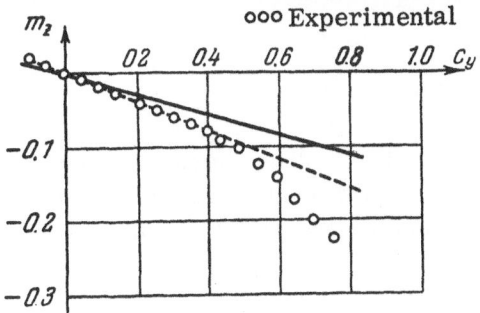

Fig. 11.13. A comparison of theoretical and experimental results for a conical annular wing with $\lambda = 1.0$.

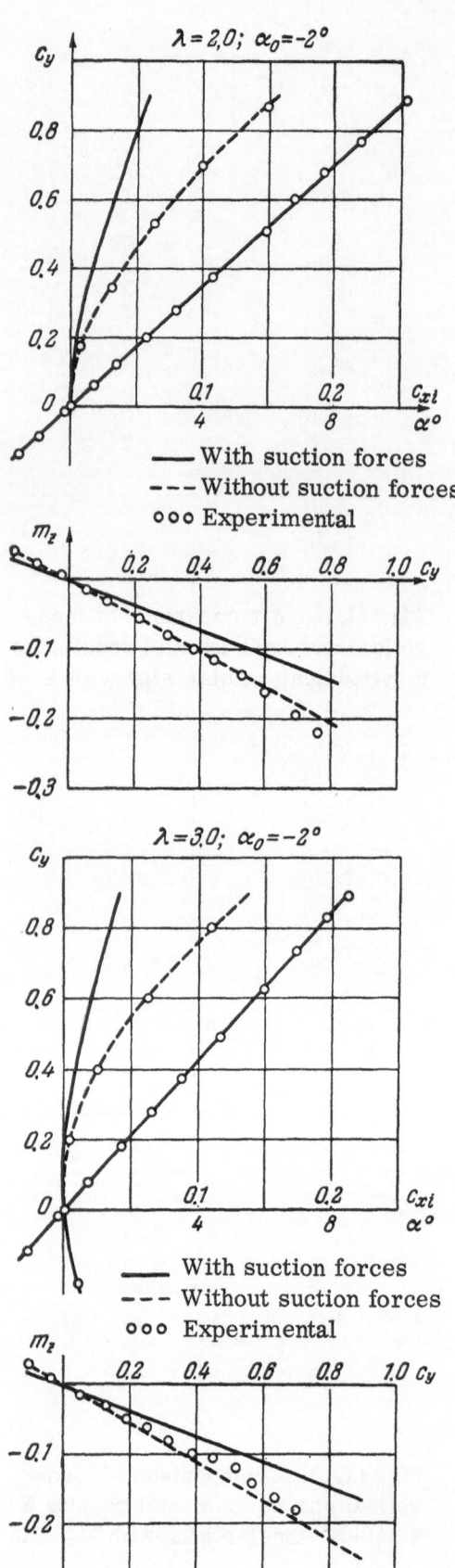

Fig. 11.14. A comparison of theoretical and experimental results for a conical annular wing with $\lambda = 2.0$.

Fig. 11.15. A comparison of theoretical and experimental results for a conical annular wing with $\lambda = 3.0$.

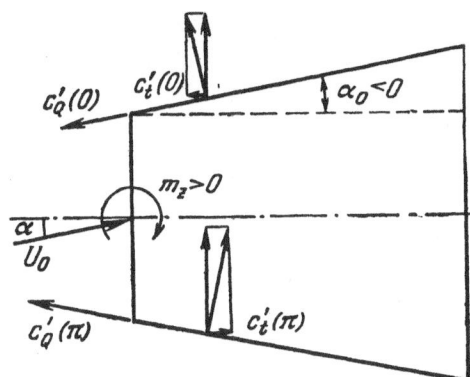

Fig. 11.16. The role of tangential and suction forces in the generation of moment in the case of a conical annular wing.

Suction forces also modify the longitudinal moment of a wing; this follows from (4.43). For $\alpha_0 < 0$ and $\alpha_0 > 0$, it follows from Fig. 11.16 that the angle of attack of the lower section is larger than that of the upper section. Hence the suction forces will be greater on the lower section, and these forces generate supplementary positive moment.

The opposite effect will naturally result from taper of the opposite sign.

It is clear from Figs. 11.13-11.15 that the total effect of the suction forces is practically never felt, and experimental results for moderate angles of attack are in good agreement with theoretical results obtained without taking account of suction forces.

CHAPTER XII

Monoplane Wings

§ 1. Some Problems in Methods of Calculation

A detailed investigation into the development of a rational scheme for the calculation of rotational-derivative coefficients of monoplane wings was carried out by E. M. Moiseev.

It was established that the set of algebraic equations to which the problem reduces has a very important property — it is very stable. This is because the elements of the principal diagonal of the coefficient matrix exceed the off-diagonal elements in absolute value, since the highest velocities are generated by the vortex directly in front of the points at which calculations are performed.

The choice of the necessary number n of vortices on each chord and N of vortices along the half-span for each attached vortex filament was also considered. If the calculations are assumed to be correct, an increase in n and N will raise the accuracy of the results. In practice we can select values of n and N such that any increase beyond these values will have no effect on the numerical results from the point of view of the required accuracy. Such an analysis is conveniently carried out graphically with n (for N = const) and N (for n = const) as the arguments and rotational-derivative coefficients as the functions. The fact that these graphs will be somewhat formal, since n and N can only take integral values, has already been noted.

We now analyze the effect of the values of n and N on the coefficients of four wings — two swept-back wings (with low and high elongation) and two straight wings (also with low and high elongation). The geometrical parameters of these wings are given in the following table.

Wing number	χ_n	λ	η
1	0	1.0	1
2	60°	1.5	5
3	45°	5.0	2
4	0	5.0	1

The results obtained for these wings are presented graphically in Figs. 12.1–12.6, and these graphs can be used to decide on a rational vortex scheme for the wings.

It is clear from Figs. 12.1–12.4 that in the calculation of the derivatives c_y^α and c_z^α there is no need to have a large number of vortices along the chord (2–4 are usually sufficient), but a fairly large number of vortices are needed along the half-span (N = 15–20).

We conclude from Fig. 12.5 and 12.6 that a similar but even stronger conclusion can be drawn concerning the calculation of $m_z^{\omega z}$.

171

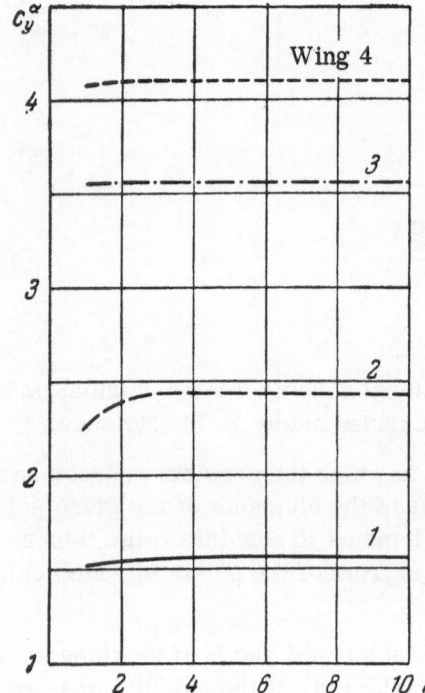

Fig. 12.1. The effect of the number of vortices n along the chord on the value of c_y^α (N = const).

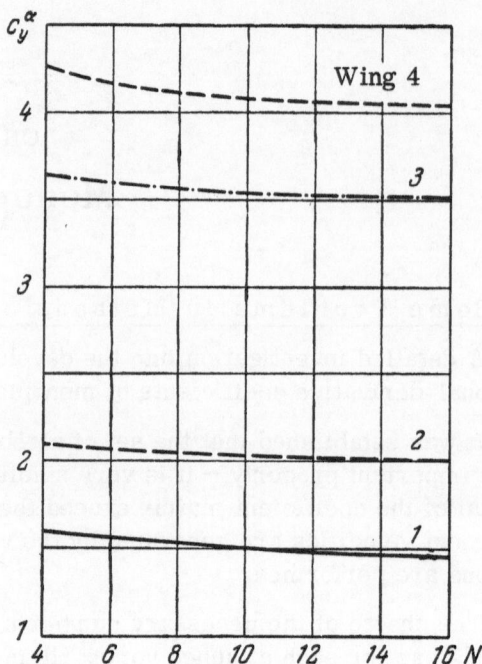

Fig. 12.2. The effect of the number of vortices N along the half-span (n = const).

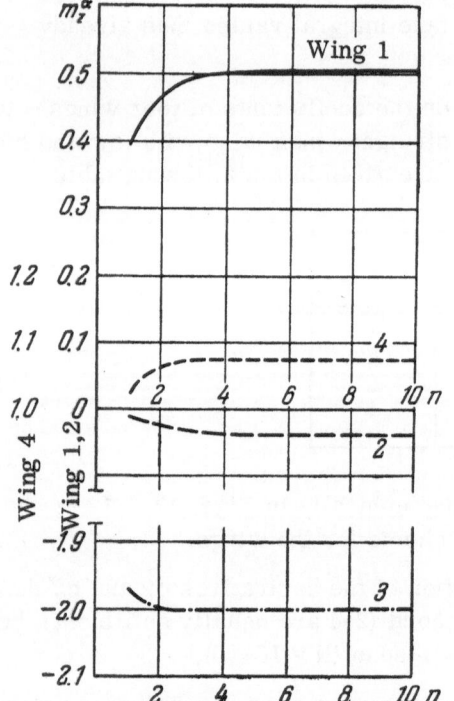

Fig. 12.3. The effect of the number of vortices n along the chord on the value of m_z^α (N = const).

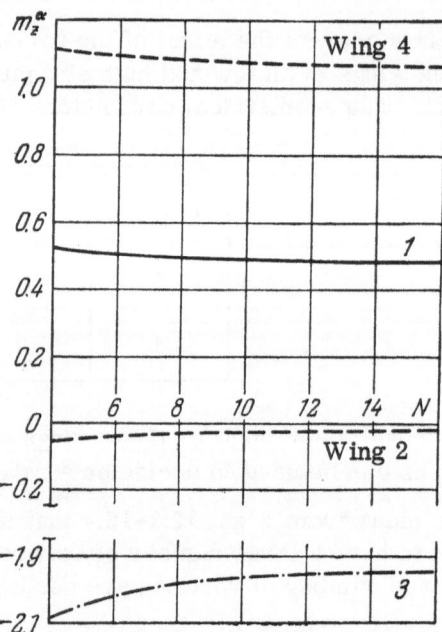

Fig. 12.4. The effect of the number of vortices N along the half-span (n = const).

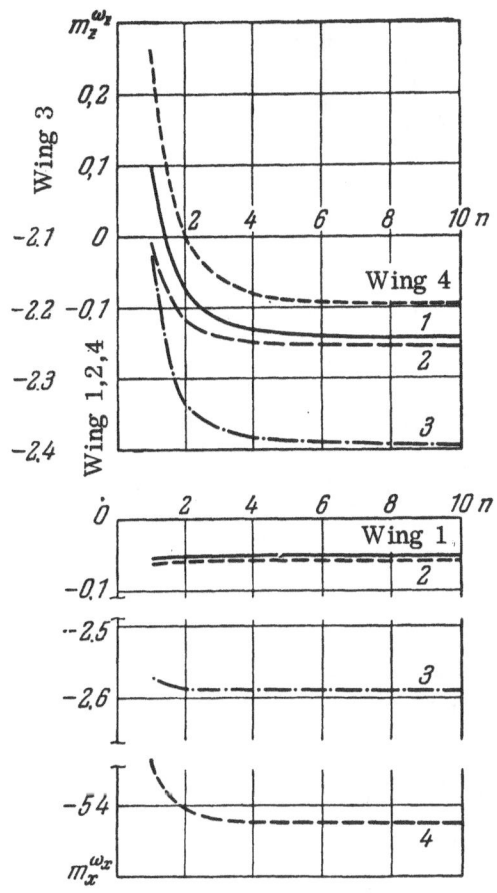

Fig. 12.5. The effect of the number of vortices n along the chord for N = const.

The most unfavorable case, in which there must be a large number of vortices both along the chord and along the span, is that of the calculation of $m_z^{\omega z}$ for a swept-back wing of high aspect ratio.

A good check on the accuracy of calculations can be obtained from the reversibility theorem. For rectangular wings with center $x_T = 0.5$ we have

$$m_z^a = c_y^{\omega z}.$$

However, in view of certain features of the equations and formulas used for the calculation of rotational-derivative coefficients, this relation for rectangular wings holds exactly for any values of n and N and it cannot be used as a check.

In systematic calculations for other wings, the equations of §3, Chapter VI, are often used to determine the errors in calculated results and can thus also be used to judge on the satisfactory choice of n and N. Figures 12.7-12.10 compare corresponding coefficients for direct and reversed triangular wings of various elongation with centering $\overline{x}_T = 0.5$. The continuous curves are for direct triangular wings (with sweepback relative to the trailing edge $\chi_1 = 0$), the dotted curves are for reversed triangular wings.

§ 2. Rotational-Derivative Coefficients and the Focus

Using the method described above, E. M. Moiseev, V. G. Tabachnikov, and the author employed an electronic computer for the systematic calculation of aerodyanmic characteristics of monoplane wings of various shapes with straight edges. We present below some of the results.

For most of the coefficients, it is convenient to use the mean aerodynamic chord of the wing as a characteristic linear dimension. This makes the dependence of the coefficients on the geometrical parameters of the wing more stable and causes it to approach the dependence for rectangular wings. It is also most often used in practice. The only exception to this rule is the coefficient $m_{x1}^{\omega x1}$, for which the characteristic dimension is related to the span l.

It is thus convenient to write the aerodynamic coefficients and the rotational-derivatives as follows:

$$\left.\begin{aligned}
c_y &= \frac{Y}{qS}, & m_{za} &= \frac{M_{za}}{qSb_a}, & m_{x1} &= \frac{M_x}{qSl}, \\
c_y &= c_y^a \alpha + c_{ya}^{\omega za} \omega_{za}, & m_{za} &= m_{za}^a \alpha + m_{za}^{\omega za} \omega_{za}, \\
m_x &= m_{x1}^{\omega x1} \omega_{x1}, & \omega_{za} &= \frac{\Omega_z b_a}{U_0}, & \omega_{x1} &= \frac{\Omega_x l}{2U_0}.
\end{aligned}\right\} \qquad (12.1)$$

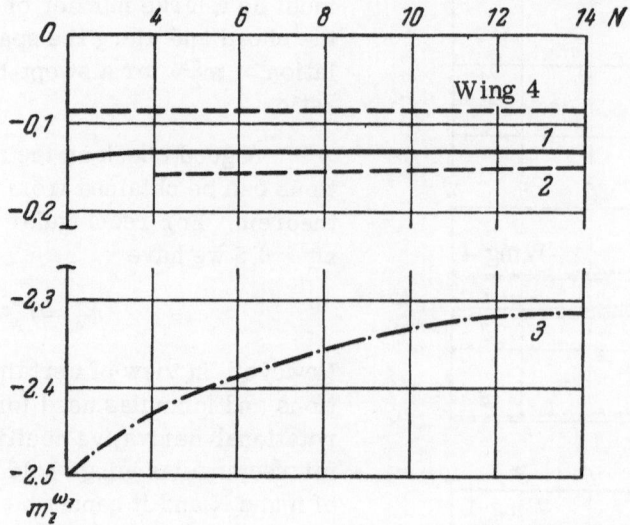

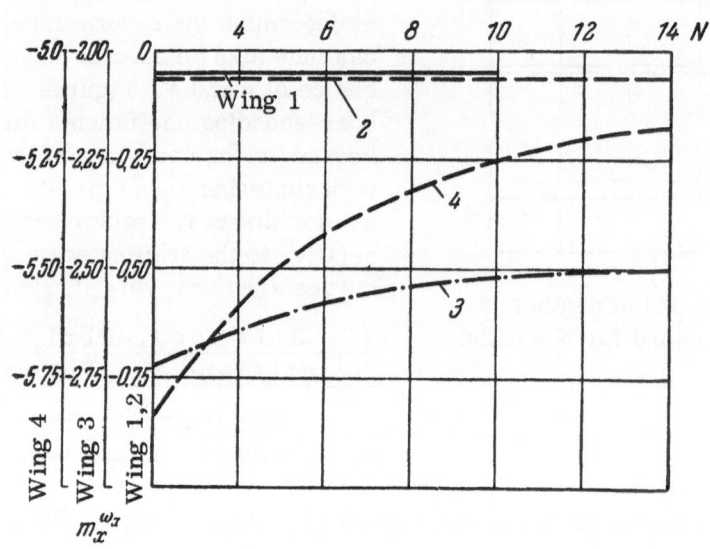

Fig. 12.6. The effect of the number of vortices N along
the half-span for n = const.

Here S is the area of the wing, b_a is the mean aerodynamic chord, l is the span, and M_{za} is the aerodynamic moment relative to the Oz axis, which passes through the leading edge of the mean aerodynamic chord.

If we need to convert to other characteristic dimensions, we can use the method described in §6, Chapter IV. For example the relation between $m_{x1}^{\omega x_1}$ and $m_x^{\omega x}$ is

$$m_{x1}^{\omega x_1} = m_x^{\omega x} \frac{8}{\lambda^2} \frac{1}{\left(1 + \frac{1}{\eta}\right)^2}.$$

(12.2)

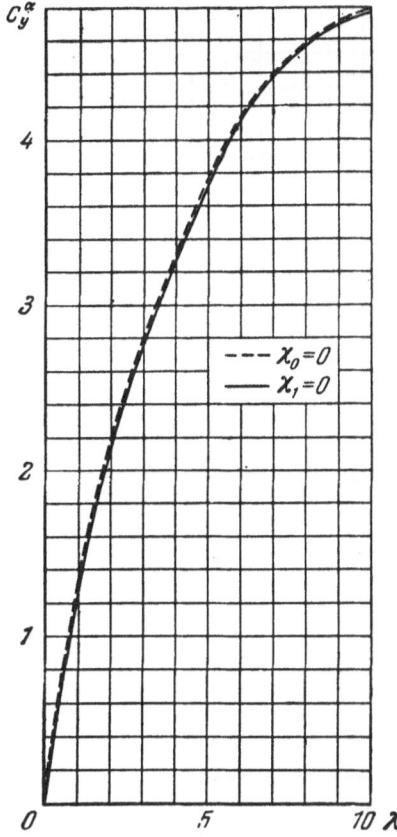

Fig. 12.7. A comparison of direct and reversed triangular wings.

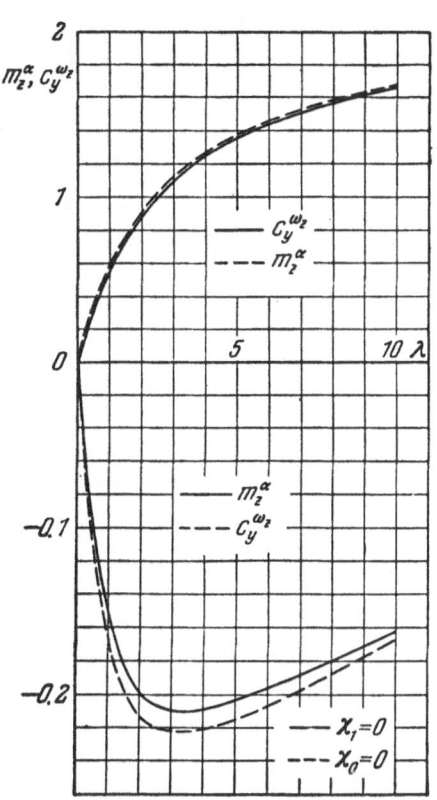

Fig. 12.8. A comparison of direct and reversed triangular wings.

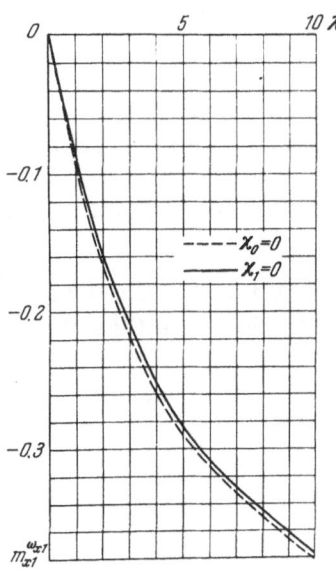

Fig. 12.9. A comparison of direct and reversed triangular wings.

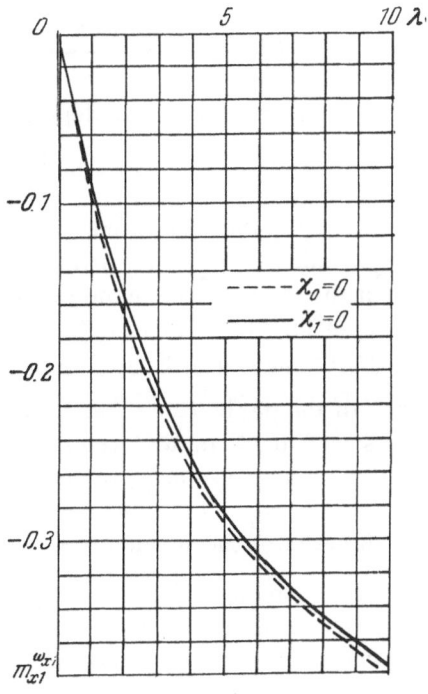

Fig. 12.10. A comparison of direct and reversed triangular wings.

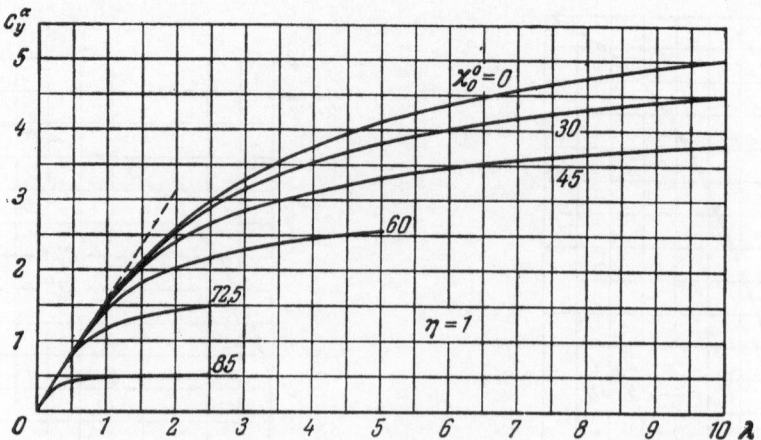

Fig. 12.11. The dependence of c_y^α on the aspect ratio of monoplane wings for $\eta = 1$.

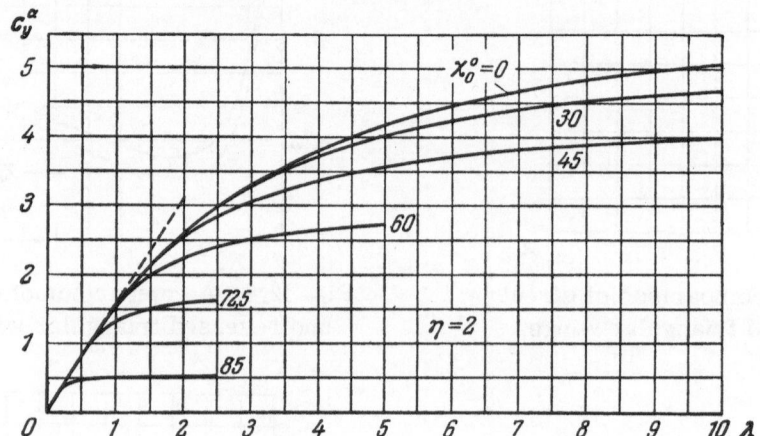

Fig. 12.12. The dependence of c_y^α on the aspect ratio of monoplane wings for $\eta = 2$.

To convert to other positions for the Oz axis parallel to the original direction, we use (4.75) for $c_y^{\omega z}$, m_z^α and $m_z^{\omega z}$, recalling that the distance h_a from the leading edge of the wing to the nose of the mean aerodynamic chord can be obtained from (4.63). For this change of the Oz axis the coefficients c_y^α and $m_{x_1}^{\omega x}$ remain unchanged. The mean aerodynamic chords for several wings are shown in Figs. 1.3-1.5.

Values of the derivative c_y^α as a function of the elongation λ for various relative reductions η and sweep-back angles χ_0 relative to the leading edge are plotted in Figs. 12.11-12.14. We do not give results for m_{za}^α; instead of this Figs. 12.15-12.18 contain graphs of the dimensionless coordinate $\bar{x}_F$ — the distance from the leading edge of the mean aerodynamic chord to the focus F in units of the mean aerodynamic chord.

The corresponding relations for the rotational-derivative coefficients $c_{ya}^{\omega za}$, $m_{za}^{\omega za}$ and $m_{x_1}^{\omega x_1}$ are plotted in Figs. 12.19-12.22, 12.23-12.26, and 12.27-12.30 as functions of the geometrical parameters of monoplane wings.

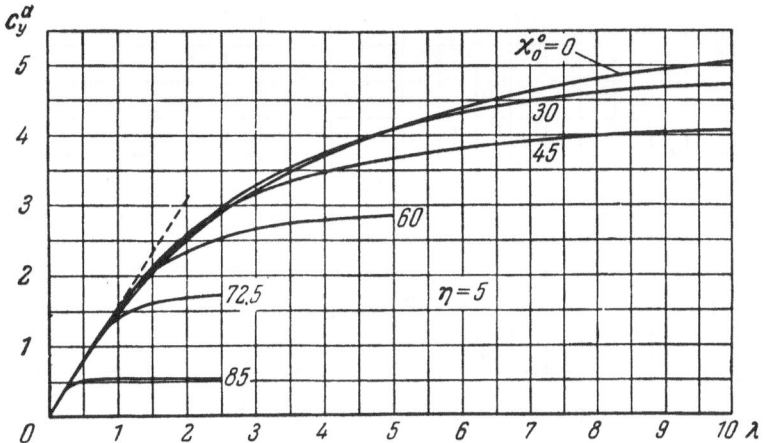

Fig. 12.13. The dependence of c_y^α on the aspect ratio of monoplane wings for $\eta = 5$.

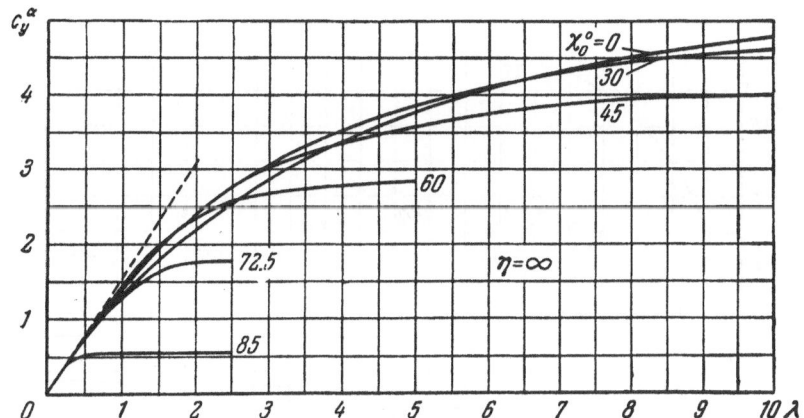

Fig. 12.14. The dependence of c_y^α on the aspect ratio of monoplane wings for $\eta = \infty$.

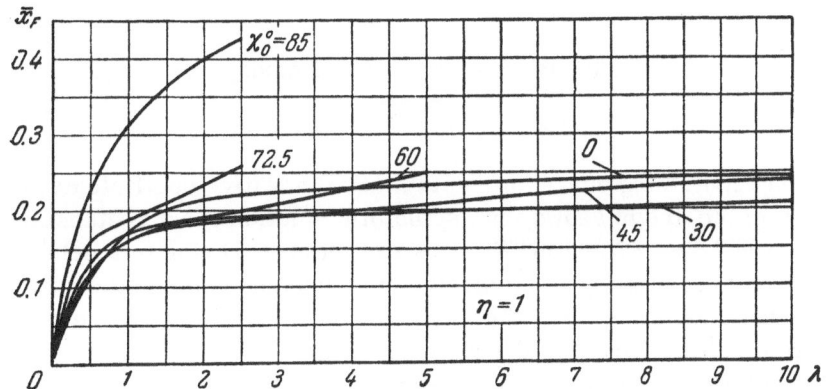

Fig. 12.15. The position of the focus of monoplane wings for $\eta = 1$.

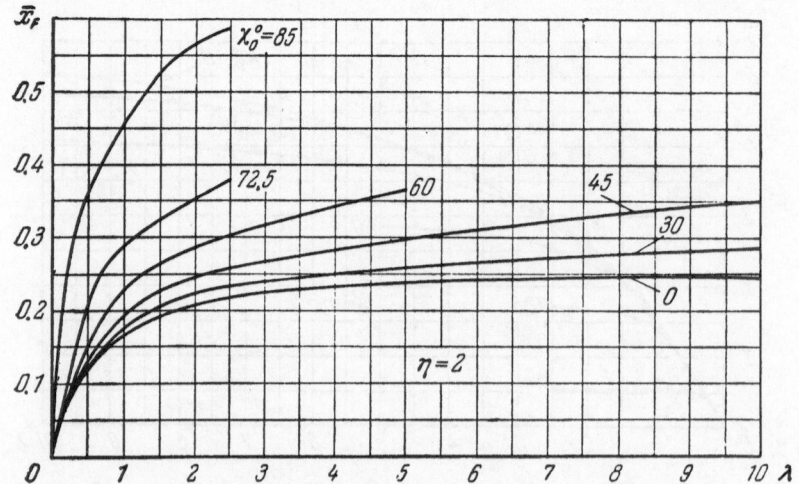

Fig. 12.16. The position of the focus of monoplane wings for
$\eta = 2$.

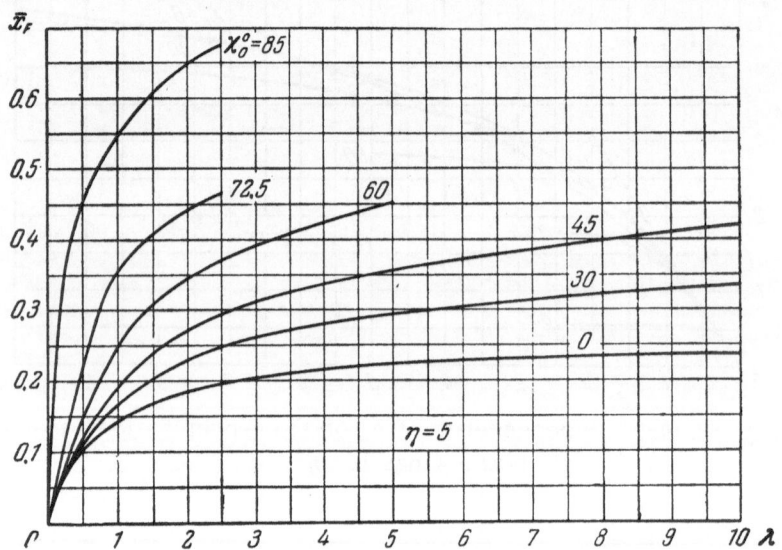

Fig. 12.17. The position of the focus of monoplane wings for
$\eta = 5$.

On all graphs containing curves for rotational-derivative coefficients, the dotted straight line emanating from the origin shows the asymptotic solution for a rectangular wing of very small elongation, $c_{ya}^{\omega za} = \pi\lambda/2$ and $m_{za}^{\omega za} = -\pi\lambda/4$, corresponding to a centering $\overline{x}_T = 0$. The results given are both interesting in themselves and useful in some other directions. For example, when we know $c_y^{\omega z}$ for a wing, then we can find m_z^{α} by using the reversibility theorem, and then the position of the focus for the reversed wing. Thus Figs. 12.19-12.22 can be used to determine the position of the focus for a series of wings with reversed (negative) sweepback relative to the leading edge. The results shown in Figs. 12.11-12.30 can also be used in analyses of the influence of geometrical parameters on the aerodynamic characteristics of wings; in the explanation of certain properties of the aerodynamics of wings of various shapes; etc.

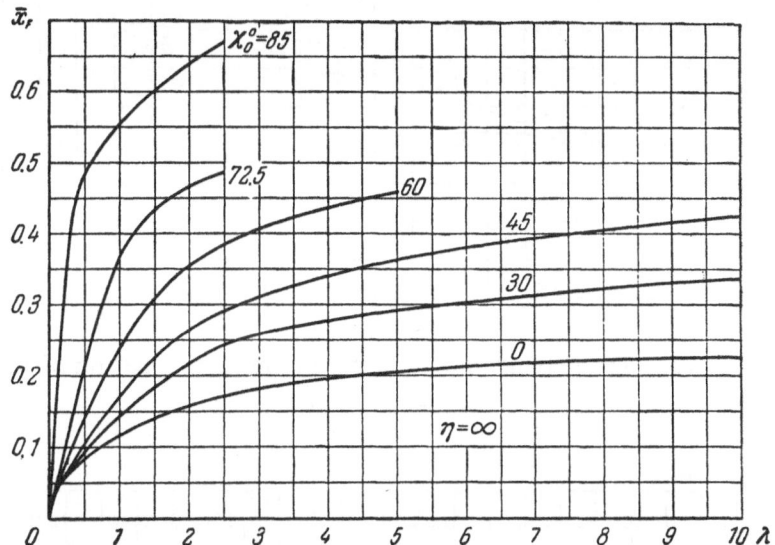

Fig. 12.18. The position of the focus of monoplane wings for $\eta = \infty$.

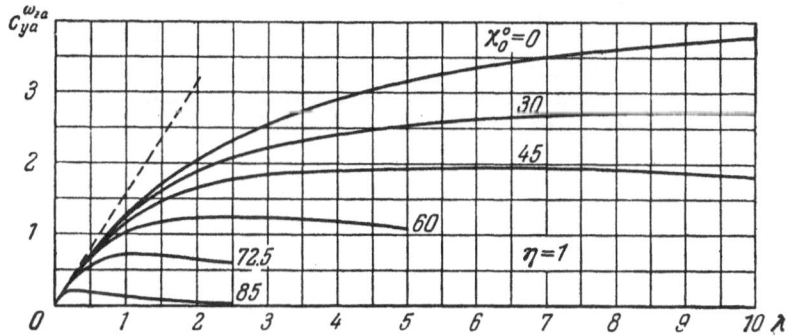

Fig. 12.19. The dependence of $c_{ya}^{\omega za}$ on the aspect ratio of monoplane wings for $\eta = 1$.

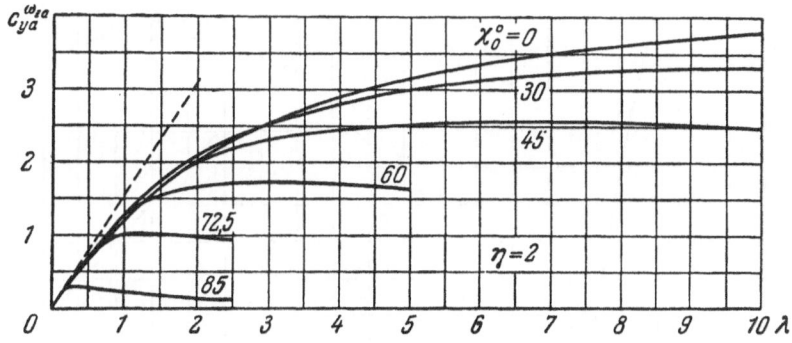

Fig. 12.20. The dependence of $c_{ya}^{\omega za}$ on the aspect ratio of monoplane wings for $\eta = 2$.

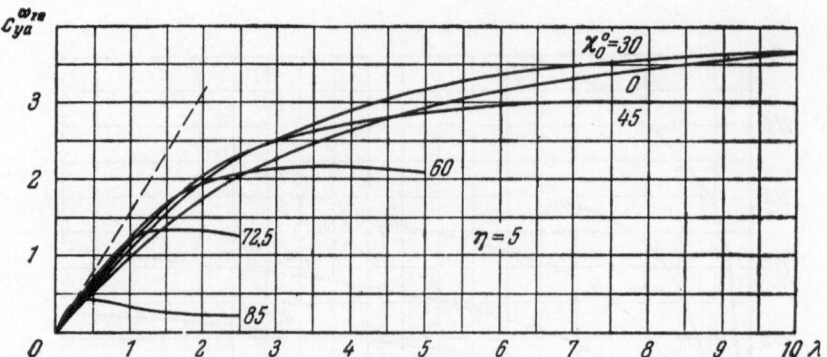

Fig. 12.21. The dependence of $c_{ya}^{\omega za}$ on the aspect ration of monoplane wings for $\eta = 5$.

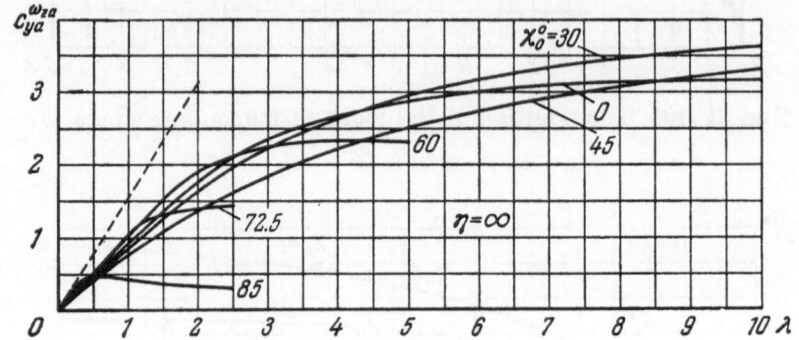

Fig. 12.22. The dependence of $c_{ya}^{\omega za}$ on the aspect ratio of monoplane wings for $\eta = \infty$.

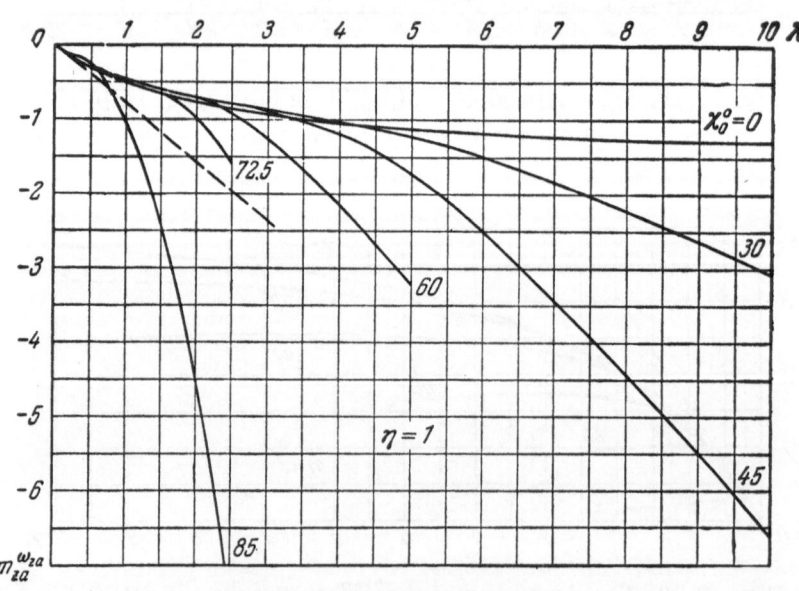

Fig. 12.23. The dependence of $m_{za}^{\omega za}$ on the aspect ratio of monoplane wings for $\eta = 1$.

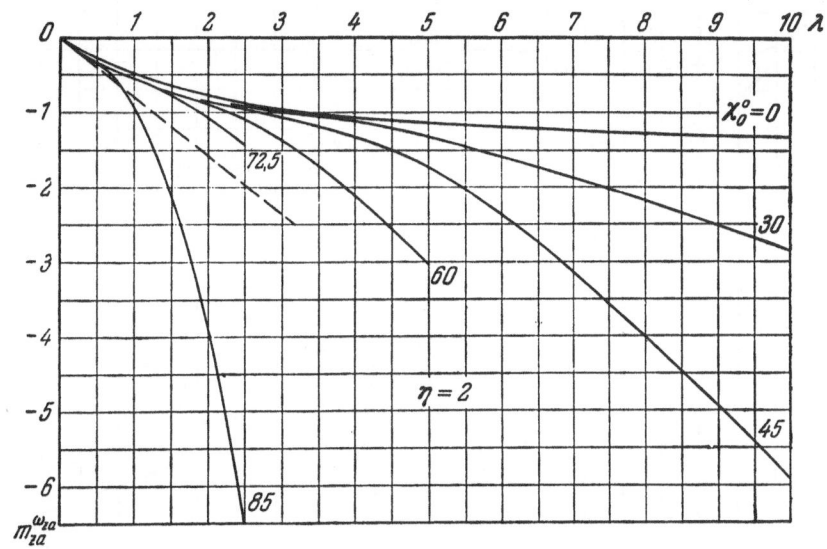

Fig. 12.24. The dependence of $m_{za}^{\omega za}$ on the aspect ratio of monoplane wings for $\eta = 2$.

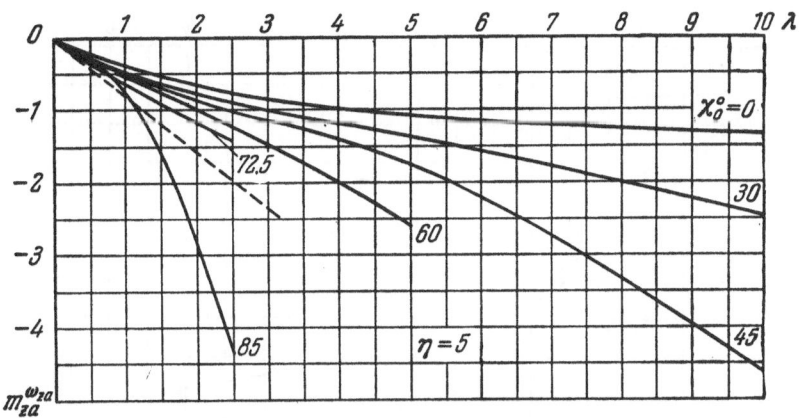

Fig. 12.25. The dependence of $m_{za}^{\omega za}$ on the aspect ratio of monoplane wings for $\eta = 5$.

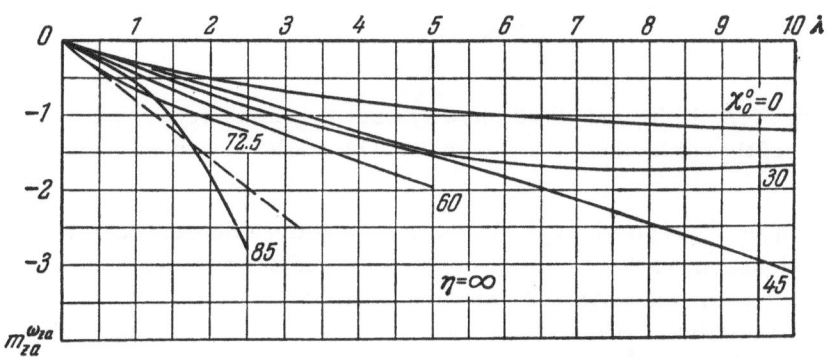

Fig. 12.26. The dependence of $m_{za}^{\omega za}$ on the aspect ratio of monoplane wings for $\eta = \infty$.

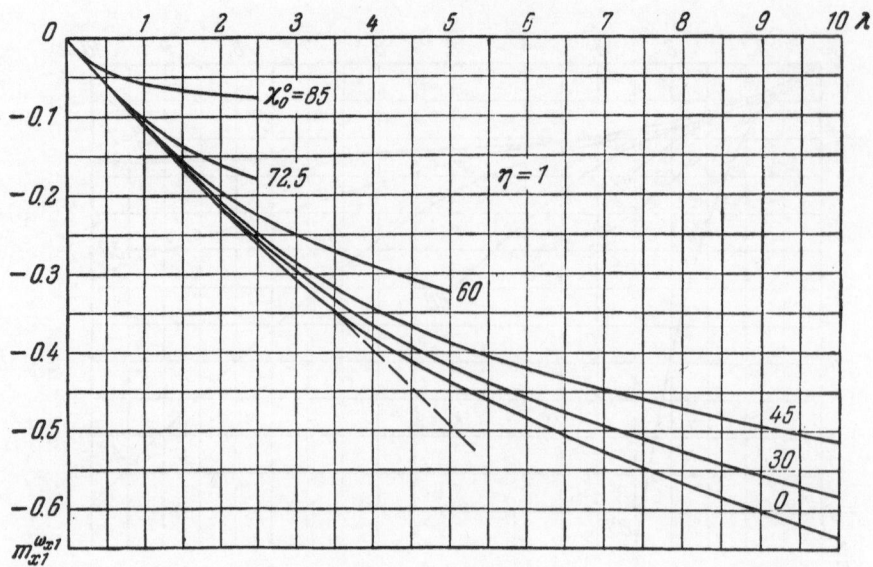

Fig. 12.27. The dependence of $m_{x1}^{\omega x_1}$ on the aspect ratio of monoplane wings for $\eta = 1$.

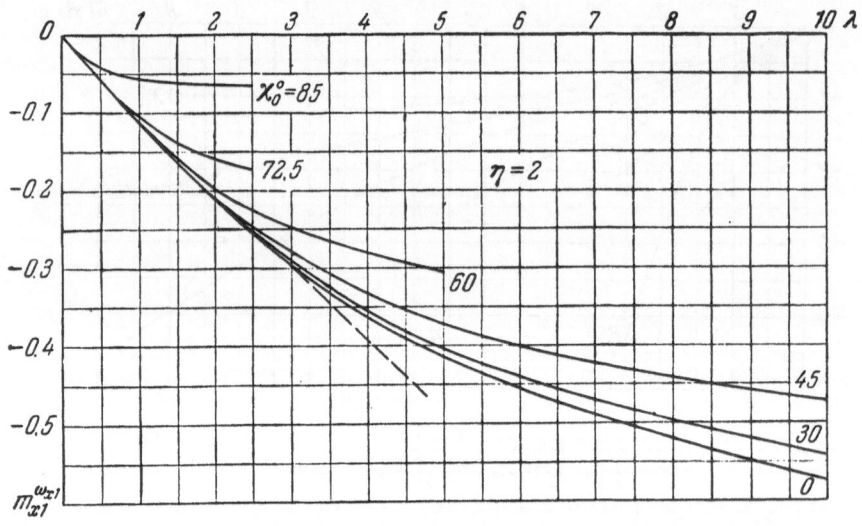

Fig. 12.28. The dependence of $m_{x_1}^{\omega x_1}$ on the aspect ratio of monoplane wings for $\eta = 2$.

For example, Figs. 12.31 and 12.32 show typical relations between c_y^α, which characterizes the lift properties of a wing, and the sweepback angle χ_0 and the reciprocal of the taper ratio $1/\eta$.

An increase in the sweepback angle leads to a considerable decrease in the derivative c_y^α for a wing of high aspect ratio. The lift–drag ratio, and for very high aspect ratios the lift itself, are given by the scheme for a side-slipping wing (§3, Chapter IX). For low aspect ratio, the angle χ_0 has practically no effect. This can be due to the fact that for small λ even a large change in χ_0 has a negligible effect on the shape of a wing in plan.

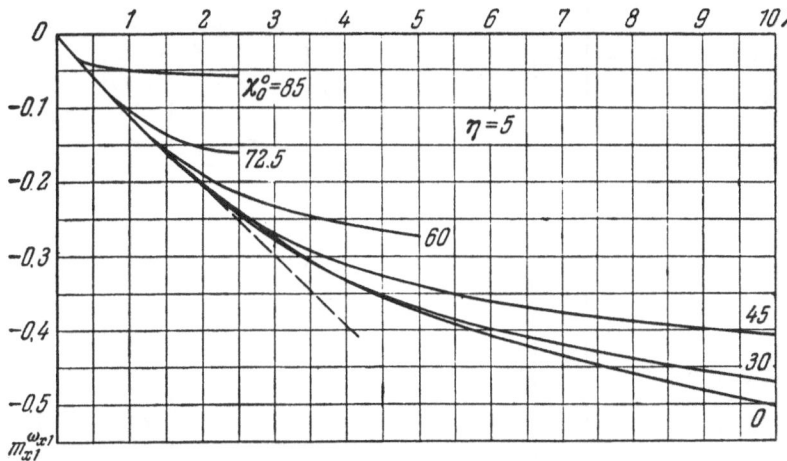

Fig. 12.29. The dependence of $m_{x1}^{\omega x1}$ on the aspect ratio of monoplane wings for $\eta = 5$.

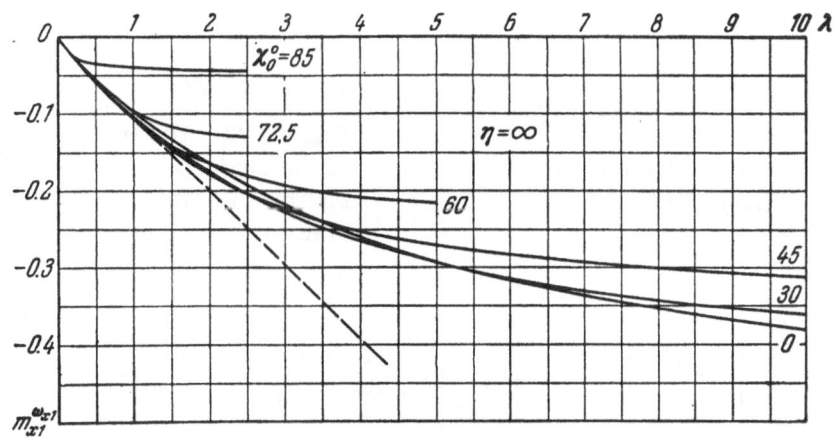

Fig. 12.30. The dependence of $m_{x_1}^{\omega x1}$ on the aspect ratio of monoplane wings for $\eta = \infty$.

Changes in the taper for constant χ_0 and λ have only a small effect on the lift properties of a wing, especially for small λ. This is also easily understood if we note that the main contribution to the lift is concentrated close to the leading edge, especially for small values of λ. An increase in the taper with $\chi_0 = $ const is equivalent to the elimination of an edge section of the wing, and such a section contributes very little lift. A similar explanation can be given of the behavior of the dependence of c_y^α on λ.

Close to the ends of a wing with a positive angle of attack there is a flow upwards from the lower surface, and the pressure is raised on the upper surface. The lower the aspect ratio, the greater will be the magnitude of the end flow, and this will lower the lift.

§ 3. Induced Drag and Suction Force

The induced drag with suction force and the suction-force coefficient were determined by B. K. Skripach. We give below some of the results he obtained.

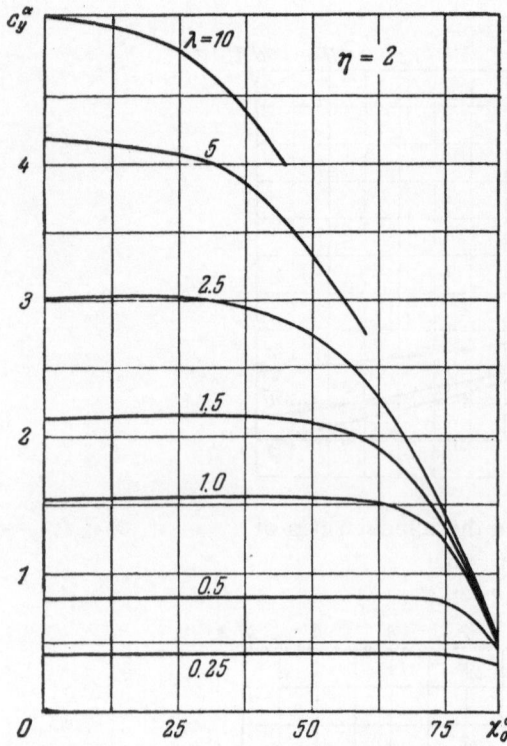

Fig. 12.31. The influence of the sweep-back angle on the lift properties of a monoplane wing.

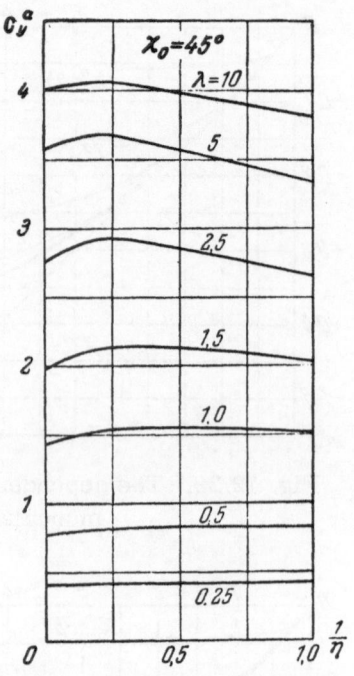

Fig. 12.32. The influence of the reciprocal of the taper on the lift properties of a monoplane wing.

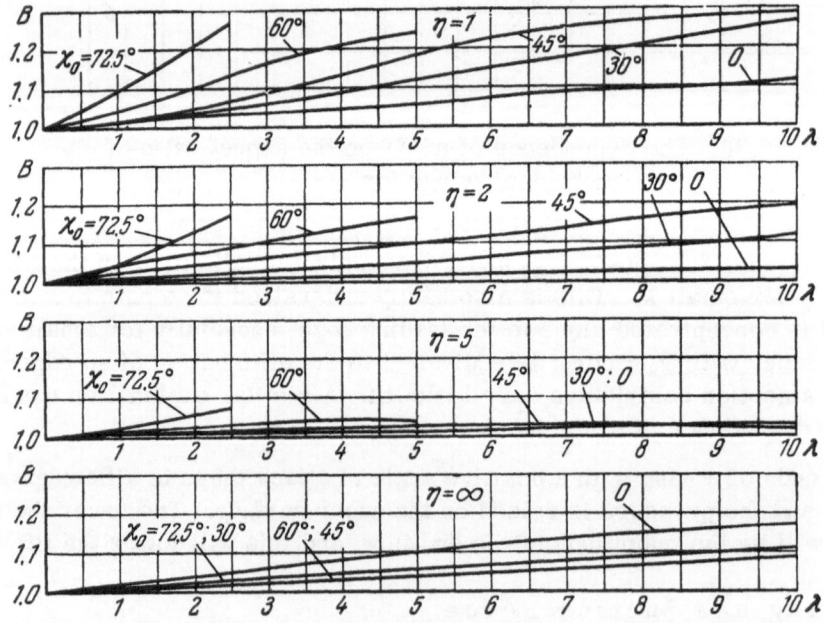

Fig. 12.33. The factor B of induced-drag with suction force on monoplane wings.

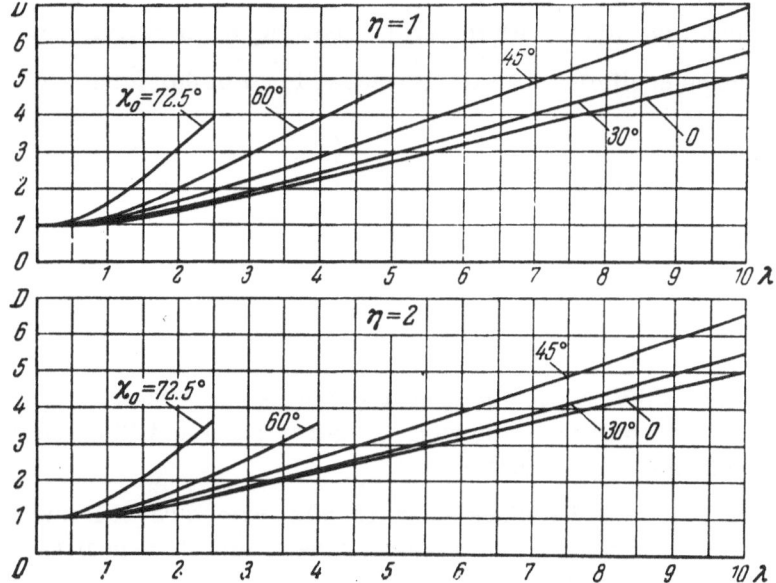

Fig. 12.34. The suction-force factor D of monoplane wings
($\eta = 1$; 2).

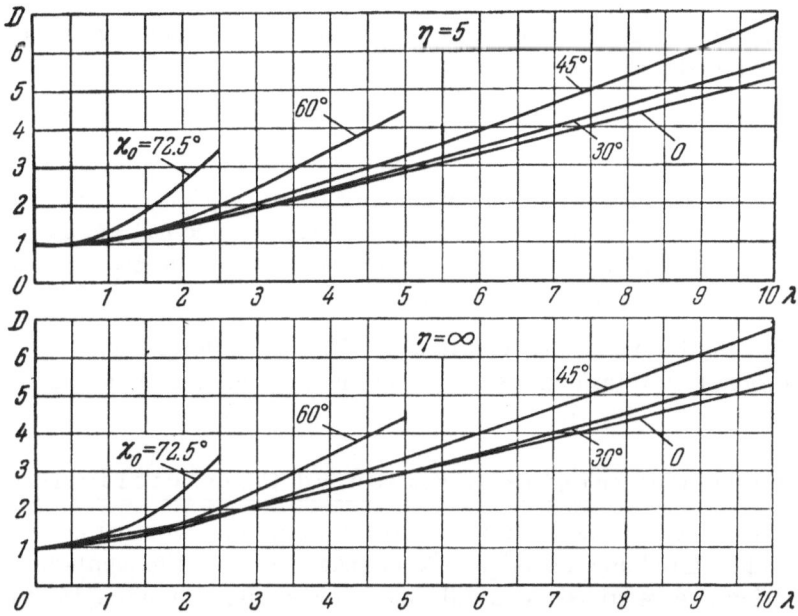

Fig. 12.35. The suction-force factor D of monoplane wings
($\eta = 5$; ∞).

We recall that from §1, Chapter IV the coefficient of the induced drag with suction force can be given by the formula

$$c_{xi} = B \frac{c_y^2}{\pi \lambda}.$$ (12.3)

Here B is a factor which depends in an incompressible medium on the shape of the wing in plan. When the circulation varies with the span according to an elliptic law we have B = 1.

The suction-force coefficient can be expressed in a similar form

$$c_Q = D \, \frac{c_y^2}{\pi \lambda},$$ (12.4)

where from (4.19) we have

$$D = \frac{\pi \lambda}{c_v^\alpha} - B.$$ (12.5)

Figure 12.33 shows the relation between the induced-drag factor with the suction force B and the geometrical parameters of the wing. Figures 12.34 and 12.35 show the corresponding relation for the suction-force factor D.

An analysis of the effect of various geometrical parameters on B and D easily shows that the limiting relations obtained for a side slipping wing of infinite aspect ratio in §3, Chapter IX hold more accurately when λ is larger, etc.

We consider one further interesting fact. When $\lambda \rightarrow 0$ the factors B and D tend to one. This means that the induced drag without suction force is double the induced drag with suction force for a wing of very small aspect ratio. In other words the induced drag is reduced by one half by the suction force in the case under consideration.

From the information given we can find the over-all aerodynamic characteristics of monoplane wings for any steady motion.

The rotation of a thin plate-wing about the Oy axis does not disturb the medium, and so we can investigate:

1) rectilinear motion of a monoplane wing (α = const, $\omega_x = \omega_z = 0$);
2) motion in a circle in the Oxy plane for $\alpha = 0$, $\omega_x = 0$ and ω_z = const;
3) translational motion of a wing with U_0 = const, when $\alpha = 0$, $\omega_z = 0$ and ω_x = const;
4) any combination of the above motions.

We recall that the information given is not sufficient to determine the action of the medium when translational or rotational oscillations are superimposed on translational motion with velocity U_0 = const [31, 32].

§4. Some Results Concerning Distributed Characteristics

In many applications it is necessary to know not only the over-all aerodynamic characteristics, but also the distributed aerodynamic characteristics of monoplane wings. These latter characteristics are required in calculations concerning the strength of a wing and in some aerodynamic calculations: for example, in the calculation of downwash behind a wing, etc. These characteristics can also be used in conjunction with consequences of the reversibility theorem for the determination of the over-all characteristics of deformed wings and the efficiency of a mechanism.

In this connection it is useful to have some representation of the nature of the distribution of the aerodynamic load along the chord and span of a wing.

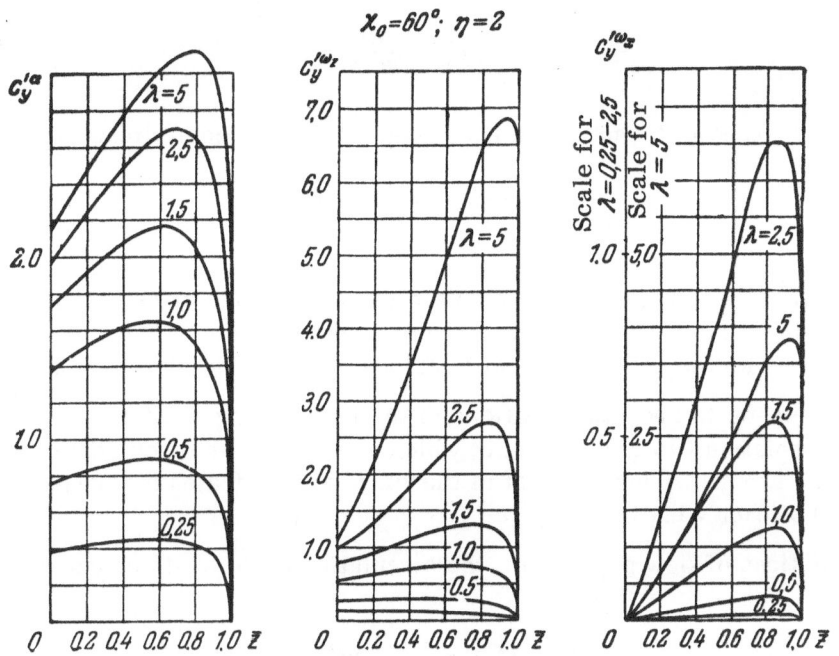

Fig. 12.36. The effect of the aspect ratio of a wing on the rotational-derivative coefficients of sections.

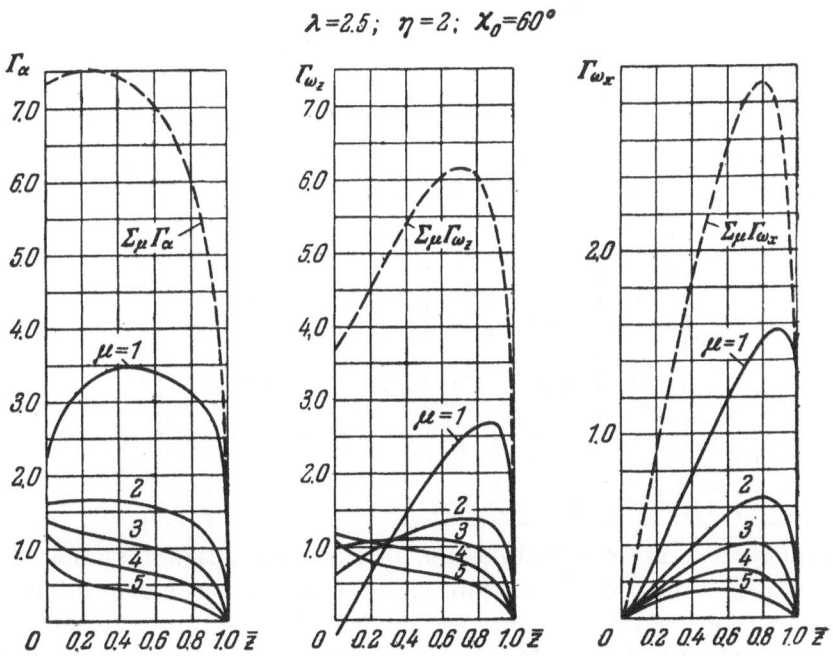

Fig. 12.37. The dimensionless circulation of various vortex filaments of a monoplane wing.

We recall that the lift coefficient of a wing section is given by the formula

$$c_y' = \frac{dY}{qb' \, dz}.$$ (12.6)

It is usually expressed in terms of the rotational-derivative coefficients of the sections:

$$c_y' = c_y'^o \alpha + c_y'^{\omega_z} \omega_z + c_y'^{\omega_x} \omega_x.$$ (12.7)

The coefficients in the expression on the right-hand side of (12.7) determine the lift properties of wing sections for various types of steady motion. Figure 12.36 shows how these coefficients vary with the span of wings of various aspect ratios for $\chi_0 = 60°$ and $\eta = 2$.

It is also useful to know the positions of the foci of wing sections for steady translational motion without rotation. In Figs. 1.3-1.5, which are drawn for three series of wings, the foci of wing sections are shown and the foci of the wings are indicated by crosses. The mean aerodynamic chord of each wing is also given.

In calculations of flow past wings, the circulation of the μ-th bound vortex filament is

$$\Gamma_{+\mu} = \frac{U_0 l}{14} (\Gamma_{\sigma\mu}\alpha + \Gamma_{\omega_z\mu}\omega_z + \Gamma_{\omega_x\mu}\omega_x).$$ (12.8)

The filaments are numbered from leading edge to trailing edge. The variation for each of the attached vortex filaments along the span for a swept-back wing ($\lambda = 2.5$, $\eta = 2$, $\chi_0 = 60°$) is shown graphically in Fig. 12.37. The calculations were carried out for n = 5, i.e., the wing was replaced by five vortex filaments for each section.

Such graphs can also be used to determine the variation of circulation along the chord of each section and then, by an application of Zhukovskii's theorem "in the small," to determine the variation of the aerodynamic load. It is useful here to know that for n = 5, for example, dimensionless distances h_μ'/b' of vortex filaments from the leading edge of the section in each section can be obtained from the accompanying table.

μ	1	2	3	4	5
$\dfrac{h_\mu'}{b'}$	1/20	4/20	9/20	13/20	17/20

§5. The Effect of the Mach Number

When systematic data are available on the aerodynamic characteristics of monoplane wings of various shapes in plan in an incompressible medium, these characteristics can easily be calculated for high subsonic velocities. The method of doing this is described in Chapter V. In finding the characteristics close to $M_\infty = 1$, it is expedient to use asymptotic formulas like those used in §5, Chapter IX.

The subscript "com" indicates, as usual, the value of a quantity for high subsonic velocities.

Figures 12.38-12.42 show graphically the dependence of the rotational-derivative coefficients and the dimensionless coordinate of the focus $\bar{x}_{F\text{com}}$ on M_∞ for rectangular wings of

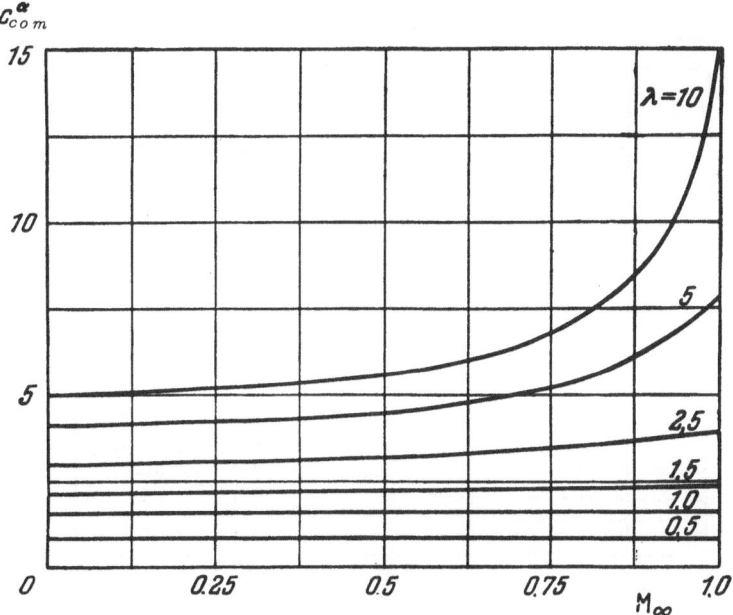

Fig. 12.38. The influence of the value of $\mathbf{M}_\infty$ on c_y^α for rectangular wings.

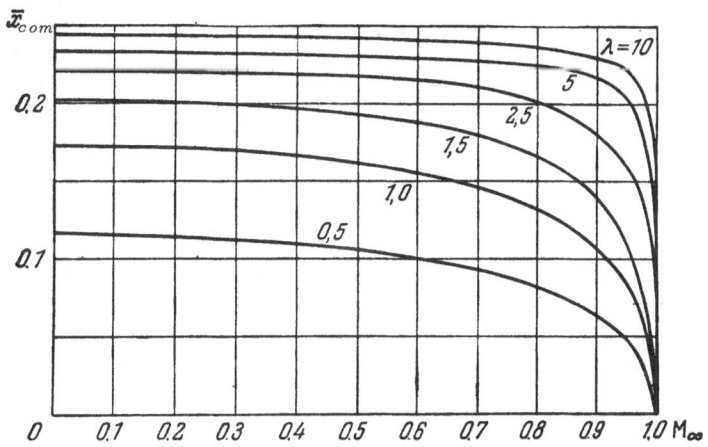

Fig. 12.39. The influence of the value of $\mathbf{M}_\infty$ on the position of the foci of rectangular wings.

various aspect ratios. The coefficients $c_y^{\omega z}$ and $m_z^{\omega z}$ were calculated for the center $\bar{x}_T = 0.5$. For $\mathbf{M}_\infty \to 1$, all the rotational-derivative coefficients were calculated by using asymptotic formulas valid for $\lambda \to 0$.

This dependence is qualitatively similar to that of the characteristics of annular wings. When $\mathbf{M}_\infty$ increases, all the rotational-derivative coefficients increase in absolute value, and they increase more rapidly when the aspect ratio is greater. The focus moves forward and, for $\mathbf{M}_\infty \to 1$, it tends theoretically towards the leading edge. When the aspect ratio decreases, the aerodynamic characteristics become more stable relative to variations in $\mathbf{M}_\infty$.

All these facts have the same explanation: A decrease in aspect ratio leads to a weakening of disturbances generated by the wing (if other conditions remain unchanged). Hence the compressibility of air begins to make itself felt for large $\mathbf{M}_\infty$.

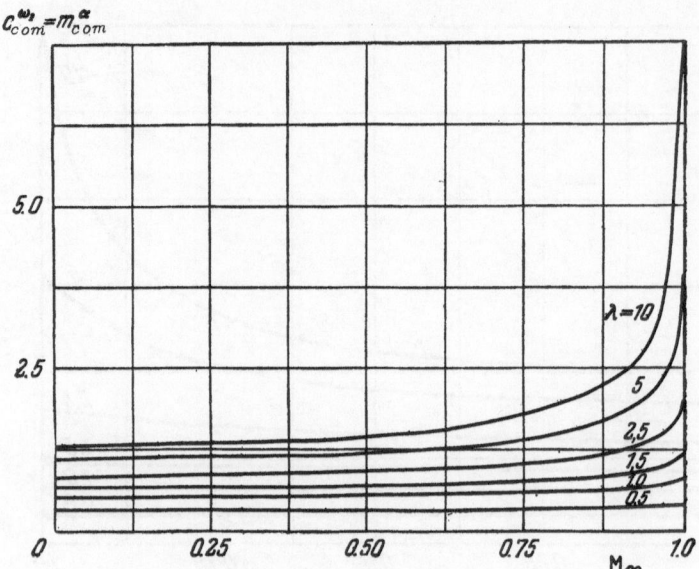

Fig. 12.40. The influence of the value of M_∞ on the coefficient $m_z^\alpha = c_y^{\omega z}$ for rectangular wings.

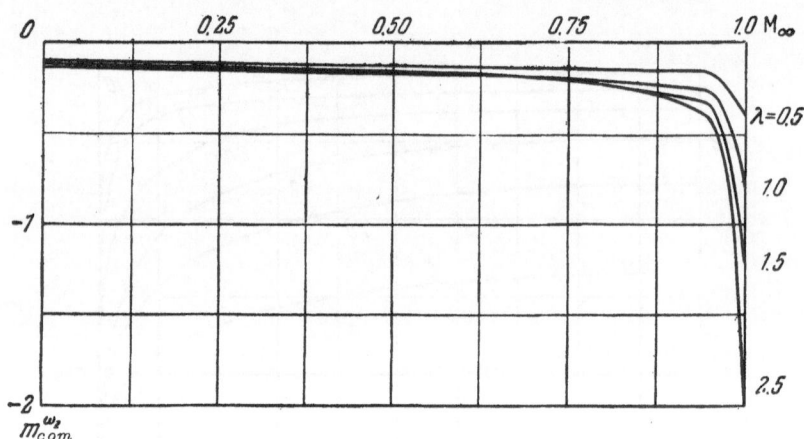

Fig. 12.41. The influence of the value of M_∞ on the coefficient $m_z^{\omega z}$ for rectangular wings.

Figure 12.43 illustrates the influence of the value of M_∞ on the induced drag with suction force taken into account and on the suction force itself.

The factor of induced drag with suction force B depends weakly on M_∞. This is also a consequence of (4.16), (4.18), (5.23), and (5.24). In fact B depends on the variation of the circulation along the span, which can be expressed by the relation

$$\Gamma_\pi(\bar{z}) = \Gamma_a(0) f(\bar{z}), \tag{12.9}$$

and on the derivative c_y^α. An analysis of these formulas easily shows that the factor $\sqrt{1 - M_\infty^2}$ cancels out. There remains only the dependence related to a change in the ratio $[\Gamma_{\alpha M}(0)]/c_{yM}^\alpha$ due to variations with respect to M_∞ of the geometrical parameters λ_M, χ_{0M}, and η_M of the wing under consideration.

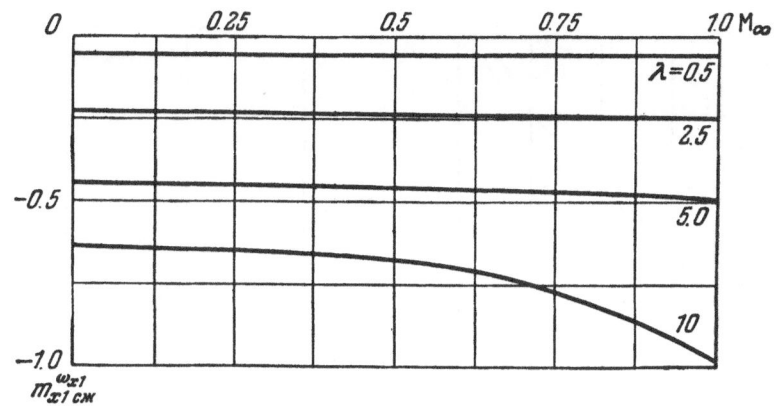

Fig. 12.42. The influence of the value of $\mathbf{M}_\infty$ on the coefficient $m_{x1}^{\omega x1}$ for rectangular wings.

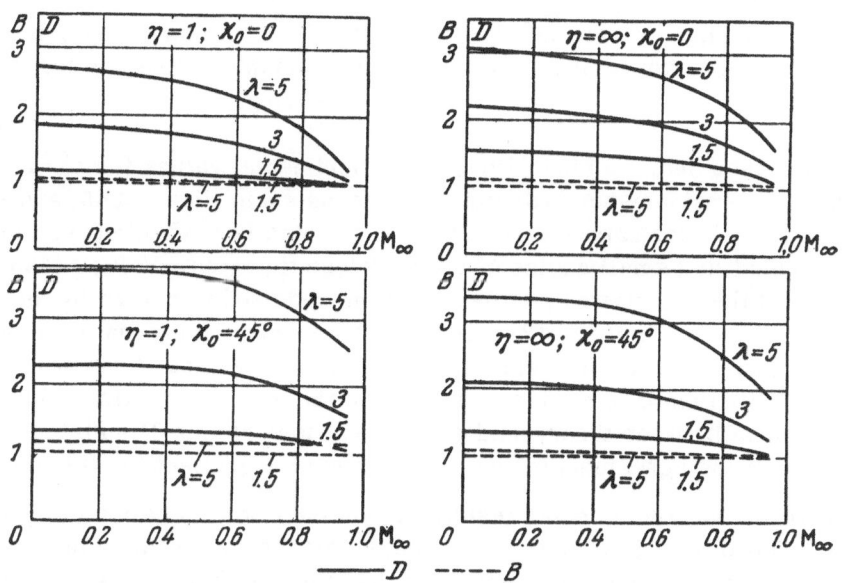

Fig. 12.43. The influence of the value of $\mathbf{M}_\infty$ on the induced-drag factor and on the suction force.

Hence from (12.5) we find that the variation of D with respect to $\mathbf{M}_\infty$ will be mainly determined by the function $c_{y\,com}^\alpha\,(\mathbf{M}_\infty)$. Thus for wings of small aspect ratio this relation will be unimportant, and for large λ the factor D decreases with increasing $\mathbf{M}_\infty$.

§6. A Comparison of Theoretical and Experimental Results

There is a large amount of experimental data available on monoplane wings, and this causes some difficulty in making a reasonable choice of material and in the analysis of this material, especially since information is sometimes lacking concerning the conditions of experiment, the models used, and in particular, the accuracy of the results. However, by using a large amount of the most reliable experimental data and making comparisons with theoretical results, it is possible to determine the correctness of the theory.

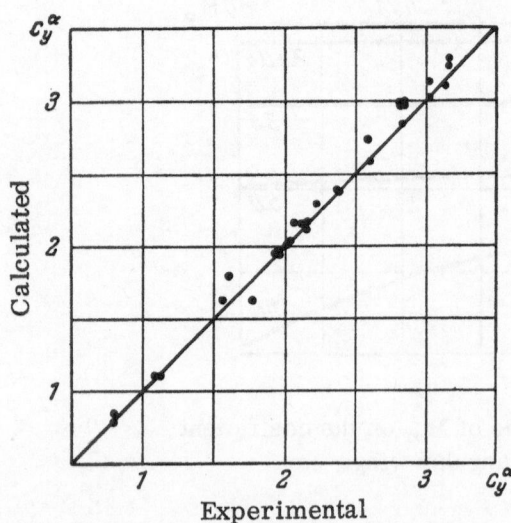
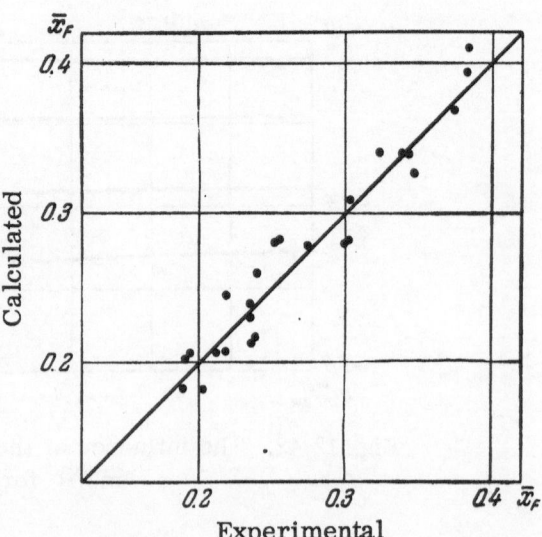

Fig. 12.44. A comparison of calculated and experimental values for various monoplane wings.

Fig. 12.45. A comparison of calculated and experimental values for various monoplane wings.

Such a comparison between theoretical and experimental values for c_y^α is shown in Fig. 12.44, for the dimensionless coordinate of the focus $\overline{x}_F$ in Figs. 12.45, and for $m_{x_1}^{\omega_{x_1}}$ in Fig. 12.46. Values of the derivative were obtained graphically from the experimentally determined function c_y^α for small angles of attack (close to $\alpha = 0$). Values of m_z^α, were obtained similarly, and then $\overline{x}_F$ was calculated. The method used to determine experimental values of $m_{x_1}^{\omega_{x_1}}$ is described in [32]. Each point in the accompanying diagrams was obtained for a single wing. Thus results for twenty or thirty wings of various shapes in plan are compared. The straight line inclined at an angle of 45° to the horizontal axis is the locus of points at which the agreement between theory and experiment is perfect. The further a point is from this line, the worse is the agreement between theory and experiment for this point.

The theoretical and experimental results are on the whole in satisfactory agreement. The best agreement is found for lift properties – here the dispersion of the points is very small (Fig. 12.44). Greater dispersion is observed for the moment characteristics (Figs. 12.45 and (12.46). This is related to a large degree to the fact that these characteristics are more delicate. They cannot be determined as accurately as those in Fig. 12.44 and the reliability of the experimental results is thus not as high.

Experimental difficulties must be overcome in the determination of the rotational-derivative coefficients $c_y^{\omega_z}$ and $m_z^{\omega_z}$ [32]. Accurate experiments with a rectangular wing of elongation $\lambda = 1.0$ with relative thickness $\overline{c} = 12\%$ were carried out by I. B. Fedorova, and the results of these experiments are compared with theoretical results in Fig. 12.47, in which values of the dimensionless parameter ω_z characterizing the distortion of the model are given. We see that the linear theory is on the whole in satisfactory agreement with the results of these experiments.

Figures 12.48-12.51 show results for distributed characteristics of monoplane wings. The first three of these graphs show the dependence on $\overline{z}$ of the ratios of the derivatives $c_y'^\alpha$ for sections to the derivative c_y^α for swept-back wings. The fourth figure shows the lift coefficient c_y' of a rectangular wing ($\lambda = 1.0$) for a few different angles of attack. Experimental results are in good agreement with theoretical results for two of the wings with leading-edge sweepback angle $\chi_0 = 45°$, and $\lambda = 5.0$ and $\lambda = 2.5$.

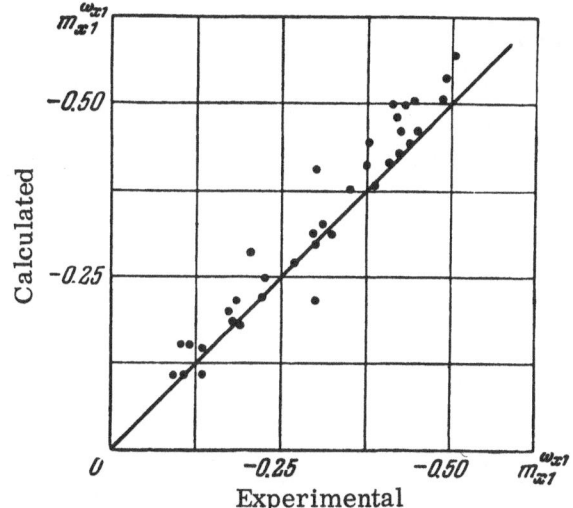

Fig. 12.46. A comparison of calculated and experimental results for various monoplane wings.

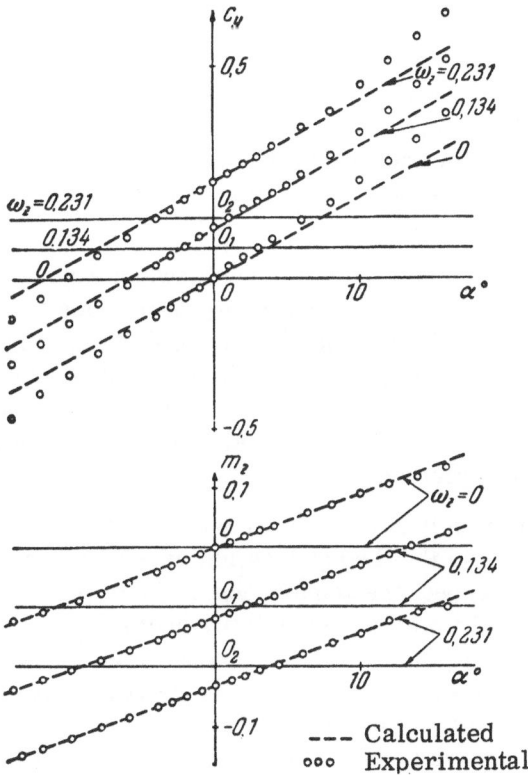

Fig. 12.47. A comparison of calculated and experimental results for various curved models of a rectangular wing with $\lambda = 1.0$.

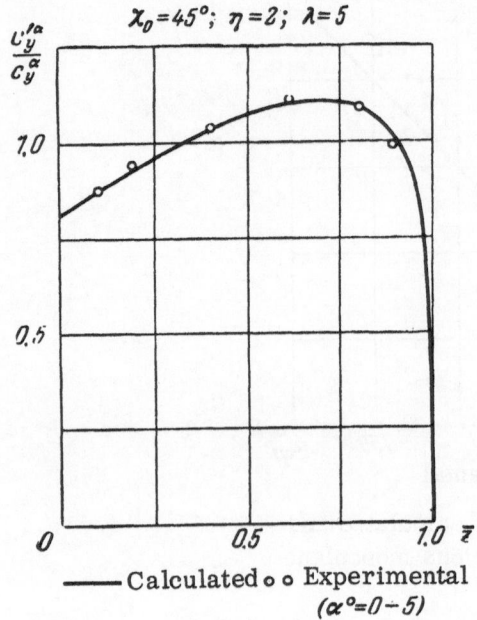
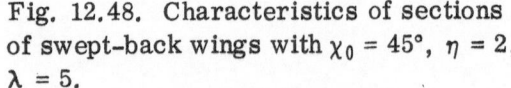
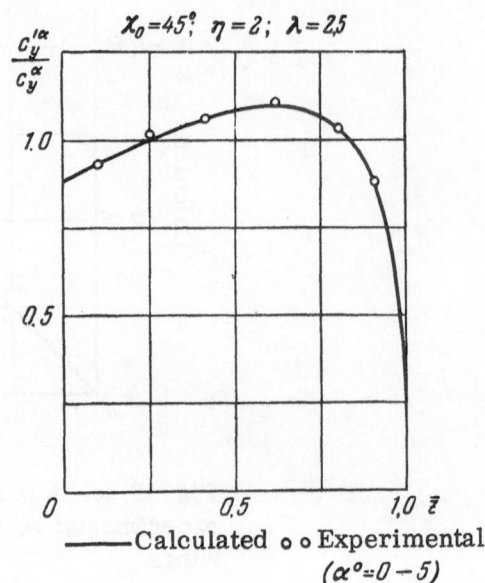

——Calculated o o Experimental
$(\alpha°=0-5)$

——Calculated o o Experimental
$(\alpha°=0-5)$

Fig. 12.48. Characteristics of sections of swept-back wings with $\chi_0 = 45°$, $\eta = 2$, $\lambda = 5$.

Fig. 12.49. Characteristics of sections of swept-back wings with $\chi_0 = 45°$, $\eta = 2$, $\lambda = 2.5$.

In two limiting cases — for very large taper η and for η close to 1 — the difference between results obtained from the linear theory and experimental results is appreciable. If η is very large, a collapse in the flow occurs at the end of the wing, and the experimental results for c_y^{ι} (or $c_y^{\iota\alpha}$) are lower than the theoretical results. It is clear from Figs. 12.50 that for the wing under consideration with a taper $\eta = 5$ this tendency can be observed in the section $\bar{z} = 0.9$.

Another type and another cause of divergence between experimental and theoretical results occurs for a wing of small aspect ratio with $\eta = 1$ or close to 1. As an example we consider a wing with elongation $\lambda = 1.0$ and thickness ratio $\bar{c} = 18\%$ (Fig. 12.51). Here the experimental values of c_y^{ι} at the end of the wing exceed the theoretical values, and with increasing angle of attack this divergence increases. This explains why, for a wing of low aspect ratio, the difference between the theoretical and experimental functional relation $c_y(\alpha)$ is of the type shown in Fig. 12.52 for large angles of attack.

The basic cause of this result is the influence of the end vortex filaments; these filaments are of finite dimension and the theory does not take this into account. When the elongation or the contraction is increased, the importance of these filaments decreases, and the difference between theoretical and experimental results decreases. It is interesting to note that an increase of the thickness ratio $\bar{c}$ weakens the above effect, as can be seen in Fig. 12.52. Hence the end overflow for a thin wing generates a large end vortex and a rarefaction on the upper surface of the wing. It has already been noted that the shape of the leading edge of a wing plays an important role in the generation of suction force. If the flow over the leading edge is smooth, without breakaway, then suction force will be generated and the induced drag will be appreciably decreased.

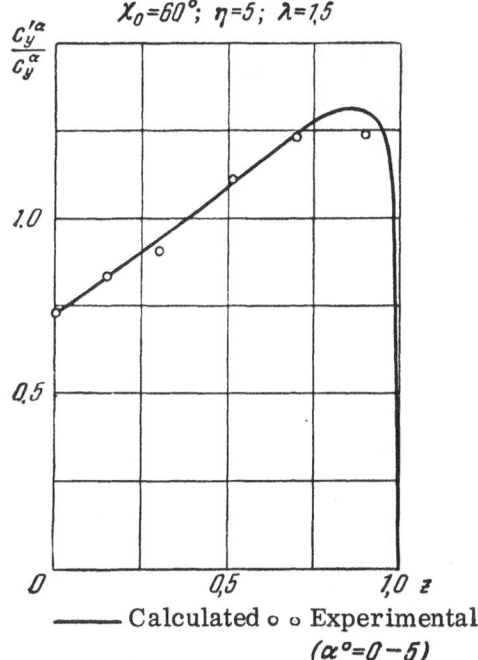

— Calculated ○ ○ Experimental
($\alpha° = 0-5$)

Fig. 12.50. Characteristics of sections of a swept-back wing with $\chi_0 = 60°$, $\eta = 5$, $\lambda = 1.5$.

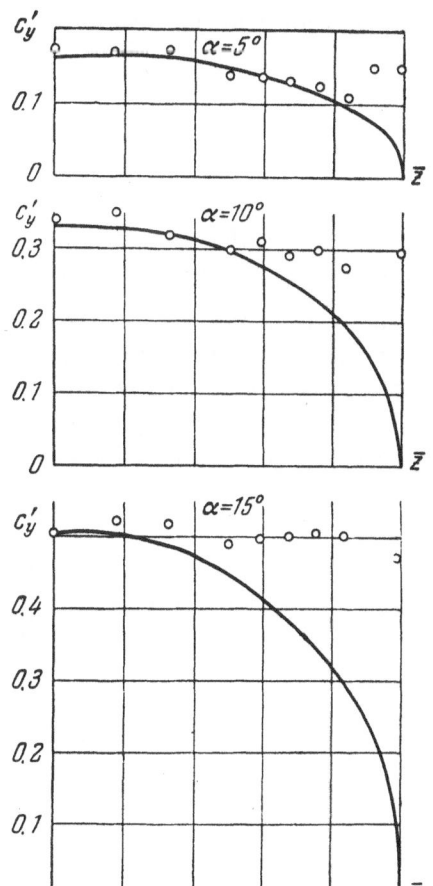

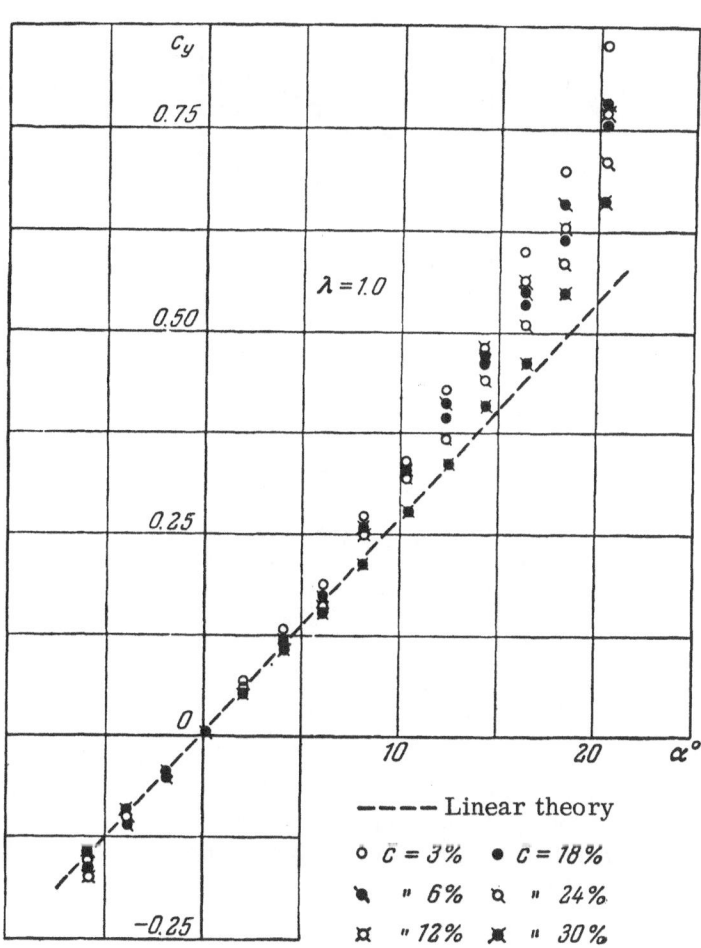

Fig. 12.52. The effect of the thickness ratio of a rectangular wing with $\lambda = 1.0$ on lift characteristics.

Fig. 12.51. Lift coefficient of sections of a rectangular wing ($\lambda = 1.0$).

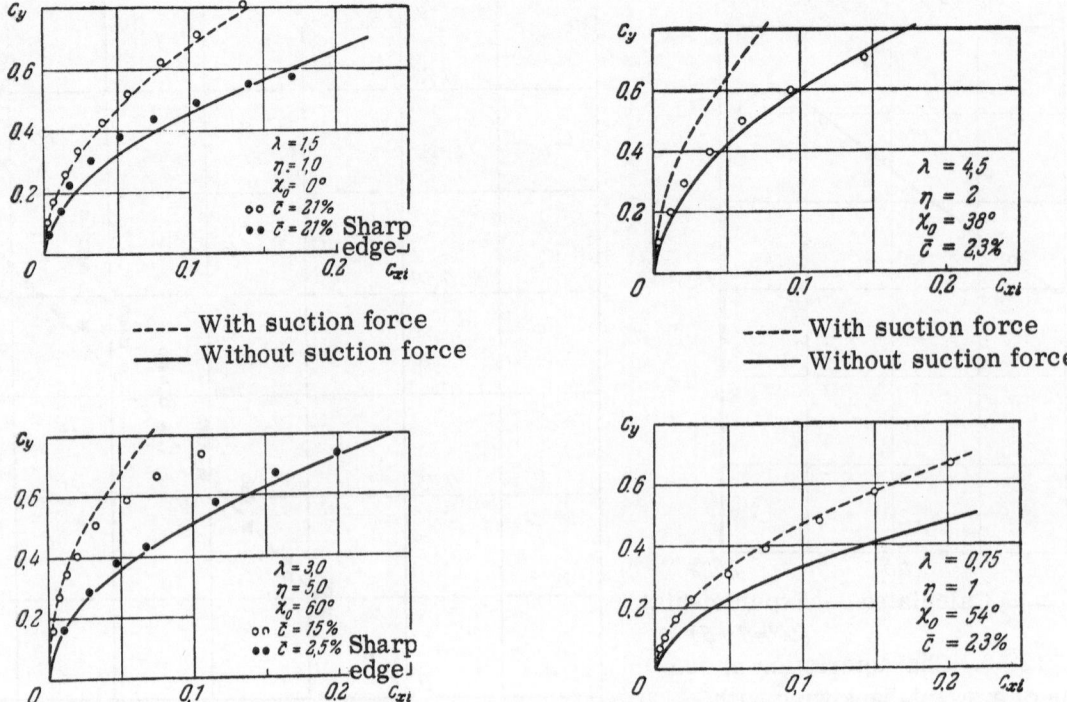

----- With suction force
—— Without suction force

----- With suction force
—— Without suction force

Fig. 12.53. The influence of the shape of the leading edge and the relative thickness of a wing on the suction force.

Fig. 12.54. The influence of the three-dimensional character of the flow on the suction force.

It is clear from Fig. 12.53 that if the leading edge is sharp, then there will be no suction force either on a thick wing ($\overline{c}$ = 21%) or on a thin wing ($\overline{c}$ = 2.5%). On a profiled wing of considerable thickness, however, the suction force is almost completely realized (see also [23]).

The relative thickness is not the only factor affecting the magnitude of the suction force. Figure 12.54 shows that although there is no suction force on a wing of low relative thickness ($\overline{c}$ = 2.3%) and high elongation (λ = 4.5), the suction force is completely realized on a wing of the same relative thickness and low elongation (λ = 0.75) with high sweepback. The decisive role here is obviously played by the intrinsic three-dimensional nature of the flow, which eliminates local separation at the leading edges.

CHAPTER XIII

CONTROLS ATTACHED TO MONOPLANE WINGS. THE EFFECT OF INTERFACE

§ 1. Some Problems in Methods of Calculation

The basic position described above concerning methods of calculating aerodynamic characteristics of isolated wings remains in force in the consideration of the motion of a wing close to an interface.

We will note the facts that must be taken into account in the calculation of the efficiency of controls of wings on the basis of the reversibility theorem.

The extra lift and moments produced by deflections of flaps or ailerons are calculated by using Eqs. (6.44), in which the integration is taken over that section of the reversed wing occupied by the control on the direct wing. Since flaps and ailerons are close to the trailing edge of a wing, the integration of the load will be over a relatively small part of the reversed wing close to its leading edge.

Among other results that can be obtained from these considerations, we obtain a simple practical method for finding qualitatively where it is best to locate the supplementary control of a wing. It is sufficient to consider the load-distribution $p_{\omega z-}$, and $p_{\omega x-}$ on a reversed wing close to the leading edge and to find the conditions under which the integrals (6.44) have their largest values. If necessary, the positions of flaps and ailerons can then be determined numerically.

There are two methods of ensuring that results obtained from Eqs. (6.44) are sufficiently accurate: We either use the discrete-vortex method and employ enough bound vortices to yield sufficient accuracy in the sums that replace the integrals; or we use information on discrete bound vortices, convert to distributed aerodynamic loads, and then calculate the integrals (6.44).

Another method considered by O. N. Sokolova is less laborious, and so we will describe it in more detail.

The pressure difference Δp on the wing is expressed in terms of the strength of the bound vortex layer by using (4.1). From (4.1), (4.2), and (4.5) we have

$$p_{\alpha} = 2\gamma_z^{\omega}, \qquad p_{\omega_z} = 2\gamma_z^{\omega z}, \qquad p_{\omega_x} = 2\gamma_z^{\omega x}. \tag{13.1}$$

Hence to obtain the desired result it is sufficient to convert from discrete vortices to a disturbuted vortex layer.

Let the circulation Γ_{+i} of the bound vortices on the reversed wing be known. Using γ_z^{α} as an example, we will show how to find the distribution of the bound vortex layer on an arbi-

197

trary section z_{k_i} = const. We write the required function in the form

$$\gamma_z^a = \sum \frac{\alpha_r (1-t)^r}{\sqrt{t}}, \qquad t = \frac{x^* - x}{b_{k_l}}. \tag{13.2}$$

Here the α_r are unknown coefficients constant for each section z_{k_i} = const, b_{k_i} is the chord of this section, and x^* is the coordinate of the leading edge of the reversed wing. The representation (13.2) automatically satisfies the Chaplygin-Zhukovskii condition on the trailing edge since here $\gamma_Z^\alpha = 0$ for t = 1. Moreover there is a singularity of the type $1/\sqrt{t}$ at the leading edge.

In the general case, the conditions for the equivalence of a distributed vortex layer and discrete bound vortices can be written for each section z_{k_i} as follows:

$$\left. \begin{aligned} \int_{x_0^*}^{x^*} \gamma_{z+}(x, z_{k_l}) \, dx &= \sum_{\mu=1}^n {}_{k_l}\Gamma_{+i}, \\ \int_{x_0^*}^{x^*} \gamma_{z+}(x, z_{k_l}) x \, dx &= \sum_{\mu=1}^n {}_{k_l}\Gamma_{+i}x_i, \\ \int_{x_0^*}^{x^*} \gamma_{z+}(x, z_{k_l}) x^2 dx &= \sum_{\mu=1}^n {}_{k_l}\Gamma_{+i}x_i^2, \\ \cdots \cdots \cdots \cdots \cdots & \end{aligned} \right\} \tag{13.3}$$

Here x_0^* is the coordinate of the trailing edge of the section of the reversed wing and x^* the corresponding coordinate of the trailing edge; the remaining notation is the same as that used in (8.43).

We use the relations

$$\left. \begin{aligned} \gamma_{z+} &= U_0(\gamma_z^a \alpha + \gamma_z^\omega z \omega_z + \gamma_z^\omega x \omega_x), \\ \Gamma_{+i} &= U_0 l_i (\Gamma_{\alpha l} \alpha + \Gamma_{\omega_z i} \omega_z + \Gamma_{\omega_x i} \omega_x), \end{aligned} \right\} \tag{13.4}$$

where γ_z^α, $\gamma_Z^{\omega z}$, and $\gamma_Z^{\omega x}$ are expressed in the form (13.2). Substituting all these expressions in (13.3), we obtain a set of algebraic equations for the coefficients α_r. These coefficients will naturally be different both for different functions and for different wing sections.

We give all the formulas for determining the α_r for the case when the summation with respect to r in (13.2) is from 1 to 3. Calculations show that this number of terms in (13.2) is usually sufficient.

If we use three terms in (13.2), then the equations for γ_Z^α are

$$\left. \begin{aligned} \frac{4}{3}a_1 + \frac{16}{15}a_2 + \frac{32}{35}a_3 &= \frac{1}{N}\frac{l}{2b_{k_l}}\sum_{\mu=1}^n {}_{k_l}\Gamma_{\alpha l}, \\ \frac{4}{15}a_1 + \frac{16}{105}a_2 + \frac{32}{315}a_3 &= \frac{1}{N}\frac{l}{2b_{k_l}}\sum_{\mu=1}^n {}_{k_l}\Gamma_{\alpha l}t_\mu, \\ \frac{4}{35}a_1 + \frac{16}{315}a_2 + \frac{32}{1155}a_3 &= \frac{1}{N}\frac{l}{2b_{k_l}}\sum_{\mu=1}^n {}_{k_l}\Gamma_{\alpha l}t_\mu^2. \end{aligned} \right\} \tag{13.5}$$

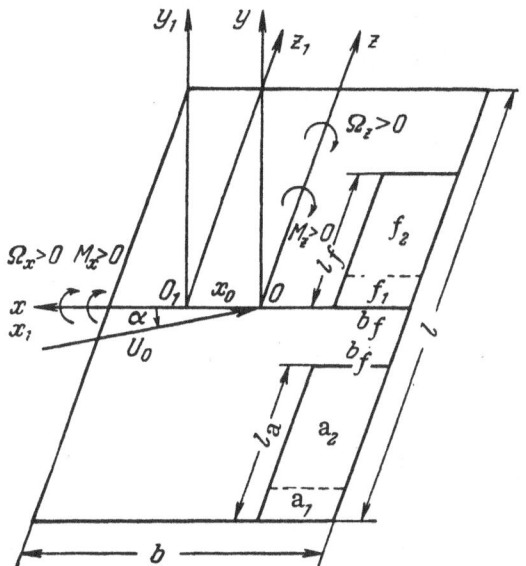

Fig. 13.1. A wing with attached control surface.

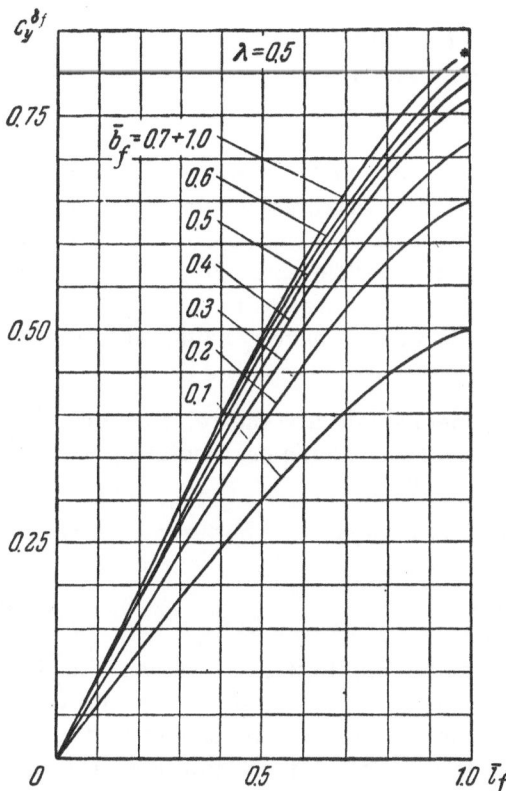

Fig. 13.2. The influence of the relative dimensions of a flap on $c_y^{\delta_f}$ for a rectangular wing with $\lambda = 0.5$.

Here N is the number of discrete vortices on a half span, l is the span, and

$$t_\mu = \frac{\mu - \frac{3}{4}}{n}. \tag{13.6}$$

To find the coefficients of the other functions $\gamma_z^{\omega z}$ and $\gamma_z^{\omega x}$ in (13.5) we must replace $\Gamma_{\alpha i}$ by $\Gamma_{\omega_z i}$ or $\Gamma_{\omega_x i}$.

§2. The Efficiency of Controls Attached to Rectangular Wings

The deflection of flaps and ailerons leads to a supplementary lift ΔY and supplementary moments ΔM_x and ΔM_z. The corresponding coefficients are referred to the characteristic dimensions of the wing, and we write

$$\Delta c_y = \frac{\Delta Y}{qS}, \qquad \Delta m_z = \frac{\Delta M_z}{qSb}, \qquad \Delta m_x = \frac{\Delta M_x}{qSb}. \tag{13.7}$$

If the wing is symmetric in plan and the flaps and ailerons are symmetrically located relative to the wing, then

$$\Delta c_y = c_y^{\delta_f}\delta_f, \qquad \Delta m_z = m_z^{\delta_f}\delta_f, \qquad \Delta m_x = m_x^{\delta_a}\delta_a. \tag{13.8}$$

Here δ_f and δ_a are the deflection angles of the flaps and ailerons, and we assume that both flaps are deflected in the same direction, while the deflections of the ailerons are identical in magnitude but opposite in sign.

We consider a rectangular wing of arbitrary aspect ratio in a standard coordinate system Oxyz fixed to the wing with origin at the midpoint of the root chord (Fig. 13.1). Positive directions of the moments and angular velocities and the positive direction of the angle of attack are shown in the diagram.

If the origin is translated from O to O_1 with no change in the direction of the axes, the coefficients $c_y^{\delta_f}$ and $m_x^{\delta_a}$ remain unchanged. If the distance from O to O_1 is x_0, then $m_z^{\delta_f}$ is transformed according to the formula

$$m_{z1}^{\delta_f} = m_z^{\delta_f} - c_y^{\delta_f}\xi_0, \qquad \xi_0 = \frac{x_0}{b}. \tag{13.9}$$

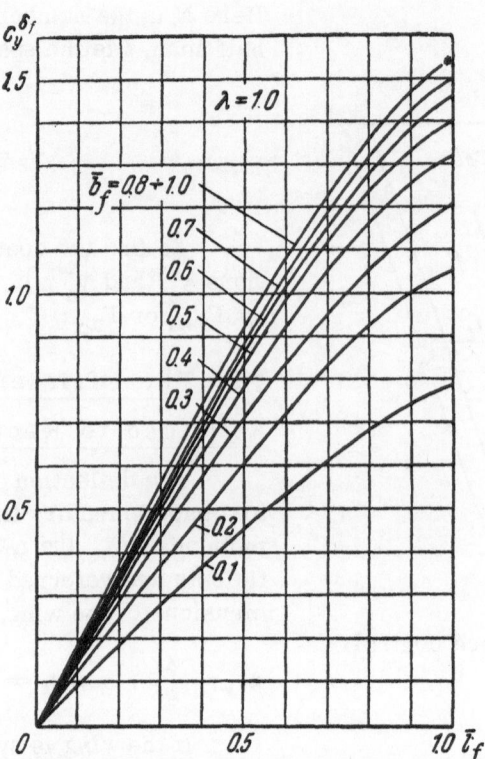

Fig. 13.3. The influence of the relative dimensions of a flap on $c_y^{\delta f}$ for a rectangular wing with $\lambda = 1.0$.

We consider flaps and ailerons that are rectangular in plan and assume that all flaps begin at the root chord and all ailerons at the tip chord. It follows from the linearity of Eqs. (6.44) that the coefficient $m_x^{\delta a}$ of the aileron $a_1 + a_2$ is equal to the sum of these coefficients for each of the ailerons a_1 and a_2 separately (Fig. 13.1). Similarly the coefficients $c_y^{\delta f}$ and $m_z^{\delta} f$ of the flap $f_1 + f_2$ can be obtained by adding the corresponding coefficients of the flaps f_1 and f_2. This property permits us to convert from the flaps and ailerons we have considered to the more general case with flaps and ailerons f_2 and a_2.

The dimensionless geometrical parameters of the mechanism are the relative chord $\bar{l}_f = l_f/(l/2)$ of a flap or $\bar{l}_a = l_a/(l/2)$ of an aileron (Fig. 13.1). These parameters can clearly vary from 0 to 1.

It is useful to remember that when $\bar{b}_f \to 1$ and $\bar{l}_f \to 1$ we have

$$c_y^{\delta_f} = c_y^{\alpha}, \qquad m_z^{\delta_f} = m_z^{\alpha}. \tag{13.10}$$

Using the results of O. N. Sokolova, we have plotted in Figs. 13.2–13.19 values of $c_y^{\delta f}$, $m_z^{\delta f}$, and $m_x^{\delta a}$, which characterize the efficiency of controls on a rectangular wing with various aspect ratios λ. These coefficients are shown as functions of the relative span of the flap or aileron for various values of $\bar{b}_f$ or $\bar{b}_a$. The stars indicate values obtained from Eqs. (13.10).

The results given by the graphs yield much qualitative and quantitative information concerning the operation of wing controls. For example it is plain from Figs. 13.2–13.7 and 13.14–13.19 that an increase in the relative chord with fixed relative span is not accompanied by a proportional improvement in the lift properties or the banking moment. The reason for this is

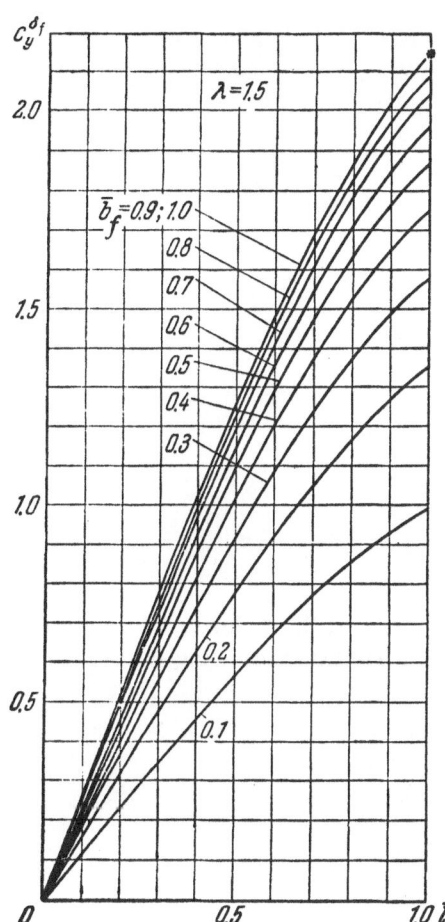

Fig. 13.4. The effect of the relative dimensions of flaps on $c_y^{\delta f}$ for a rectangular wing with $\lambda = 1.5$.

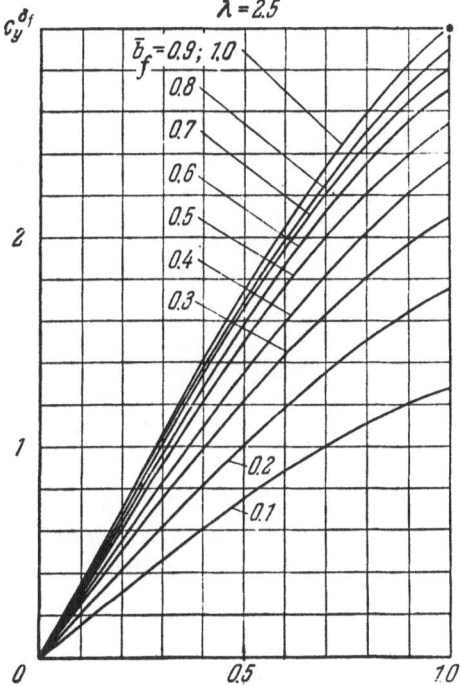

Fig. 13.5. The effect of the relative dimensions of flaps on $c_y^{\delta f}$ for a rectangular wing with $\lambda = 2.5$.

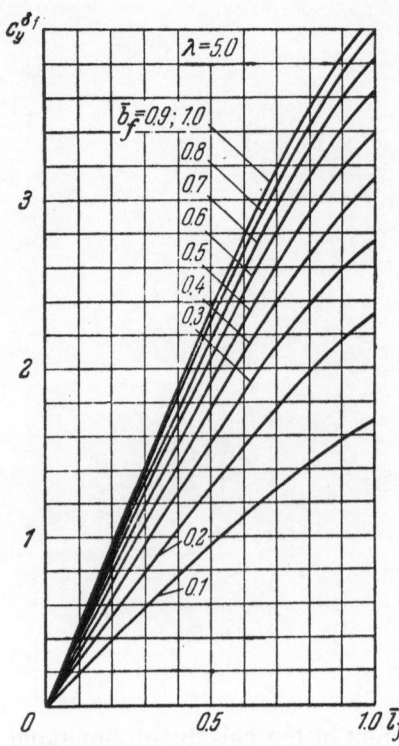

Fig. 13.6. The effect of the relative dimensions of flaps on $c_y^{\delta}f$ for a rectangular wing with $\lambda = 5.0$.

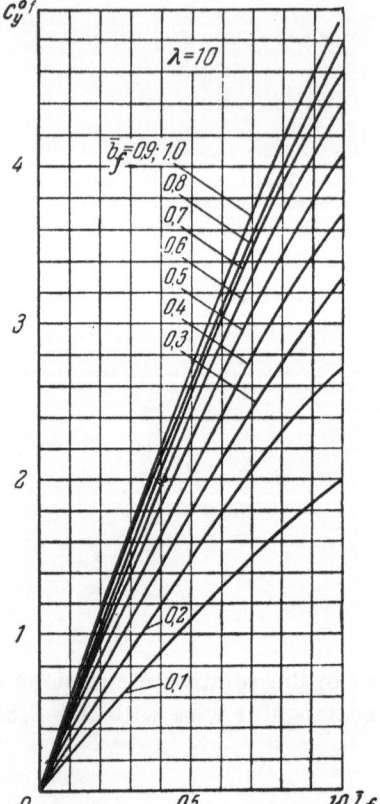

Fig. 13.17. The effect of the relative dimensions of flaps on $c_y^{\delta}f$ for a rectangular wing with $\lambda = 10$.

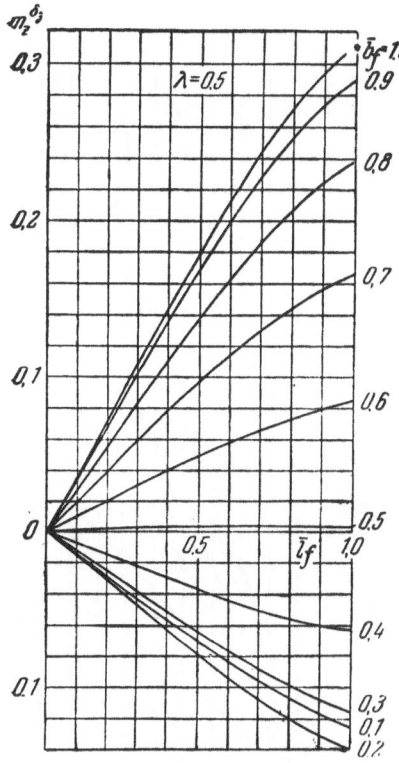

Fig. 13.8. The effect of the relative dimensions of flaps on $m_z^{\delta_f}$ for a rectangular wing with $\lambda = 0.5$.

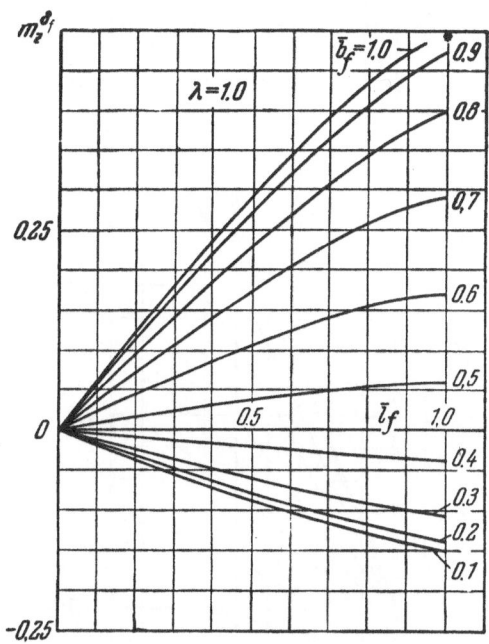

Fig. 13.9. The effect of the relative dimensions of flaps on $m_z^{\delta_f}$ for a rectangular wing $\lambda = 1.0$.

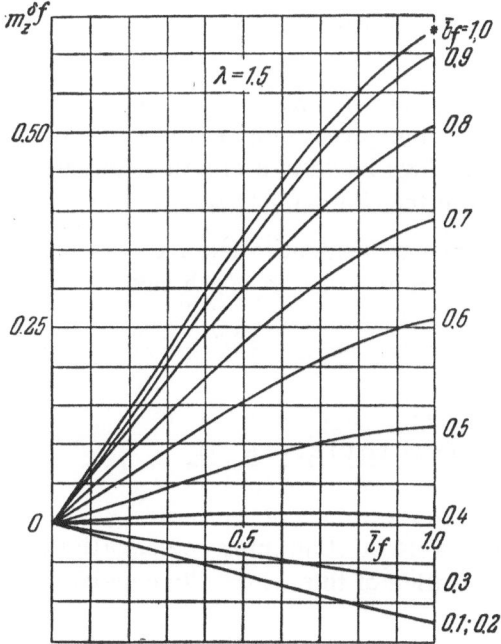

Fig. 13.10. The effect of the relative dimensions of flaps on $m_z^{\delta_f}$ for a rectangular wing with $\lambda = 1.5$.

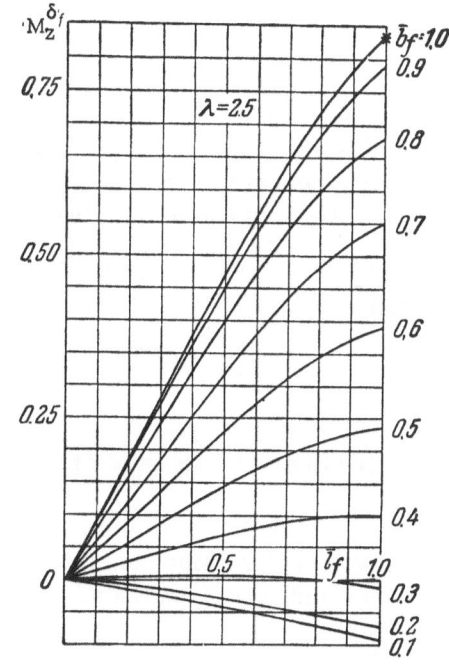

Fig. 13.11. The effect of the relative dimensions of flaps on $m_z^{\delta_f}$ for a rectangular wing with $\lambda = 2.5$.

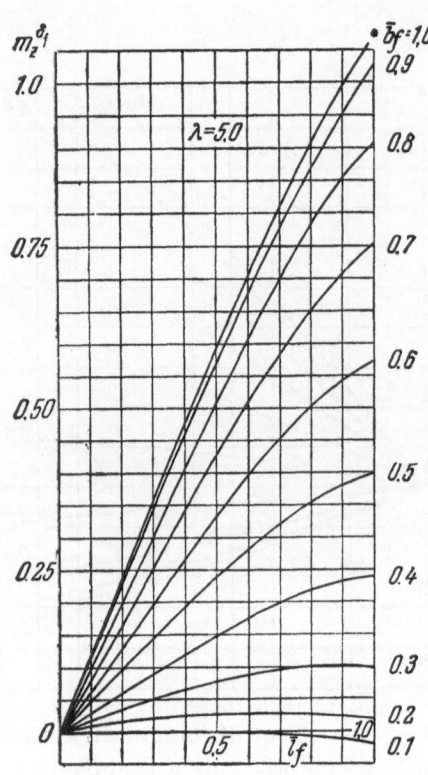
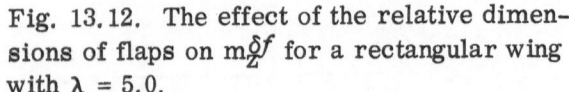

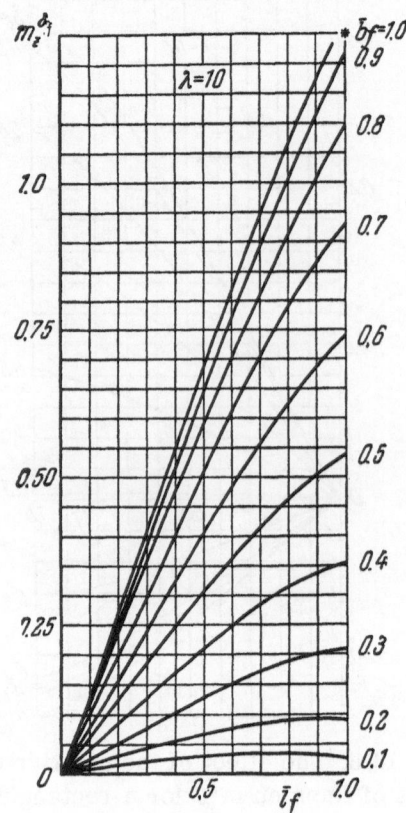

Fig. 13.12. The effect of the relative dimensions of flaps on $m_z^{\delta f}$ for a rectangular wing with $\lambda = 5.0$.

Fig. 13.13. The effect of the relative dimensions of flaps on $m_z^{\delta f}$ for a rectangular wing with $\lambda = 10$.

that a deflection of a flap or aileron at subsonic velocities causes disturbances over the whole wing. The wing itself therefore plays an important part in the generation of ΔY, ΔM_X, and also even ΔM_Z. An increase of the relative area of flaps and wings leads to a simultaneous decrease of the area of the wing and a lowering of recorded useful effect for ΔY and ΔM_X.

We note that in practical applications it is often more convenient to use the coefficients m_{X1} and $m_{X1}^{\delta f}$ and to use l as a characteristic dimension instead of b in which case

$$m_{x1}^{\delta a} = \frac{b}{l}\, m_x^{\delta a}.$$ (13.11)

§3. The Influence of Interfaces on the Characteristics of Rectangular Wings

We now consider some concrete data concerning the effect of an interface parallel to the plane of a monoplane wing on the aerodynamic characteristics of this wing. We consider wings of various aspect ratio.

We recall that at a solid boundary (like the earth's surface), which acts like a screen, the normal components of the disturbed velocity must be damped out. At a free surface (such as the surface of the sea) the pressure varies continuously, and so the tangential component of the disturbed velocity vanishes (Fig. 8.13).

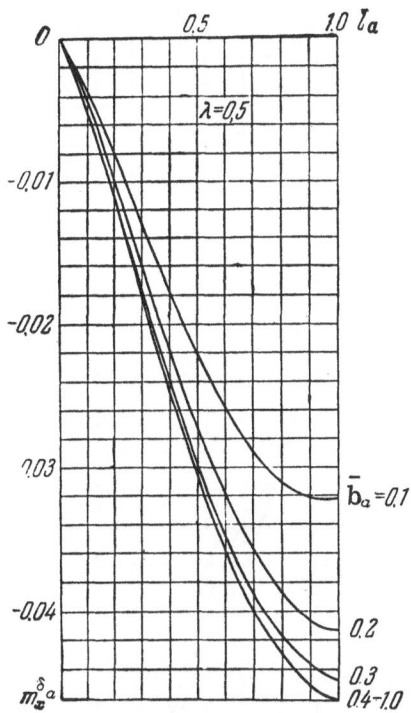
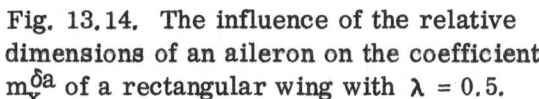
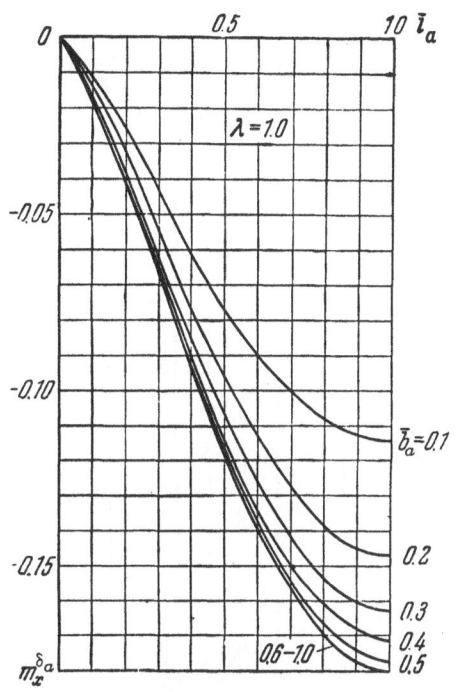

Fig. 13.14. The influence of the relative dimensions of an aileron on the coefficient $m_x^{\delta a}$ of a rectangular wing with $\lambda = 0.5$.

Fig. 13.15. The influence of the relative dimensions of an aileron on the coefficient $m_x^{\delta a}$ of a rectangular wing with $\lambda = 1.0$.

Hence in the case of a free surface we actually solve the problem of a biplane of finite span without stagger with the planes identically oriented, while in the case of a solid surface we consider planes identical in size but with angles of attack of opposite signs. In the first case the pressure on the upper surface of the lower plane is lowered, and the pressure on the lower surface of the upper plane is raised (when the angle of attack is positive). Hence the interference between the planes clearly lowers the lift of each of the planes. Conversely, in the second case the lift is increased.

The rotational-derivative coefficients of a rectangular wing moving parallel to a surface of separation are

$$
\left.
\begin{aligned}
c_y &= \frac{Y}{qS}, & m_z &= \frac{M_z}{qSb}, & m_{x1} &= \frac{M_x}{qSl}, \\
c_y &= c_y^\alpha \alpha + c_y^{\omega_z} \omega_z, & m_z &= m_z^\alpha \alpha + m_z^{\omega_z} \omega_z, \\
m_{x1} &= m_{x1}^{\omega_{x1}} \omega_{x1}, & \omega_{x1} &= \frac{\Omega_x l}{2U_0}, & \omega_z &= \frac{\Omega_z b}{U_0}.
\end{aligned}
\right\}
\tag{13.12}
$$

The coordinate system is standard with origin at the mid-point of the root chord.

In addition to the aspect ratio λ of the wing we now have another parameter on which the coefficients (13.12) depend. We use $\bar{h} = h/l$ for this parameter, where h is the distance from the wing to the interface and l is the span.

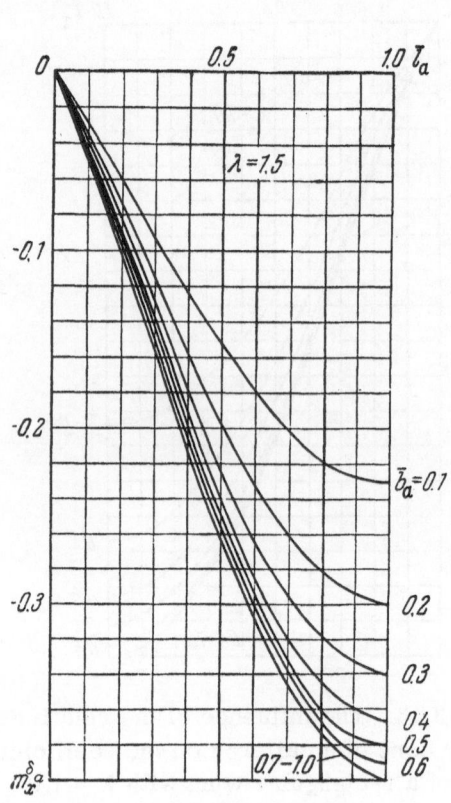

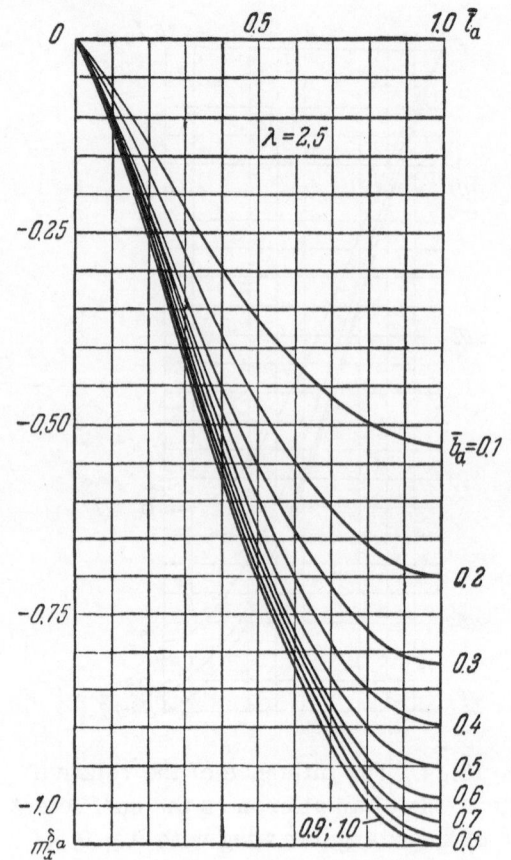

Fig. 13.16. The influence of the relative dimensions of an aileron on the coefficient $m_x^{\delta_a}$ of a rectangular wing with $\lambda = 1.5$

Fig. 13.17. The influence of the relative dimensions of an aileron on the coefficient $m_x^{\delta_a}$ of a rectangular wing with $\lambda = 2.5$.

Calculations of rotational-derivative coefficients of rectangular wings close to an interface were carried out by M. D. Patlasova, Ya. E. Polonsk, and G. A. Yakovlev. Figures 13.20-13.23 contain graphs of these coefficients as functions of the aspect ratio λ for various relative distances $\bar{h}$ from the interface.

It is clear from Fig. 13.20 that the lift properties of a wing with fixed angle of attack are improved close to a solid surface (the surface of the earth) and are worsened close to a free surface (water). The same type of behavior takes place for the same reasons in connection with the damping of bank (Fig. 13.23) — close to the earth it is increased and close to water it is decreased. The coefficient $m_z^{\omega_z}$ (Fig. 13.22) shows a more complex dependence for the center $\bar{x}_T = 0.5$ under consideration. For low aspect ratio its behavior is similar to the above, but for aspect ratios exceeding $\lambda = 2$ it reacts in the opposite way.

We note that an interface has only a weak effect for relative distances $\bar{h} > 0.5$, and all the rotational-derivative coefficients tend to their limiting values for an isolated wing when $\bar{h} \to \infty$.

All the data given and the conclusions arrived at were obtained from the linear theory and they refer, in particular, to small angles of attack. Naturally not all phenomena connected with the motion of a part of the control close to a surface of separation can be investigated in this way. For large angles of attack or for very small distances from an interface other new factors may play a fundamental part.

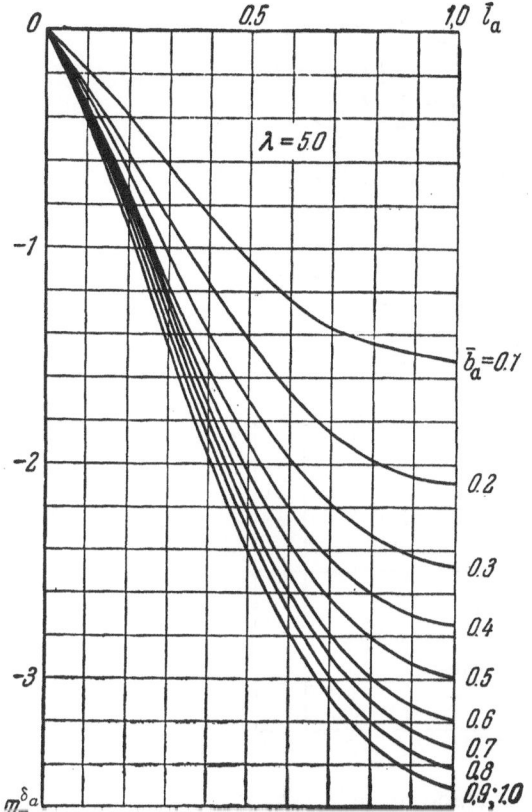
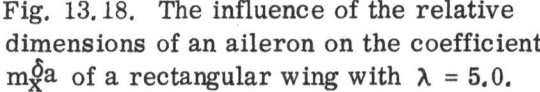
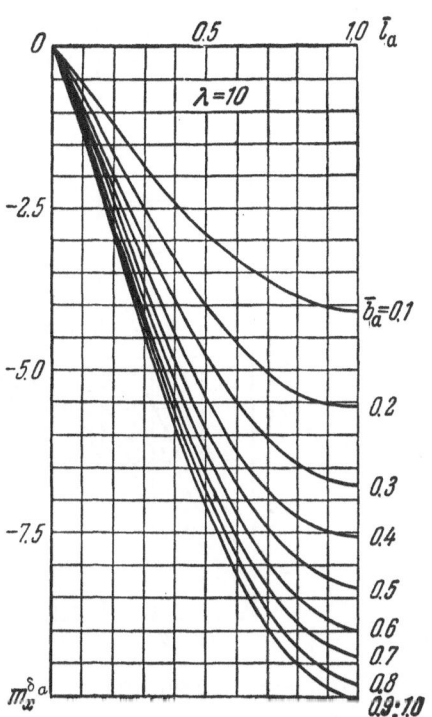

Fig. 13.18. The influence of the relative dimensions of an aileron on the coefficient $m_x^{\delta}a$ of a rectangular wing with $\lambda = 5.0$.

Fig. 13.19. The influence of the relative dimensions of an áileron on the coefficient $m_x^{\delta}a$ of a rectangular wing with $\lambda = 10$.

§ 4. The Influence of Interfaces on the Efficiency of Controls

Attached to a Rectangular Wing

Consequences of the reversibility theorem yield a rather simple method of determining the over-all effect due to deformation of the wing or to deflections of flaps or ailerons for a wing moving close to an interface. It is only necessary to know the distributed characteristics for a reversed wing taking account of the effects of the boundary of the flow. Since the characteristics of a reversed rectangular wing with center $\bar{x}_T = 0.5$ are the same as the characteristics of a direct wing, no new problems need be solved in the case under consideration.

We investigate this type of control in the same way as we investigated an isolated rectangular wing (§2 of the present chapter). We use the same coefficients, but we indicate the rotational–derivative coefficients of an isolated wing ($\bar{h} \to \infty$) by the subscript "∞" ($c_{y\infty}^{\delta}f$, $m_{x\infty}^{\delta}f$, and $m_{x\infty}^{\delta}f$). It is most convenient to analyze the ratio of these characteristics for a wing moving parallel to a surface of separation to the same characteristics for an isolated wing. In contrast to §2, it is more convenient here to consider $m_z^{\delta}f$ and $m_z^{\delta}f_{\infty}$ with the center $\bar{x}_T = 0$ (the Oz axis coincides with the leading edge of the wing). The remaining coefficients do not depend on the center x_T.

Calculations performed by G. A. Yakovlev yield the following results.

The ratios $c_y^{\delta}f / c_y^{\delta}f$, $m_z^{\delta}f / m_z^{\delta}f_{\infty}$, and $m_x^{\delta}a / m_{x\infty}^{\delta}a$ are practically independent of the relative span $\bar{l}_f$ of flaps or $\bar{l}_f$ of ailerons. Hence the influence of these parameters need not be

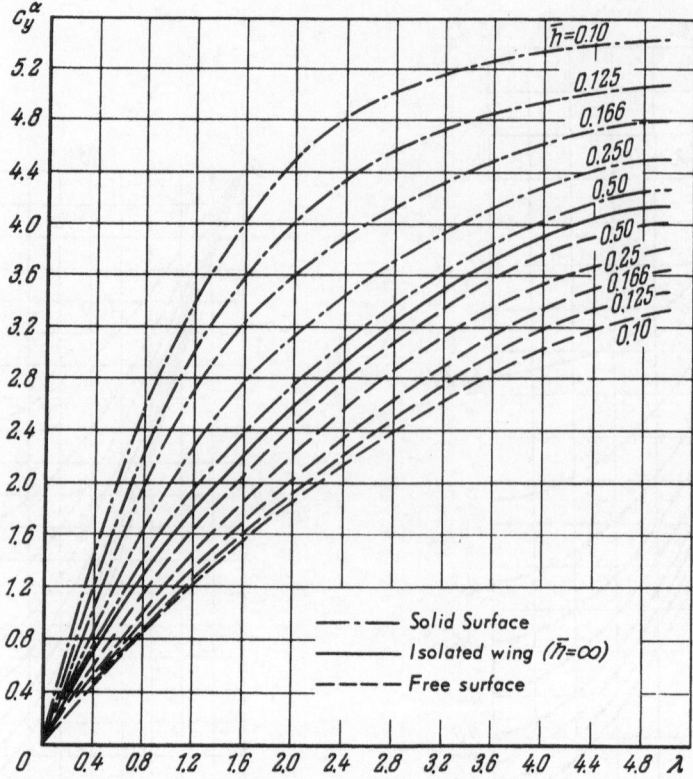

Fig. 13.20. The influence of an interface on the coefficients c_y^{α} of rectangular wings.

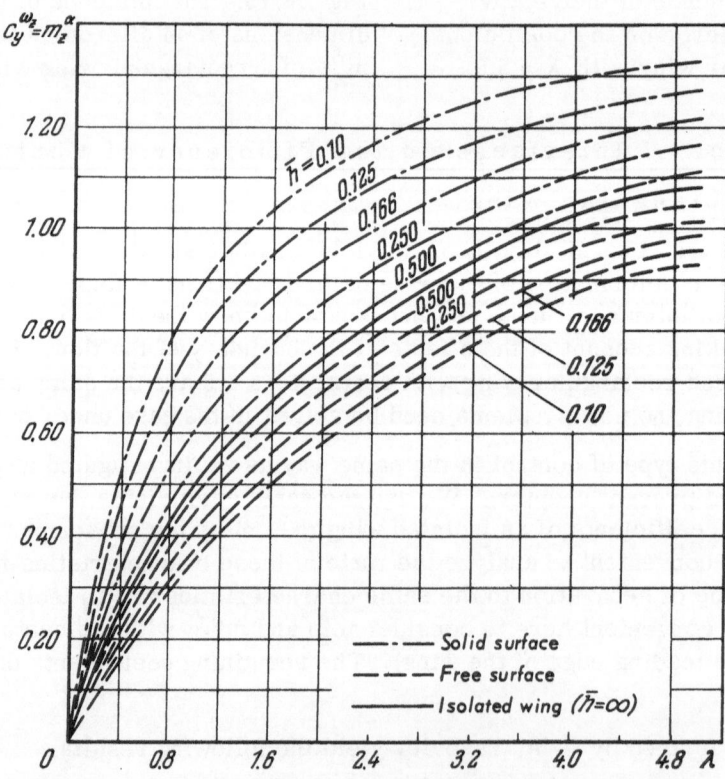

Fig. 13.21. The influence of an interface on the coefficients $m_z^{\alpha} = c_z^{\omega_z}$ of rectangular wings.

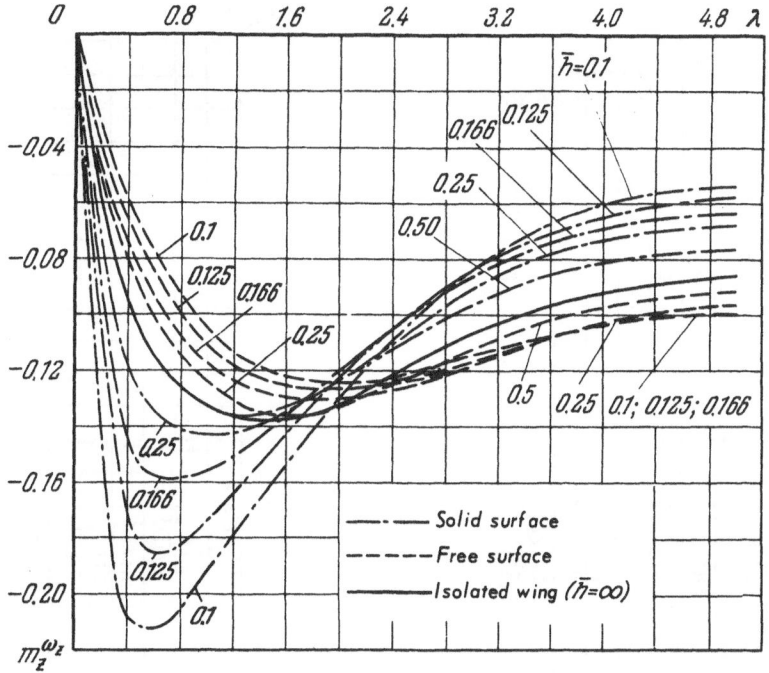

Fig. 13.22. The influence of an interface on the coefficients $m_z^{\omega_z}$ of rectangular wings.

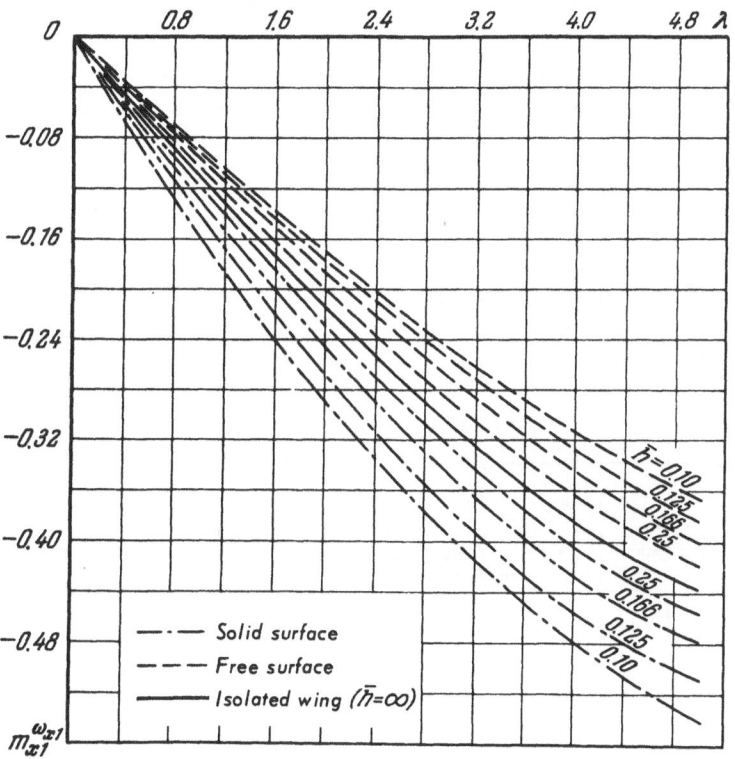

Fig. 13.23. The influence of an interface on the coefficients $m_{x_1}^{\omega_{x_1}}$ of rectangular wings.

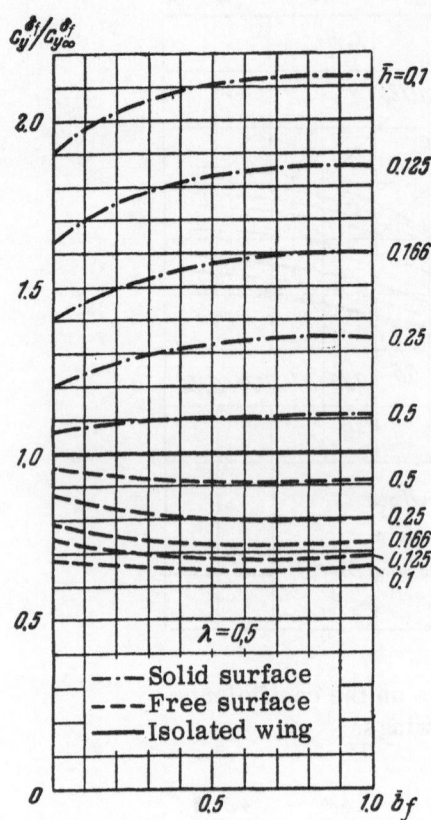

Fig. 13.24. The influence of an interface on the characteristics of flaps on a rectangular wing with $\lambda = 0.5$.

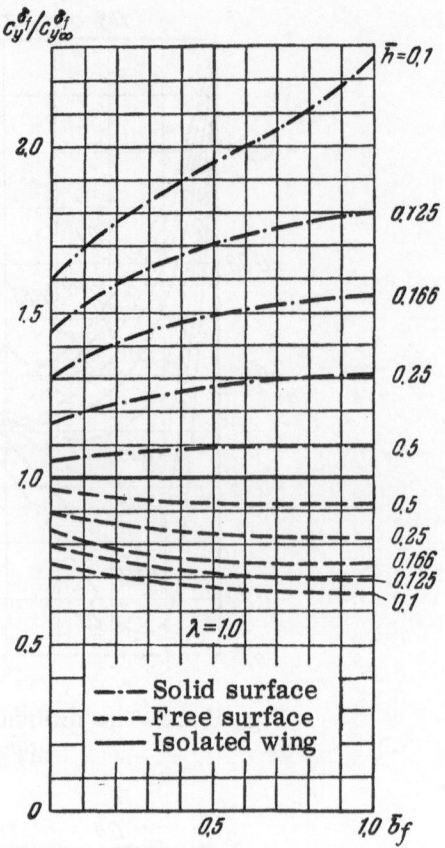

Fig. 13.25. The influence of an interface on the characteristics of flaps on a rectangular wing with $\lambda = 1.0$.

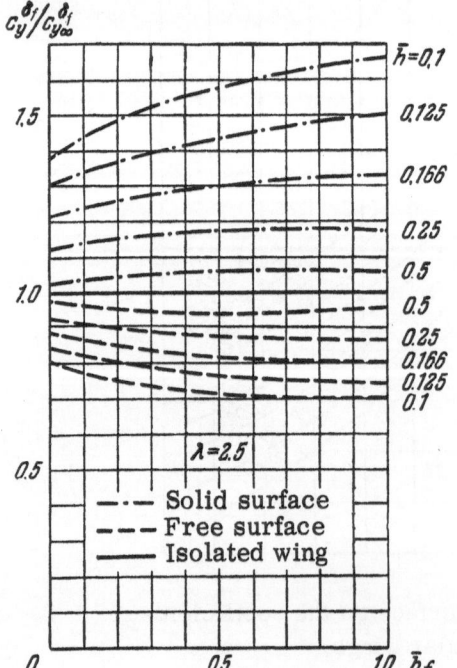

Fig. 13.26. The influence of an interface on the characteristics of flaps on a rectangular wing with $\lambda = 2.5$.

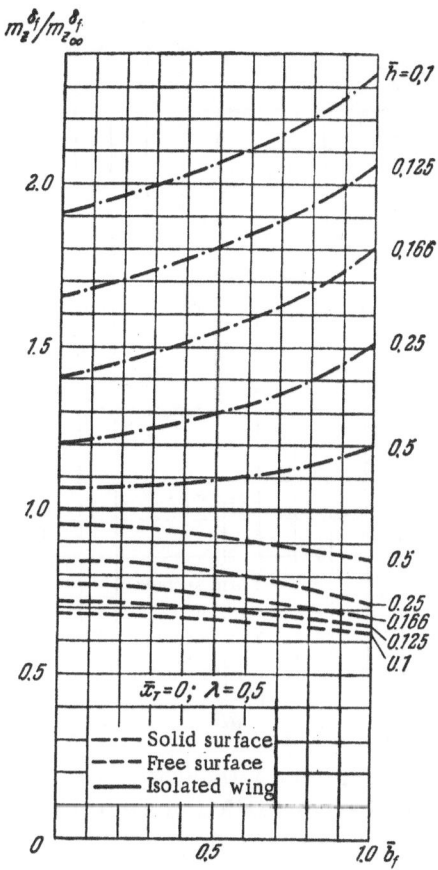

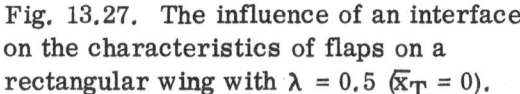

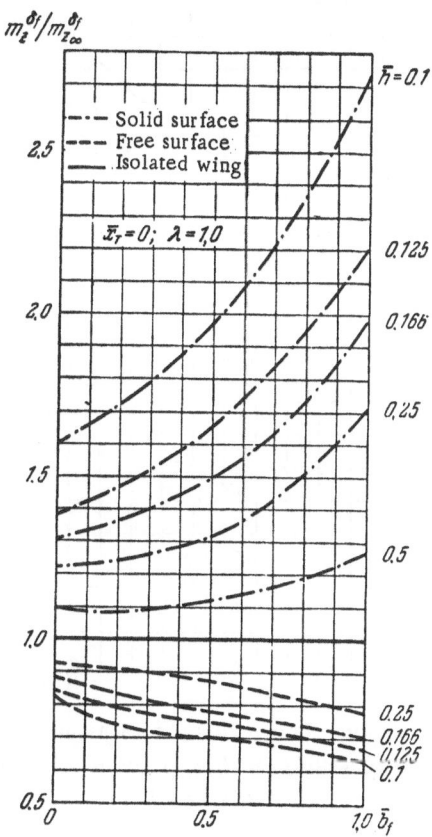

Fig. 13.27. The influence of an interface on the characteristics of flaps on a rectangular wing with $\lambda = 0.5$ ($\bar{x}_T = 0$).

Fig. 13.28. The influence of an interface on the characteristics of flaps on a rectangular wing with $\lambda = 1.0$ ($\bar{x}_T = 0$).

considered. This does not mean that the coefficients $c_y^{\delta f}$, $m_z^{\delta f}$, or $m_x^{\delta a}$ are independent of $\bar{l}_f$ or $\bar{l}_a$.

In Figs 13.24–13.32 we show graphs of these ratios as functions of the dimensionless chord $\bar{b}_f$ of a flap and $\bar{b}_a$ of an aileron for various values of the relative distance $\bar{h}$ from the interface. The results are given for rectangular wings of three aspect ratios: $\lambda = 0.5$, $\lambda = 1.0$; and $\lambda = 2.5$.

The accompanying graphs make it clear that the influence of the magnitude of the dimensionless chord on the ratios is comparatively negligible. This means that the coefficients $c_y^{\delta f}$, $m_z^{\delta f}$ and $m_x^{\delta a}$ will depend on $\bar{b}_f$ and $\bar{b}_a$ in approximately the same way as the corresponding characteristics of isolated wings.

When a wing is moving close to the earth's surface the lift properties of the wing and the attached controls are improved. A free surface leads to the opposite effect. The causes of these effects were considered in the previous section.

The above effects are naturally more strongly felt for small relative distances $\bar{h}$. When $\bar{h} > 0.5$ they become negligible.

We recall that all the above conclusions and results were obtained from linear theory, which holds for small angles of deflection of flaps and ailerons and in smooth flow past both wings and attached mechanisms.

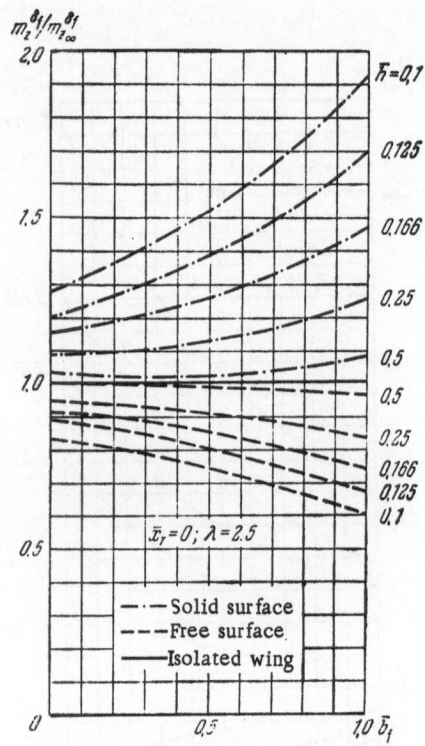

Fig. 13.29. The influence of an interface on the characteristics of flaps on a rectangular wing with $\omega = 2.5$ ($\bar{x}_T = 0$).

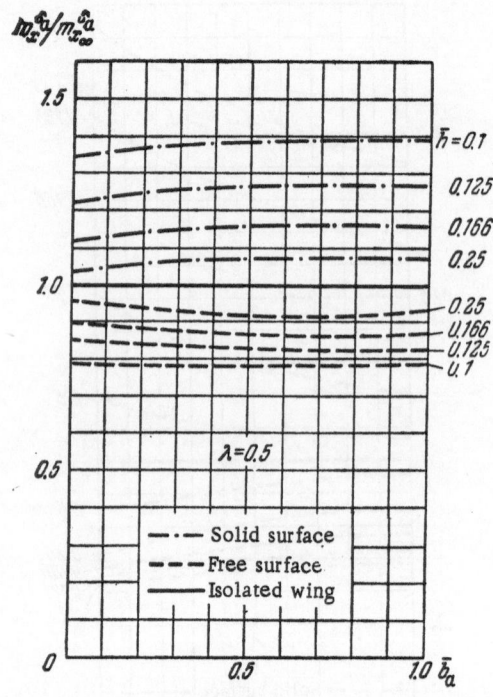

Fig. 13.30. The influence of an interface on the characteristics of ailerons on a rectangular wing with $\lambda = 0.5$.

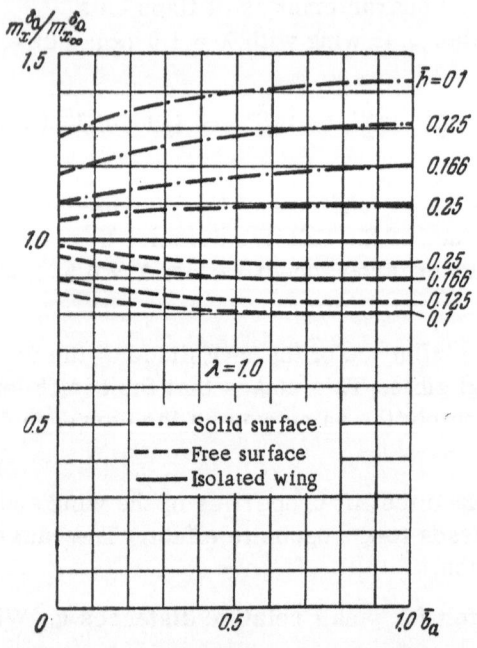

Fig. 13.31. The influence of an interface on the characteristics of ailerons on a rectangular wing with $\lambda = 1.0$.

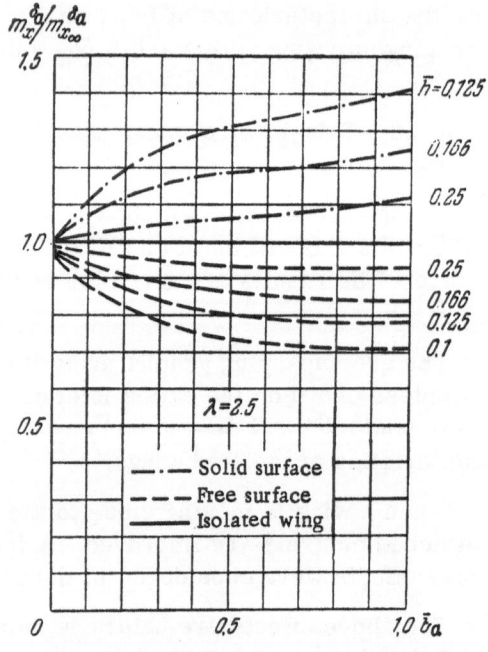

Fig. 13.32. The influence of an interface on the characteristics of ailerons on a rectangular wing with $\lambda = 2.5$.

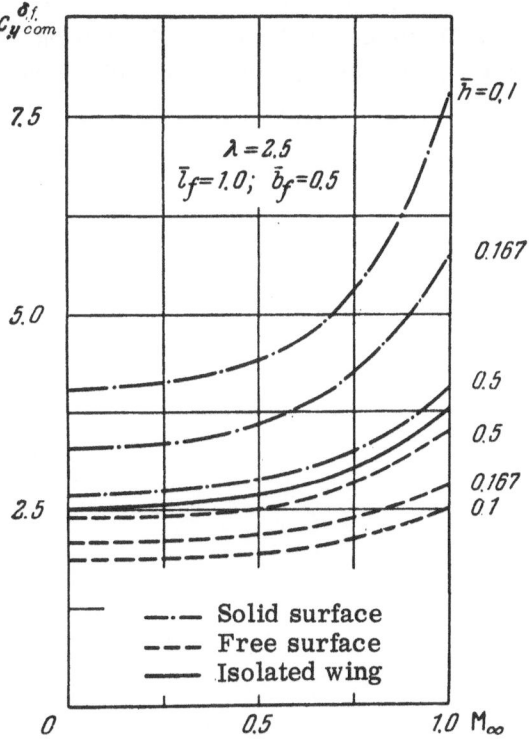

Fig. 13.33. The influence of the value of $\mathbf{M_\infty}$ on the efficiency of flaps on a rectangular wing for $\lambda = 2.5$.

§5. The Effect of the Mach Number

The rule for the conversion of aerodynamic characteristics to high subsonic velocities was established in Chapter V for isolated wings without attached controls. However, it is easily generalized to apply to the cases considered above.

We easily see that (5.19) and (5.20) are still valid, and we have

$$c_{com}^{\delta_f} = \frac{c_{yM}^{\delta_f}}{\sqrt{1 - M_\infty^2}}, \quad m_{com}^{\delta_f} = \frac{m_{zM}^{\delta_f}}{\sqrt{1 - M_\infty^2}}, \quad m_{com}^{\delta_a} = \frac{m_{x1M}^{\delta_a}}{\sqrt{1 - M_\infty^2}}. \tag{13.13}$$

For the determination of the aerodynamic parameters of a transformed wing in an incompressible medium (all the coefficients have the subscript "M"), we use the geometrical parameters (5.7) as arguments. Here we have the extra dimensionless parameters

$$\bar{h}_M = \frac{h_M}{l_M}, \quad \bar{b}_{fM} = \frac{b_{fM}}{b_M}, \quad \bar{b}_{aM} = \frac{b_{aM}}{b_M}, \quad \bar{l}_{fM} = \frac{2l_{fM}}{l_M}, \quad \bar{l}_{aM} = \frac{2l_{aM}}{l_M}. \tag{13.14}$$

Since the transformation (5.2) changes linear dimensions only along the Ox axis and leaves dimensions in the directions of the Oy and Oz axes unchanged, we have

$$\bar{h}_M = \bar{h}, \quad \bar{b}_{fM} = \bar{b}_f, \quad \bar{b}_{aM} = \bar{b}_a, \quad \bar{l}_{fM} = \bar{l}_f, \quad \bar{l}_{aM} = \bar{l}_a. \tag{13.15}$$

As an example we show graphically in Fig. 13.33 results illustrating the influence of the compressibility of air on the coefficient $c_y^{\delta f}$ for a rectangular wing with aspect ratio $\lambda = 2.5$, both for an isolated wing and for a wing moving close to an interface.

CHAPTER XIV

APPARENT MASSES OF WINGS

§1. Some Problems in Methods of Calculation

E. P. Kapustina carried out an extensive investigation into the rational development of methods of calculating apparent-mass coefficients of annular and monoplane wings. She also carried out systematic calculations on an electronic computer and obtained a large amount of data on the coefficients.

It was established that the algebraic equations to which the problem reduces are very stable, this being due to the predominant role of terms close to the diagonal of the coefficient matrix.

The convergence of the solution with increasing number of vortices was investigated by numerical experiments; results of calculations using different values of m were compared. The results for an annular wing with aspect ratio $\lambda = 1$ are given in the accompanying table.

$2m$	k_{22}	Δ_1 %	k_{66}	Δ_1 %	k_{26}
4	0.691		0.0173		0
		8.9		27.6	
6	0.759		0.0242		0
		3.4		10.7	
8	0.785		0.0271		0
		2.2		6.5	
10	0.803		0.0290		0
		1.5		4.4	
12	0.815		0.0303		0
		1.3		3.2	
14	0.826		0.0313		0
		0.8		1.5	
16	0.833		0.0318		0

The problem was solved in the standard coordinate system with center $\bar{x}_T = 0.5$ with various numbers 2m of bound annular vortices (from 4 to 16). The percentage variation Δ_1 of the apparent-mass coefficient when the value of m increases to its next larger value is given between the main lines of data in the table. For all m the apparent-mass coefficient k_{26} is zero for the center used.

215

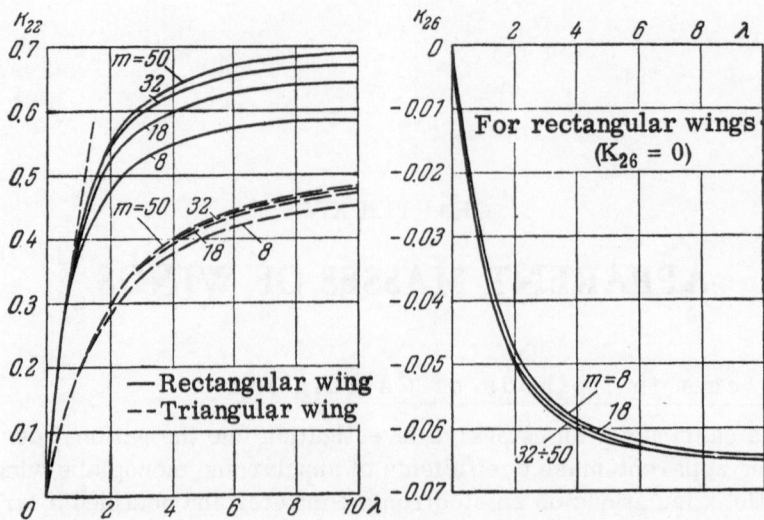

Fig. 14.1. The effect of the number of vortices m on the apparent-mass coefficient of monoplane wings.

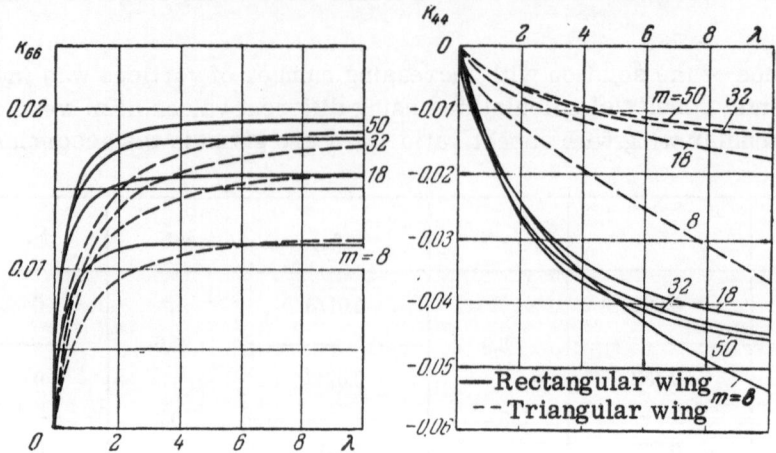

Fig. 14.2. The effect of the number of vortices m on the apparent-mass coefficient of monoplane wings.

Figures 14.1 and 14.2 show graphically the values of the apparent-mass coefficients k_{22}, k_{26}, k_{44} and k_{66} for rectangular and triangular wings of various aspect ratios. Various values of m = nN for the half-span were used in the calculations (m varies from 8 to 50).

The dotted straight line emanating from the origin in Fig. 14.1 is obtained from Eq. (9.9).

We note one fact concerning the distribution of discrete vortices over a monoplane wing which must be remembered when we are calculating apparent masses. If, in flow with circulation past a wing, the directions parallel to the span and chord are not equivalent and the nature of the flow over the leading, trailing, and side edges is qualitatively different, then the picture is different. The flow over all edges is identical. In the case under consideration the Chaplygin-Zhukovskii hypothesis will not hold, infinite rarefactions will occur on each of the edges, etc.

Hence in the determination of k_{22} for translational oscillations of a square wing in the direction of the Oy axis, which is perpendicular to the plane of the wing, conditions are identical on all sides of the wing. It is therefore logical to use the same number of bound vortices along the chord and along the span. In asymmetrical motion, of course, even in the case of a rectangular wing, we may need to use more vortices in one direction than in the other. Here, however, we do not obtain any qualitative difference between the flow over the edges of a wing such as was obtained in the calculation of the rotational-derivative coefficients.

§2. Annular Wings

We consider a cylindrical annular wing of arbitrary elongation λ in a standard coordinate system with $\overline{x}_T = 0.5$ (Figs. 8.10 and 8.11).

We use the chord b and the area of the rectangular wing with equivalent over-all dimensions $S = 2rb$ as characteristic dimensions and we write the dimensional apparent-mass coefficients in the form:

$$k_{22} = \frac{\lambda_{22}}{\rho Sb}, \qquad k_{26} = \frac{\lambda_{26}}{\rho Sb^2}, \qquad k_{66} = \frac{\lambda_{66}}{\rho Sb^3}. \tag{14.1}$$

For annular wings of infinite aspect ratio the apparent-mass coefficients are $\pi/2$ times their values for plates of infinite span (§2, Chapter IX). Hence for an annular wing of infinite aspect ratio we have

$$k_{22} = \frac{\pi^2}{8}, \qquad k_{66} = \frac{\pi^2}{256}, \qquad k_{26} = 0. \tag{14.2}$$

Figure 14.3 shows graphically the values of k_{22} and k_{66} for cylindrical annular wings as functions of the aspect ratio λ. The dotted curve shows the limiting values for $\lambda = \infty$ obtained from (14.2). For center $\overline{x}_T = 0.5$, the coefficient k_{26} is identically zero for annular wings of any aspect ratio; this can be proved from considerations of symmetry without any calculation.

The qualitative effect of elongation on the apparent-mass coefficient is similar to that encountered in connection with the derivative c_y^α. This similarity is not due to chance, since there is much in common to the two cases. The smaller the aspect ratio, the greater will be

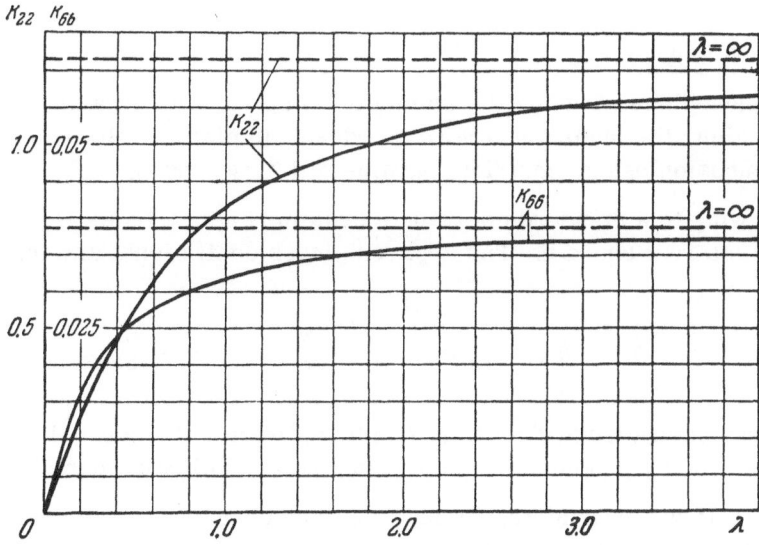

Fig. 14.3. The apparent-mass coefficient of annular wings.

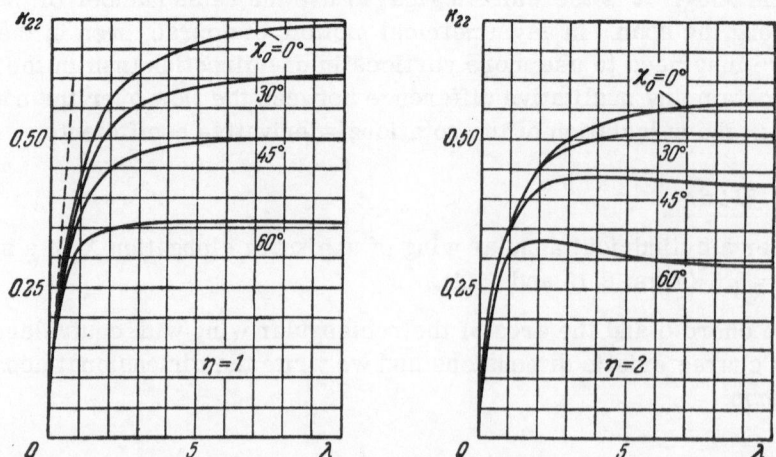

Fig. 14.4. The apparent-mass coefficients k_{22} of monoplane wings.

the unfavorable interaction between sections of an annular wing corresponding to the same meridional half-plane (for example, between the lower surface of the upper section $\varphi = 0$ and the upper surface of the lower surface $\varphi = \pi$). In the case under consideration, only when there is no circulatory flow will there be no free vortices and consequently no resulting effects.

In a qualitative analysis of the relations discussed above and with the aim of understanding them better, we use the following relation between the apparent mass and the hydrodynamic action of the medium on an annular wing. From (1.49) we conclude that k_{22} is proportional to the lift Y due to translational oscillations of the wing along the Oy axis; k_{66} is proportional to the moment M_z due to nonuniform rotation of the wing about the Oz axis; k_{26} is proportional to the moment M_z due to translational oscillations of the wing along the Oy axis.

We recall that the circulation-free flow scheme permits the modeling of phenomena occurring when the mean velocity of motion U_0 is close to zero [38, 39, 40].

Hence the results obtained can be used in the determination of the over-all action of an incompressible medium on annular wings during the evolution of these actions, in gusts of wind, etc.

§3. Monoplane Wings

We consider monoplane wing-plates, symmetric in plan, with straight edges, and with constant sweepback along the leading and trailing edges. In plan the shapes of such wings are described by the dimensionless geometrical parameters λ, χ_0 and η.

We relate the wing to a standard coordinate system (Figs. 4.10 and 4.12) with the origin O at the midpoint of the root chord b. The apparent-mass coefficients are introduced in the form

$$k_{22} = \frac{\lambda_{22}}{\rho Sb}, \quad k_{26} = \frac{\lambda_{26}}{\rho Sb^2}, \quad k_{44} = \frac{\lambda_{44}}{\rho Sb^3}, \quad k_{66} = \frac{\lambda_{66}}{\rho Sb^3}. \tag{14.3}$$

If we know these coefficients, then Eqs. (1.46) easily give the over-all action of an incompressible medium on a monoplane wing carrying out an arbitrary maneuver in situ.

In Fig. 14.4.-14.12 we present graphs showing the dependence of the above coefficients on the aspect ratio λ for various sweepback angles χ_0 of the leading edge and taper ratios η.

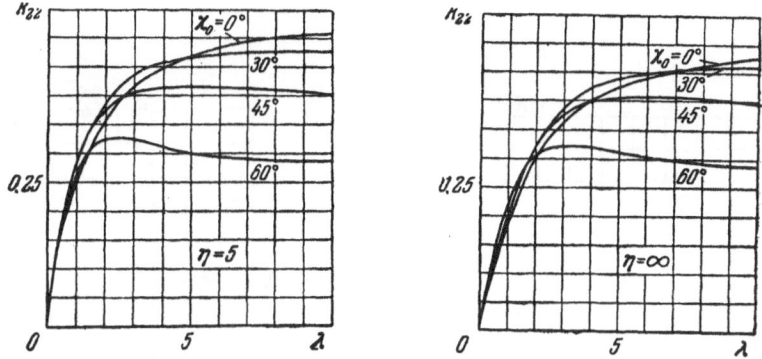

Fig. 14.5. The apparent-mass coefficients k_{22} of monoplane wings.

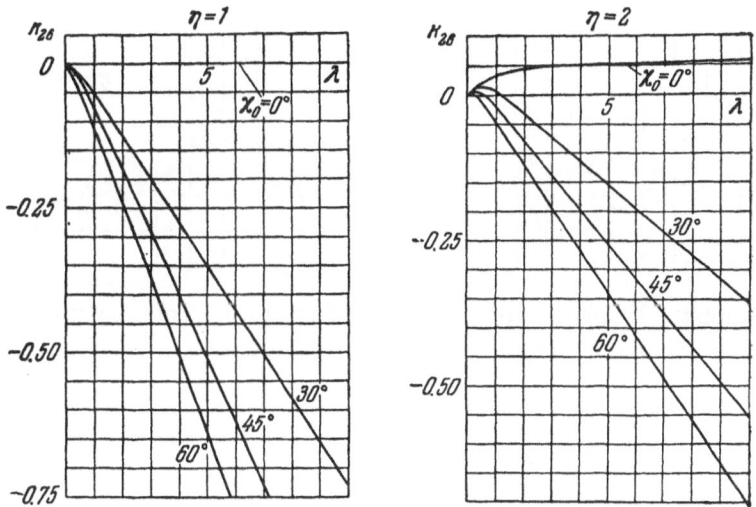

Fig. 14.6. The apparent-mass coefficients k_{26} of monoplane wings.

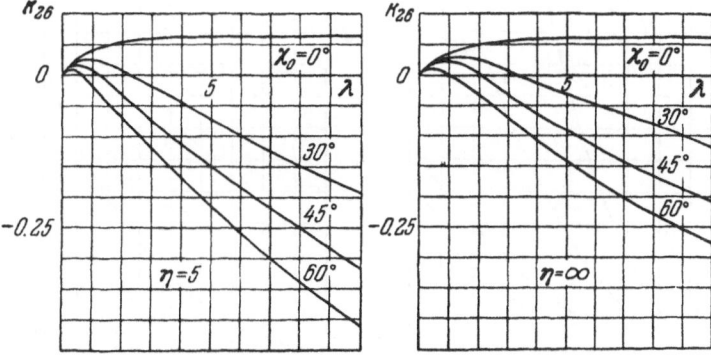

Fig. 14.7. The apparent-mass coefficients k_{26} of monoplane wings.

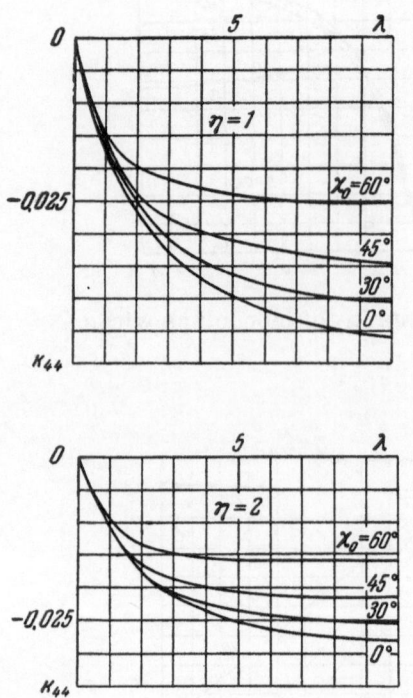

Fig. 14.8. The apparent-mass coefficients k_{44} of monoplane wings.

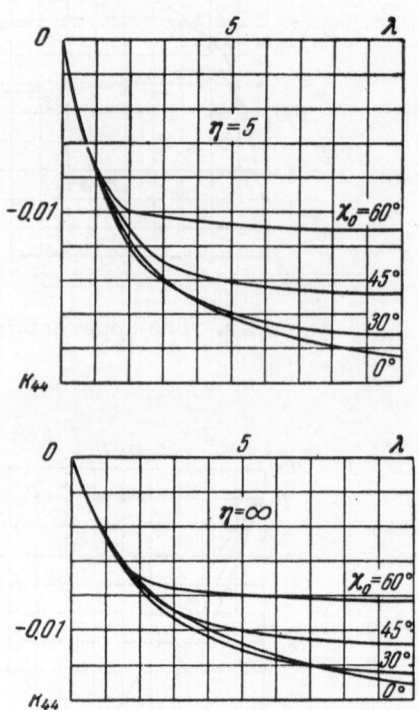

Fig. 14.9. The apparent-mass coefficients k_{44} of monoplane wings.

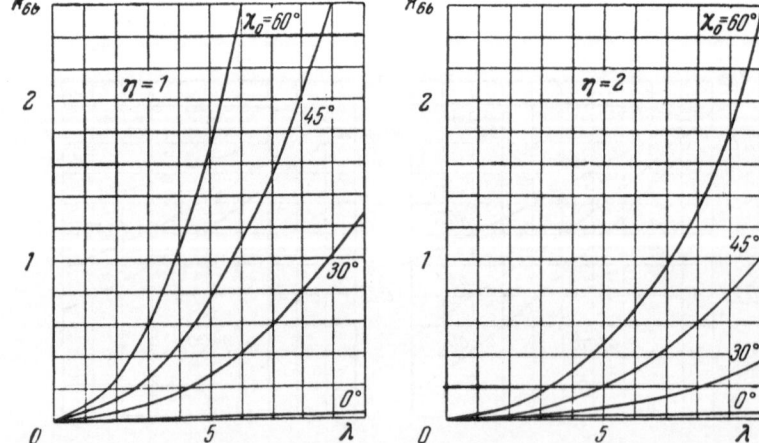

Fig. 14.10. The apparent-mass coefficients k_{66} of monoplane wings.

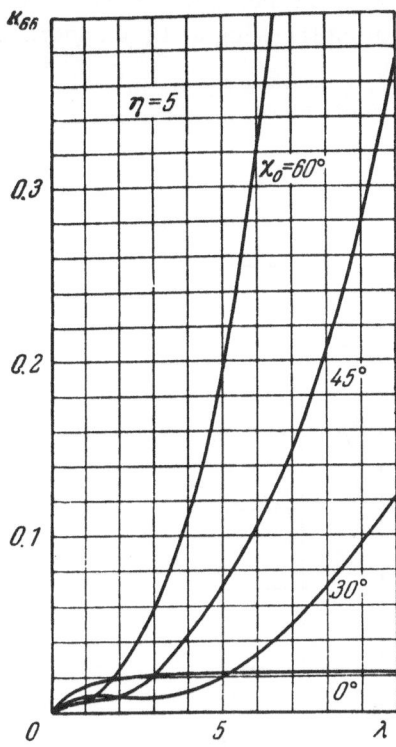

Fig. 14.11. The apparent-mass coefficients k_{66} of monoplane wings.

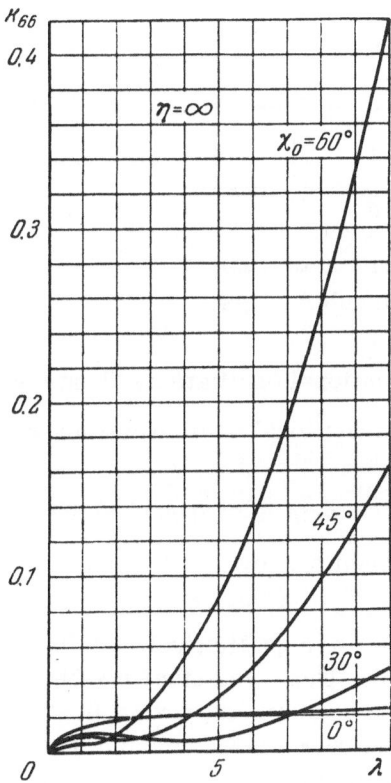

Fig. 14.12. The apparent-mass coefficients k_{66} of monoplane wings.

It is useful for the better understanding of the results, as in the previous case, to obtain a relation between physically intuitive effects of the medium on the wing and the apparent-mass coefficients. Using (1.46) we can assert the following:

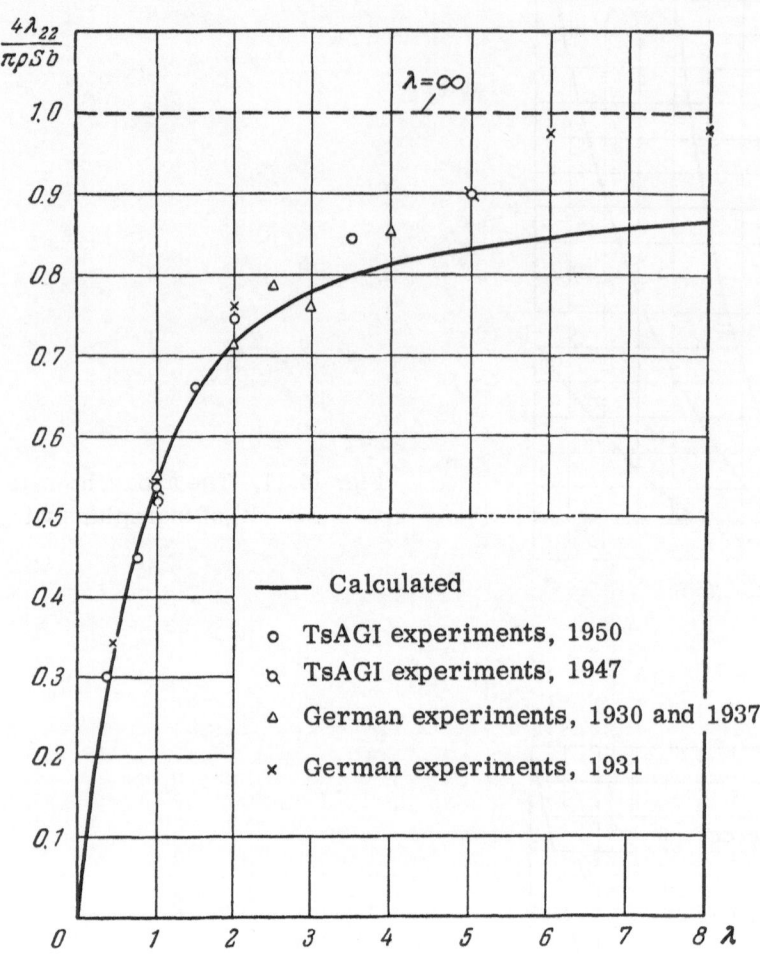

Fig. 14.13. A comparison of experimental and calculated values of the apparent masses of rectangular wings.

The coefficient k_{22} determines the lift Y due to translational oscillations along the Oy axis;

The coefficient k_{66} characterizes the moment M_z for rotational oscillations about the Oz axis;

The moment M_x for rotational oscillations about the Ox axis is expressible in terms of k_{44};

Finally, the coefficient k_{26} is proportional to the moment M_z for translational oscillations along the Oy axis.

Many of the relations shown in accompanying diagrams are qualitatively similar to those that were obtained for the rotational-derivative coefficients. These and other coefficients have properties in common, since the actions of the medium on a body are expressed in similar form in terms of them. There is, however, an essential difference between them, since they correspond to essentially different flow schemes. In particular there are no vortex trails in

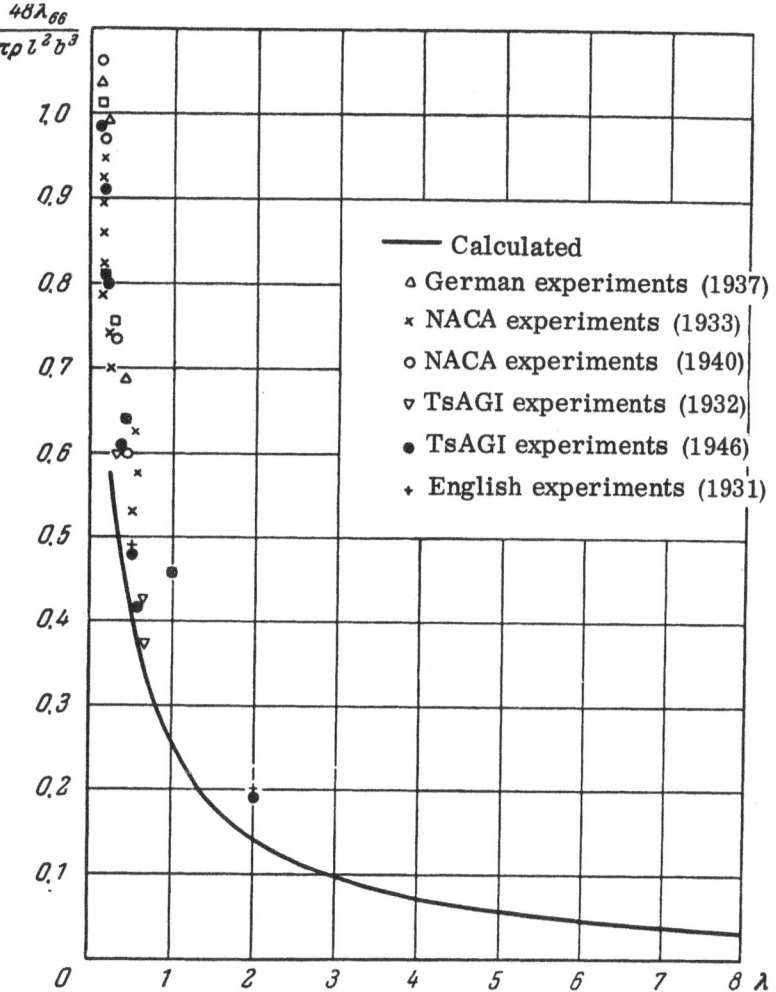

Fig. 14.14. A comparison of experimental and calculated values
of the apparent mass of rectangular wings.

one case (for the apparent mass), while there are vortex trails in the other.

As we have already noted, the circulation-free flow scheme corresponds to phenomena that occur when the velocity U_0 is close to zero.

Hence results concerning the apparent mass can be used in the calculation of the action of an incompressible medium on monoplane wings during maneuvers in situ, in wind gusts, etc.

§4. A Comparison of Theoretical and Experimental Results

Various methods of experimental determination of apparent-mass coefficients are related to investigations of oscillations of bodies in situ in a continuous medium (air, water, etc.) and in a vacuum. There is a description of these methods in [38].

In the present section we compare some of the calculated results given above with experimental results obtained by various authors between 1930 and 1950 and collected in the article referred to.

Figure 14.13 shows values of the apparent mass λ_{22} of rectangular wings of various aspect ratios. The indicated parameters are referred to the apparent mass of a plate of infinite span, and so these parameters tend to one as λ increases.

The work referred to above also gives experimental data for the moment M_z acting on a rectangular plate when it performs rotational oscillations about the Oz axis. In Fig. 14.14 we compare calculated and experimental data on the dimensionless coefficient, proportional to k_{66} considered in [38].

It is clear from Figs. 14.13 and 14.14 that even though the phenomenon is rather complex and the scheme for calculating and apparent-mass coefficient rather simple, the agreement between experimental and calculated results is in general satisfactory.

LITERATURE CITED

1. N. E. Zhukovskii, Bound Vortices, Collected Works, Vol. IV, Gostekhizdat (1949).
2. N. E. Zhukovskii, Vortex Theory of Screw Propellers, Collected Works, Vol. IV Gostekhizdat (1949).
3. N. E. Zhukovskii, The Theoretical Basis of Aeronautics, Collected Works, Vol. V, Gostekhizdat (1950).
4. N. E. Zhukovskii, A Modification of the Kirchhoff Method for the Determination of the Two-Dimensional Motion of a Liquid with a Constant Velocity Given on an Unknown Streamline, Collected Works, Vol. II, Gostekhizdat (1948).
5. S. A. Chaplygin, Results of Theoretical Investigations of the Motion of an Airplane, Collected Works, Vol. II, Gostekhizdat, (1948).
6. S. A. Chaplygin, The Pressure of Plane-parallel Flow over an Obstacle (Airplane Theory), Collected Works, Vol. II, Gostekhizdat (1948).
7. S. A. Chaplygin, The Theory of the Latticed Wing, Collected Works, Vol. II Gostekhizdat (1948).
8. S. A. Chaplygin, The Effect of Plane-parallel Airflow over a Moving Cylindrical Body, Collected Works, Vol. II, Gostekhizdat, (1948).
9. L. Prandtl, Hydro-Aeromechanics, Dover, New York (1957).
10. L. Prandtl, Beitrag zur Theorie der tragenden Fläche, ZAMM 16, No. 6 (1939).
11. N. E. Kochin, I. A. Kibel', and N. V. Roze, Theoretical Hydrodynamics, Ch. I., Fizmatgiz, (1963).
12. N. E. Kochin, The Hydrodynamic Theory of Cascades, Gostekhizdat (1949).
13. V. V. Golubev, The Theory of Airplane Wings of Finite Span, Trudy Tsent. Aero. Gidr. Inst., No. 108 (1931).
14. V. V. Golubev, Lectures on the Theory of Wings, Gostekhizdat (1949).
15. A. I. Nekrasov, The Theory of Wings in Nonsteady Flow, Izd-vo Akad. Nauk SSSR (1947).
16. L. I. Sedov, Plane Problems of Hydrodynamics and Aerodynamics, Wiley, New York (1965).
17. B. N. Yur'ev, Experimental Aerodynamics, Ch. II, Oborongiz (1938).
18. E. Carafoli, Aerodynamics of Aircraft Wings, Izd-vo Akad. Nauk SSSR (1956).
19. S. M. Belotserkovskii, A. S. Ginevskii, and Ya. E. Polonskii, Aerodynamic Force and Moment Characteristics of a Cascade of Thin Profiles, Industrial Aerodynamics, No. 22 Oborongiz (1962).
20. W. Birnbaum, Die tragende Wirbelfläche als Hilfsmittel zur Behandlung des ebenen Problems der Tragflügeltheorie, ZAMM 3, No. 4 (1923).
21. K. Wieghardt, Uber die Auftriebsverteilung des einfachen Rechteflügels über die Tiefe, ZAMM 19, No. 5, (1939).
22. H. Blenk, Der Eindecker als tragende Wirbelfläche, ZAMM 5, No. 1 (1925).
23. M. Hansen, Messungen an Kreistragflächen, Luftfahrtforschung (1939).
24. N. E. Kochin, The Theory of Wings of Finite Span of Circular Plan Form, Collected Works, Vol. II, Izd-vo Akad. Nauk SSSR (1949).

25.   V. V. Golubev, The Theory of wings of small aspect ratio, Izv. Akad. Nauk SSSR, Otd. Tekh. Nauk, No. 3, (1947).

26.   A. A. Dorodnitsyn, A generalization of airfoil theory to the case of a wing with a curved axis and an axis not perpendicular to the flow, Prikl. Matem. i Mekh., VIII (1) (1944).

27.   P. I. Chushkin, The Calculation of Distributed Circulation for Rectangular Wings of Small Aspect Ratio, Collection of Theoretical Works on Aerodynamics, Oborongiz (1957).

28.   G. A. Kolesnikov, A Method of Calculating the Distribution of Circulation of Wings of Small Aspect Ratio, Collection of Theoretical Works on Aerodynamics, Oborongiz, (1957).

29.   V. V. Struminskii and N. K. Lebed', A Method of Calculating Circulation about the Span of a Swept-Back Wing, Collection of Theoretical Works on Aerodynamics, Oborongiz (1957).

30.   V. M. Falkner, The solution of lifting-plane problems by vortex-lattice theory, ARC Report and Memoranda, No. 2591 (1953).

31.   S. M. Belotserkovskii, Three-dimensional non-steady motion of lifting surfaces, Prikl. Matem. i Mekh. XIX (4) (1955).

32.   S. M. Belotserkovskii, The representation of nonsteady aerodynamic moments and forces by means of rotational-derivative coefficients, Izv. Akad. Nauk SSSR, Otd. Tekh. Nauk No. 7 (1956).

33.   S. M. Belotserkovskii, A. S. Ginevskii, and Ya. E. Polonskii, The Aerodynamic Forces Acting on a Profile Lattice in Nonsteady Flow, Industrial Aerodynamics No. 20, Oborongiz (1961).

34.   A. H. Flax, Reverse flow and variational theorems for lifting-surfaces in nonstationary compressible flow, JAS, No. 2 (1953).

35.   B. I. Ul'yanov, Some generalizations and consequences of the reversibility theorem for nonsteady motion, Izv. Akad. Nauk Otd. Tekh. Nauk SSSR, Mekhanika i Mashinostroenie, No. 1 (1961).

36.   S. M. Belotserkovskii, Horseshoe vortices in nonsteady motion, Prikl. Matem. i Mekh., Vol. XIX, No. 2 (1955).

37.   S. M. Belotserkovskiik Annular vortices in nonsteady motion, Prikl. Matem. i Mekh., Vol. XX, No. 2 (1956).

38.   I. S. Riman, and R. L. Kerps, Apparent masses of bodies of various forms, Trudy Tsent. Aero. Gidr. Inst., No. 635 (1947).

39.   K. K. Fedyaevskii and S. M. Belotserkovskii, Aerodynamic forces acting on buildings on the earth's surface during gusts, Izv. Akad. Nauk SSSR, Otd. Tekh. Nauk No. 6 (1954).

40.   K. K. Fedyaevskii, The approximate theoretical determination of apparent masses of rectangular plates, Prikl. Matem. i Mekh. XVI (3) (1952).

41.   L. A. Simonov and S. A. Khristianovich, The effect of compressibility on the induced velocity of wings and propellers, Prikl. Matem. i Mekh., VIII (2) (1944).

42.   S. L. Sobolev, Equations of Mathematical Physics, GITTL (1950).

43.   S. M. Belotserkovskii, Investigations in the theoretical determination of rotational-derivative coefficients (Proceedings of the All-Union Congress on Theoretical and Applied Aerodynamics), Annotation of Proceedings, Izd-vo Akad. Nauk SSSR (1960).

44.   F. R. Küssner, Kritische Bemerkungen zur dreidimensionalen Tragflächen Theorie im Unterschallgebiet, Ztschr. f. Flugwiss, 4 No. 1-2, (1956).

45.   E. L. Blokh, The center of pressure of rectangular profiles in a grid, Technical Reports MAP, No. 123 (1948).

46.   V. M. Belyakov, R. I. Kravtsova, and M. G. Rappoport, Tables of elliptical integrals, Vol. I, Izd-vo Akad. Nauk SSSR (1962).

INDEX